Advances in Face Detection and Facial Image Analysis

Michal Kawulok • M. Emre Celebi
Bogdan Smolka

Editors

Advances in Face Detection and Facial Image Analysis

 Springer

Editors
Michal Kawulok
Automatic Control, Electronics and
    Computer Science
Silesian University of Technology
Gliwice, Poland

M. Emre Celebi
Computer Science
Louisiana State University in Shreveport
Shreveport, LA, USA

Bogdan Smolka
Automatic Control, Electronics and
    Computer Science
Silesian University of Technology
Gliwice, Poland

ISBN 978-3-319-25956-7   ISBN 978-3-319-25958-1   (eBook)
DOI 10.1007/978-3-319-25958-1

Library of Congress Control Number: 2016934052

Springer Cham Heidelberg New York Dordrecht London

Printed on acid-free paper

Springer International Publishing AG Switzerland is part of Springer Science+Business Media (www.
springer.com)

Preface

Face detection and recognition is an active research topic of computer vision and pattern recognition. A number of novel methods that are emerging every year improve the accuracy of face and facial landmarks detection, thereby increasing the number and diversity of human face analysis applications. New research directions include, in particular, facial dynamics recognition underpinned with psychological background aimed, for example, at behavior analysis, deception detection, or diagnosis of psychological disorders.

The goal of this volume is to summarize the state of the art in face detection and analysis, taking into account the most promising advances in this field.

The volume opens with three chapters on face detection and lighting adjustment. In the chapter "A Deep Learning Approach to Joint Face Detection and Segmentation," Luu et al. propose an algorithm for detecting faces using multiscale combinatorial grouping, deep learning (using the GPU-based Caffe framework), and one-class support vector machines. The detected faces are then refined using the modified active shape models. The effectiveness of the proposed face detection and segmentation algorithm is demonstrated on the Multiple Biometric Grand Challenge and Labeled Faces in the Wild databases.

In the chapter "Face Detection Coupling Texture, Color and Depth Data," Nanni et al. demonstrate how to combine multiple face detectors. While it is well known that an ensemble of detectors increases the detection rate at the cost of higher false-positives, the authors discuss various filtering techniques for keeping the false-positives low. The authors also make their MATLAB code publicly available, which is a valuable contribution to the face detection community.

In the chapter "Lighting Estimation and Adjustment for Facial Images," Jiang et al. present a lighting estimation algorithm for facial images and a lighting adjustment algorithm for video sequences. For the estimation of the illumination condition of a single image, a statistical model is proposed to reconstruct the lighting subspace. For lighting adjustment of image sequences, an entropy-based optimization algorithm, which minimizes the difference between successive images, is introduced. The effectiveness of the proposed algorithms is evaluated on face detection, recognition, and tracking.

The volume continues with three chapters on facial expression recognition and modeling. In the chapter "Advances, Challenges, and Opportunities in Automatic Facial Expression Recognition," Martinez and Valstar discuss the state of the art in automatic facial expression recognition. The authors first present three approaches for modeling facial expressions that are commonly adopted by the researchers. They then review the existing techniques used in subsequent stages of the standard algorithmic pipeline applied to detect and analyze facial expressions. Finally, the authors outline the crucial challenges and opportunities in the field, taking into account various aspects concerned with psychology, computer vision, and machine learning.

In the chapter "Exaggeration Quantified: An Intensity-Based Analysis of Posed Facial Expressions," Bhaskar et al. report their study on detecting posed facial expressions. The authors outline the importance of differentiating between posed and spontaneous expressions and present a new two-stage method to solve this problem. This method first determines the expression type and then classifies it as posed or spontaneous using a support vector machine.

In the chapter "Method of Modelling Facial Action Units Using Partial Differential Equations," Ugail and Ismail et al. develop a method for modeling action units to represent human facial expressions in three dimensions using biharmonic partial differential equations. The authors represent the action units in terms of Fourier coefficients related to the boundary curves, which allows the storage of both the face and facial expressions in highly compressed form.

The volume continues with four chapters on face recognition. In the chapter "Trends in Machine and Human Face Recognition," Mandal et al. present the state of the art in automatic machine face identification, as well as the advances in the psychological research on recognizing faces by humans. The chapter contains an interesting discussion on the most important factors that contribute to the interclass variation of human faces, such as lighting direction and facial expression, that often complicate the design of face recognition systems. The authors briefly review the history of face recognition algorithms, evaluation methodologies, and benchmarks. They also outline the current trends, including hot research topics as well as emerging benchmarks. Finally, the chapter contains an inspiring discussion concerning the psychological findings on human perception of faces, which contrasts the challenges of computer vision with those that humans encounter in recognizing faces.

In the chapter "Labeled Faces in the Wild: A Survey of Papers Published on the Benchmark," Learned-Miller et al. provide a detailed summary of the exciting research on face recognition that has been conducted on their Labeled Faces in the Wild (LFW) database since its publication in 2007. It is fascinating to witness how researchers have made progress on this challenging benchmark over the years, reaching an impressive accuracy of 99.63 % in 2015. The authors also discuss emerging databases and benchmarks, which will hopefully allow the face recognition community to identify new challenges and eventually improve the state of the art.

In the chapter "Reference-Based Pose-Robust Face Recognition," Kafai et al. introduce a novel reference-based face recognition framework. This framework involves the creation of a reference-based descriptor for the probe and gallery images by comparing them to a reference set, rather than comparing the probe and gallery images directly. The proposed framework is used in conjunction with various descriptors (e.g., the local binary patterns) and compared against several state-of-the-art face recognition algorithms on various public databases.

In the chapter "On the Importance of Frame Selection for Video Face Recognition," Dhamecha et al. discuss the importance of frame selection in video face recognition, provide a brief survey of existing techniques, and present an entropy-based frame selection algorithm. The authors demonstrate the performance of the proposed algorithm on the Point-and-Shoot-Challenge database.

The volume concludes with four chapters on applications of face analysis. In the chapter "Modeling of Facial Wrinkles for Applications in Computer Vision," Batool and Chellappa address the analysis and modeling of aging human faces. The authors focus on facial wrinkles, classified as subtle discontinuities or cracks in surrounding inhomogeneous skin texture. They review image features that can be used to capture the intensity gradients caused by facial wrinkles in the context of applications such as age estimation, facial expression analysis, and localization or detection of wrinkles and their removal for facial image retouching.

In the chapter "Communication-Aid System Using Eye-Gaze and Blink Information," Abe et al. present a system based on a personal computer and home video camera, designed to track user's eye gaze and blinks allowing for human-machine interaction. The authors also describe a method for the classification of eye blink types, which enables the detection of voluntary and involuntary blinks, allowing for efficient interaction with the computer for people with severe physical disabilities. The communication-aid systems in which the proposed solutions can be utilized are also discussed.

In the chapter "The Utility of Facial Analysis Algorithms in Detecting Melancholia," Hyett et al. present an overview of the image processing/analysis algorithms for the detection and diagnosis of depressive disorders. The authors focus particularly on the differentiation of melancholia from the other non-melancholic conditions.

In the chapter "Visual Speech Feature Representations: Recent Advances," Sui et al. summarize the latest research in the field of visual speech recognition (VSR), which has emerged to provide an alternative solution to improve speech recognition performance. VSR systems can be applied in noisy environments, where the speech recognition performance can be boosted through the fusion of audio and visual modalities. The authors survey the most recent advances in the field of visual feature extraction and also discuss future research directions.

A chapter entitled "Extended Eye Landmarks Detection for Emerging Applications," by Florea et al., completes the volume. The authors investigate the problem of eye landmark localization and gaze direction recognition in the context of eye tracking applications. Various approaches are described and extensively tested on several image databases and compared against state-of-the-art methods.

We hope that this volume, focused on face detection and analysis, will demonstrate the significant progress that has occurred in this field in recent years. We also hope that the developments reported in this volume will motivate further research in this exciting field.

Gliwice, Poland   Michal Kawulok
Shreveport, LA, USA   M. Emre Celebi
Gliwice, Poland   Bogdan Smolka

Contents

A Deep Learning Approach to Joint Face Detection and Segmentation

Khoa Luu, Chenchen Zhu, Chandrasekhar Bhagavatula, T. Hoang Ngan Le, and Marios Savvides

Abstract Robust face detection and facial segmentation are crucial pre-processing steps to support facial recognition, expression analysis, pose estimation, building of 3D facial models, etc. In previous approaches, the process of face detection and facial segmentation are usually implemented as sequential, mostly separated modules. In these methods, face detection algorithms are usually first implemented so that facial regions can be located in given images. Segmentation algorithms are then carried out to find the facial boundaries and other facial features, such as the eyebrows, eyes, nose, mouth, etc. However, both of these tasks are challenging due to numerous variations of face images in the wild, e.g. facial expressions, illumination variations, occlusions, resolution, etc. In this chapter, we present a novel approach to detect human faces and segment facial features from given images simultaneously. Our proposed approach performs accurate facial feature segmentation and demonstrates its effectiveness on images from two challenging face databases, i.e. Multiple Biometric Grand Challenge (MBGC) and Labeled Faces in the Wild (LFW).

1 Introduction

The problem of facial segmentation has been intensely studied for decades with the aim of ensuring generalization of an algorithm to unseen images. However, a robust solution has not yet been developed due to a number of challenging properties of the problem. For example, facial pose varies in captured images, illumination changes over time, different cameras have different outputs for an identical object, movement of objects cause blurring of colors, skin tones vary dramatically across individuals and ethnicities, and some background objects' color is similar to human facial skin. In this chapter, we present a novel approach based on deep learning framework to simultaneously detect human faces and segment facial features from

K. Luu (✉) • C. Zhu • C. Bhagavatula • T.H.N. Le • M. Savvides
CyLab Biometrics Center, Department of Electrical & Computer Engineering,
Carnegie Mellon University, Pittsburgh, PA, USA
e-mail: kluu@andrew.cmu.edu; chenchez@andrew.cmu.edu; cbhagava@andrew.cmu.edu;
thihoanl@andrew.cmu.edu; msavvid@ri.cmu.edu

© Springer International Publishing Switzerland 2016
M. Kawulok et al. (eds.), *Advances in Face Detection and Facial Image Analysis*,
DOI 10.1007/978-3-319-25958-1_1

1

Fig. 1 Joint face detection and segmentation using our proposed algorithm on an example form the LFW database

given digital images. Instead of focusing on general object detection in natural images [1], our method aims at building a robust system for face detection and facial feature segmentation. Our proposed method takes the advantages of the Multiscale Combinatorial Grouping (MCG) algorithm [2] and the deep learning Caffe framework [3] to robustly detect human faces and locate facial features. One-class Support Vector Machines (OCSVM) [4] are developed in the later steps to verify human facial regions. Finally, the region refinement technique [1] and the Modified Active Shape Models (MASM) [5] method are used in the post-processing steps to refine the segmented regions and cluster facial features resulting in bounding boxes and segmentations as shown in Fig. 1.

Compared to FaceCut [6], an automatic facial feature extraction method, our proposed approach contains several critical improvements. Firstly, instead of using an off-the-shelf commercial face detection engine as in [6], our method robustly performs face detection and facial boundary segmentation at the same time. Secondly, instead of using a color-based GrowCut approach [7] that is unable to deal with grayscale and illuminated images, our method robustly works with both grayscale and color images captured in various resolutions. In addition, since our method is implemented using a deep learning framework, it is able to learn features from both human faces and non-facial background regions. Therefore, our method is able to robustly detect human faces and segment facial boundaries in various challenging conditions. It also is able to deal with facial images in various poses, expressions, and occlusions. Our approach is evaluated on the NIST Multiple Biometric Grand Challenge (MBGC) 2008 still face challenge database [8] to evaluate its ability to deal with varying illuminations and slight pose variation as well as on the challenging Labeled Faces in the Wild (LFW) database [9].

The rest of this chapter is organized as follows. In Sect. 2, we review prior work on graph cuts based approaches to image segmentation and face detection methods. In Sect. 3, we present our proposed approach to simultaneously detect human faces and facial boundaries from given input images. Section 4 presents experimental results obtained using our proposed approach on two challenging databases, MBGC still face challenge database and the LFW database. Finally, our conclusions on this work are presented in Sect. 5.

2 Related Work

In this section, we describe prior graph cuts-based image segmentation algorithms and previous face detection approaches.

2.1 Image Segmentation

Recent development in algorithms such as graph cuts [10], GrabCut [11], GrowCut [7] and their improvements have shown accurate segmentation results in natural images. However, when dealing with facial images, it is hard to determine the weak boundary between the face and the neck regions by using conventional graph cuts based algorithms. This can be seen in Fig. 2, which shows an incorrectly segmented face obtained using the classical GrowCut algorithm. It is also to be noted that GrowCut requires the manually marking of points in the foreground and background and that it can't separately segment facial components such as eyes, nose and mouth.

In the graph cuts algorithm [10], input images are treated as graphs and their pixels as nodes of the graphs. In order to segment an object, max-flow/min-cut algorithms are used. For the rest of this chapter we refer to this original approach to image segmentation as the GraphCuts method. The GrabCut [11] algorithm was later introduced to improve the GraphCuts method by using an iterative segmentation scheme and applying graph cuts at intermediate steps. The input provided by the user consists of a rectangular box around the object to be segmented. The segmentation process is employed using the color statistical information inside and outside the box. The image graph is re-weighted and the new segmentation in each step is refined by using graph cuts.

Grady [12] presented a new approach to image segmentation using random walks. This approach also requires initial seed labels from the user but allows the use of more than two labels. A random walker starting at each unlabeled pixel first reach a pre-labeled pixel and its analytical probabilities are calculated. Then, image segmentation is carried out by assigning each unlabeled pixel the label for which the greatest probability is calculated.

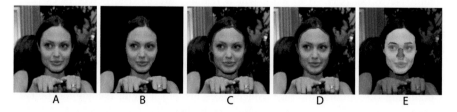

Fig. 2 Comparisons of facial segmentation algorithms: (**a**) original image, (**b**) face segmentation using color based information, (**c**) face segmentation results obtained using GrowCut, (**d**) automatic segmentation of facial features using our modified GrowCut algorithm and statistical skin information, (**e**) facial features segmentation using our proposed algorithm

2.2   Face Detection

Face detection has been a well studied area of computer vision. One of the first well performing approaches to the problem was the Viola-Jones face detector [13]. It was capable of performing real time face detection using a cascade of boosted simple Haar classifiers. The concepts of boosting and using simple features has been the basis for many different approaches [14] since the Viola-Jones face detector. These early detectors tended to work well on frontal face images but not very well on faces in different poses. As time has passed, many of these methods have been able to deal with off-angle face detection by utilizing multiple models for the various poses of the face. This increases the model size but does afford more practical uses of the methods. Some approaches have moved away from the idea of simple features but continued to use the boosted learning framework. Li and Zhang [15] used SURF cascades for general object detection but also showed good results on face detection.

More recent work on face detection has tended to focus on using different models such as a Deformable Parts Model (DPM) [16, 17]. Zhu and Ramanan's work was an interesting approach to the problem of face detection in that they combined the problems of face detection, pose estimation, and facial landmarking into one framework. By utilizing all three aspects in one framework, they were able to outperform the state-of-the-art at the time on real world images. Yu et al. [18] extended this work by incorporating group sparsity in learning which landmarks are the most salient for face detection as well as incorporating 3D models of the landmarks in order to deal with pose. Chen et al. [19] have combined ideas from both of these approaches by utilizing a cascade detection framework while synchronously localizing features on the face for alignment of the detectors. Similarly, Ghiasi and Fowlkes [20] have been able to use hierarchical DPMs not only to achieve good face detection in the presence of occlusion but also landmark localization. However, Mathias et al. [21] were able to show that both DPM models and rigid template detectors similar to the Viola-Jones detector have a lot of potential that has not been adequately explored. By retraining these models with appropriately controlled training data, they were able to create face detectors that perform similarly to other, more complex state-of-the-art face detectors.

All of these approaches to face detection were based on selecting a feature extractor beforehand. However, there has been work done in using a Convolutional Neural Networks (CNNs) to learn which features are used to detect faces [22]. Neural Networks have been around for a long time but have been experiencing a resurgence in popularity due to hardware improvements and new techniques resulting in the capability to train these networks on large amounts of training data.

Our approach is to use a CNN to extract features useful for detecting faces as well with one key difference. In all of the prior approaches, the face is detected as a bounding box around the face. Since faces are not a rigid, rectangular object, a bounding box will necessarily include more than the face. Ideally, a detector would be able to give a pixel level segmentation of the face which could be used to ensure only features from the face are extracted for any problems that use face detection. Some of the works comes close to determining a tighter bound on the face detection by finding landmarks on the face including a facial boundary. Another work similar to our approach presented in [1]. The difference is that our method aims at working on robust face detection and facial feature segmentation including subdividing the face into smaller semantic regions. In addition, since there is only one class in our defined system, i.e. the face region, instead of multiple classes, the One-Class Support Vector Machines (OCSVM) [4, 23] are a good fit in our problem. Our approach uses CNNs to determine a pixel level segmentation of the faces and use these to generate a bounding box for measuring performance against other methods.

3 Deep Learning Approach to Face Detection and Facial Segmentation

In this section, we present our proposed deep learning approach to detect human faces and segment features simultaneously. We first review the CNNs and its very fast implementation in the GPU-based Caffe framework [3]. Then, we will present our proposed approach in the second part of this section.

3.1 Deep Learning Framework

Convolutional Neural Networks are biologically inspired variants of multilayer perceptrons. The CNN method and its extensions, i.e. LeNet-5 [24], HMAX [25], etc., simulate the animal visual cortex system that contains a complex arrangement of cells sensitive to receptive fields. In this model, designed filters are treated as human visual cells in order to explore spatially local correlations in natural images. It efficiently presents the sparse connectivity and the shared weights since kernel filters are replicated over the entire image with the same parameters in each layer. The pooling step, a form of down-sampling, plays a key role in CNNs. Max-pooling is a popular pooling method for object detection and classification since max-pooling reduces computation for upper layers by eliminating non-maximal values and provides a small amount of translation invariance in each level.

Although CNNs can explore deep features, they are computationally very expensive. The algorithm runs faster when implemented in a Graphics Processing Unit (GPU). The Caffe framework [3] is a rapid deep learning implementation using CUDA C++ for GPU computation. It also supports bindings to Python/Numpy and MATLAB. It can be used as an off-the-shelf deployment of state-of-the-art models. In our proposed system, the AlexNet [26] network is implemented in Caffe and is employed to extract features from candidate regions. Given a candidate region, two 4096-dimensional feature vectors are extracted by feeding two types of region images into the AlexNet and the two vectors are concatenated to form the final feature vector. The AlexNet has 5 convolution layers and 2 fully connected layers. All the activation functions are Rectified Linear Units(ReLU). Max pooling with a stride of 2 is applied at the first, second and fifth convolution layers. Two types of region images are cropped box and region foreground. They are both warped to a fixed 227×227 pixel size.

3.2   Our Proposed Approach

In our proposed method, the MCG algorithm is first employed to extract superpixel based regions. In each region, there are two kinds of features calculated and used as inputs into the GPU-based Caffe framework presented in Sect. 3.1. The OCSVM is then applied to verify facial locations. Finally, the post-processing steps, i.e. region refinement and MASM, are employed in the final steps.

Multiscale Combinatorial Grouping [2] is considered as one of the state-of-the-art approaches for bottom-up hierarchical image segmentation and object candidate generation with both high accuracy and low computational time compared against other methods. MCG computes a segmentation hierarchy at multiple image resolutions, which are then fused into a single multiscale hierarchy. Then candidates are produced by combinatorially grouping regions from all the single scale hierarchies and from the multiscale hierarchy. The candidates are ranked based on simple features such as size and location, shape and contour strength.

Our proposed approach first employs MCG on the input image $\mathbf{X}$ to extract N face candidate regions $\mathbf{x}_i, i = 1..N$. Then, features are extracted from two types of representations of the regions, i.e. the bounding box of the region with only the foreground $\mathbf{x}_i^F$ and the ones with background $\mathbf{x}_i^B$ as shown in Fig. 3. The first representation $\mathbf{x}_i^F$, i.e. the segmented region, is used to learn the facial information. Meanwhile, the second type $\mathbf{x}_i^B$, i.e. the bounding boxes, aims at learning relative information between face and their background information. In this way, the learning features include both human faces and common backgrounds. This approach helps our proposed system robust against various challenging conditions as presented in Sect. 1.

We fuse the two representations as inputs into the CNN to extract deep features $\mathbf{x}_i^D$. In the later step, One-class Support Vector Machines are developed to verify human facial regions. Finally, the region refinement technique and the Modified Active Shape Models are used in the post-processing steps to refine the segmented regions and cluster facial features.

3.2.1   One-Class Support Vector Machines (OCSVM)

Given a set of N feature vectors $\{\mathbf{x}_1^D, \ldots, \mathbf{x}_N^D\}$, where $\mathbf{x}_i^D \in \mathbb{R}^d$ are features extracted using deep learning, the OCSVM algorithm finds a set of parameters $\{\mathbf{x}_0, r\}$, where $\mathbf{x}_0 \in \mathbb{R}^d$ denotes the center position and $r \in \mathbb{R}$ is the radius. They describe the optimal sphere that contains the training data points while allowing some slack for errors. This process can be employed by solving the following minimization problem:

$$\Lambda(\mathbf{x}_0, r, \epsilon) = r^2 + C \sum_{i=1}^{N} \epsilon_i$$

$$s.t., \quad \|\mathbf{x}_i^D - \mathbf{x}_0\|^2 \leq r^2 + \epsilon_i, \forall i, \epsilon_i \geq 0 \tag{1}$$

where $C \in \mathbb{R}^n$ denotes the tradeoff between the volume of the description and the errors, and ϵ_i represents the slack variables to describe data points outside the hyper-sphere. The quadratic programming problem in Eq. (1) can be mathematically solved by using Lagrange Multiplies and setting partial derivatives to zeros as follows:

$$L(\mathbf{x}_0, r, \epsilon, \alpha, \gamma) = r^2 + C \sum_{i=1}^{N} \epsilon_i$$

$$- \sum_i \alpha_i \{ r^2 + \epsilon_i - (\mathbf{x}_i \cdot \mathbf{x}_i - 2\mathbf{x}_0 \cdot \mathbf{x}_i + \mathbf{x}_0 \cdot \mathbf{x}_0) \}$$

$$- \sum_i \gamma_i \epsilon_i \tag{2}$$

L has to be minimized with respect to $\mathbf{x}_0, \mathbf{r}$ and ϵ, and maximized with respect to α, γ.

The distance from a point $\mathbf{z}$ to center of hyper-sphere $\mathbf{x}_0$ is defined as:

$$\|\mathbf{z} - \mathbf{x}_0\|^2 = (\mathbf{z} \cdot \mathbf{z}) - 2 \sum_i \alpha_i (\mathbf{z} - \mathbf{x}_i) + \sum_{i,j} \alpha_i \alpha_j (\mathbf{x}_i \cdot \mathbf{x}_j) \tag{3}$$

The decision rule in our One-class Support Vector Machines approach is computed using the regular ℓ_2-norm distance between the center $\mathbf{x}_0$ and a new point $\mathbf{z}$ as follows:

$$f(\mathbf{z}) = \begin{cases} 1, & \text{if} \|\mathbf{z} - \mathbf{x}_0\|^2 \leq r^2 \\ -1, & \text{otherwise} \end{cases} \tag{4}$$

In Eq. (4), when the computed ℓ_2-norm distance between $\mathbf{x}_0$ and $\mathbf{z}$ is smaller or equal to the radius r, the given point $\mathbf{z}$ will be considered as the target.

Similar to Support Vector Machines (SVMs), OCSVM can be presented in a kernel domain by substituting the inner products of $\mathbf{x}_i^D$ and $\mathbf{x}_j^D$, where $i, j = 1..N$, with a kernel function $K(\mathbf{x}_i, \mathbf{x}_j)$ to enhance the learning space:

$$K(\mathbf{x}_i, \mathbf{x}_j) = \exp(-\mathbf{x}_i - \mathbf{x}_j)^2 / s^2 \tag{5}$$

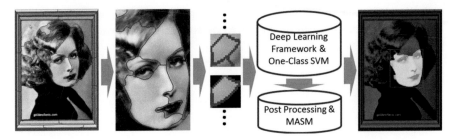

Fig. 3 Our proposed framework for face detection and facial segmentation simultaneously

where s denotes the standard deviation. Then, given a region feature $\mathbf{z} \in \mathbb{R}^d$, its distance to center $\mathbf{x}_0$ can be computed as:

$$\|\mathbf{z} - \mathbf{x}_0\|^2 = K(\mathbf{z}, \mathbf{z}) - 2 \sum_i \alpha_i K(\mathbf{z}, \mathbf{x}_i) + \sum_{i,j} \alpha_i \alpha_j K(\mathbf{x}_i, \mathbf{x}_j) \tag{6}$$

Our work aims at finding facial region detection as an one-class pattern problem. Therefore, OCSVM is employed to model deep learning features. The learned facial features boundary is more accurate when adding some negative samples during training steps as presented in [4]. Figure 3 illustrates our proposed approach to joint face detection and facial feature segmentation.

3.2.2 Post-processing Steps

Non-maximum suppression (NMS) [1] is employed on the scored candidates to refine segmented regions. Then, the Modified Active Shape Model (MASM) technique is used to refine the locations of facial feature components in the final step. The deep learning features extracted from CNNs are categorized to produce category-specific coarse mask predictions to refine the remaining candidates. Combining this mask with the original region candidates provides a further boost.

Though there are many approaches to landmark localization on face images, many of them use a 68 point landmarking scheme. Some well known methods such as Supervised Descent Method (SDM) by Xiong and De la Torre [27] and the work by Yu et al. [18] are such examples. However, with the 68 point scheme defined by these methods, segmenting certain regions of the face becomes much more difficult. Particularly the nose region as the landmarks do not give any estimate as to the boundary of the nose. For this reason, we have chosen to use the Modified Active Shape Model [5] method for the task of automatic facial landmark localization. Even though MASM may not perform as well at localization of some points, the additional points in the landmarking scheme are a necessity to the final semantic segmentation of the face region. However, MASM is a PCA based approach and does not always accurately localize landmarks in faces with shapes radically different from those exhibited by faces in the training set. Our proposed method can overcome the shortcomings of ASM based approaches in order to ensure that the facial region and individual facial features are segmented accurately as possible. Instead of using commercial face detection engines, our approach robustly finds facial locations that are used as inputs for MASMs. More importantly, the segmented facial boundaries extracted using the OCSVMs and the deep features are used to refine the facial feature regions.

4 Our Experimental Results

Our proposed models are trained on images of the CMU Multi-PIE database [28]. The OCSVM is trained on images drawn from various facial angles so that the face detection and segmentation is robust against off-angle faces. The MASM component is trained from images drawn from the frontal view set of the database.

Our approach is tested on 500 random images from the MBGC database to evaluate its robustness to varying illumination and in-plane rotation of faces. As can be seen from Table 1, which shows the mean and standard deviation of the point to point landmark fitting errors (for 79 landmarks) obtained by MASM, FaceCut and our proposed approach across all MBGC test images when compared to manually annotated images (ground truths), the results using our approach in improved fitting accuracy of the facial boundary landmarks. Our approach is also tested on images from the LFW database and was again able to accurately segment facial features in these challenging everyday images. Figure 4 shows sample detection and segmentation results using our approach on images from LFW databases respectively and demonstrates the effectiveness of the algorithm.

5 Conclusions

This paper has presented a novel approach to automatically detect and segment human facial features. The method combines the positive features of the deep learning method and the Modified Active Shape Model to ensure highly accurate

Table 1 Comparison of the mean (in pixels) and standard deviation (in pixels) of the fitting error across landmarks produced by MASM and our FaceCut approach on images from the MBGC database

Algorithms	Mean (all landmarks)	Std. dev. (all landmarks)	Mean (facial boundary landmarks)	Std. dev. (facial boundary landmarks)
MASM [5]	9.65	11.66	14.85	12.21
FaceCut [6]	9.56	11.51	14.46	11.59
Our approach	9.47	11.20	14.24	11.12

Fig. 4 The results using our proposed approach on the LFW database: the *first column*: original image, the *second column*: face detection results, the *third column*: facial features segmented

and completely automatic segmentation of facial features and the facial boundary. The effectiveness of detection and segmentation using our approach is demonstrated on unseen test images from the challenging MBGC and LFW databases.

References

1. B. Hariharan, P. Arbeláez, R. Girshick, J. Malik, Simultaneous detection and segmentation, in *European Conference on Computer Vision (ECCV)* (2014)
2. P. Arbelaez, J. Pont-Tuset, J. Barron, F. Marques, J. Malik, Multiscale combinatorial grouping, in *Proceedings of the IEEE International Conference on Computer Vision and Pattern Recognition (CVPR)* (2014)
3. Y. Jia, E. Shelhamer, J. Donahue, S. Karayev, J. Long, R. Girshick, S. Guadarrama, T. Darrell, *Caffe: Convolutional Architecture for Fast Feature Embedding* (2014). arXiv preprint arXiv:1408.5093
4. D.M.J. Tax, One-class classification, Ph.D. thesis, Delft University of Technology, June 2001
5. K. Seshadri, M. Savvides, Robust modified active shape model for automatic facial landmark annotation of frontal faces, in *Proceedings of the 3rd IEEE International Conference on Biometrics: Theory, Applications and Systems (BTAS)* (2009), pp. 319–326
6. K. Luu, T.H.N. Le, K. Seshadri, M. Savvides, Facecut - a robust approach for facial feature segmentation, in *ICIP* (2012), pp. 1841–1848
7. V. Vezhnevets, V. Konouchine, "GrowCut" - interactive multi-label N-D image segmentation by cellular automata, in *Proceedings of the International Conference on Computer Graphics and Vision (GRAPHICON)* (2005)
8. P.J. Phillips, P.J. Flynn, J.R. Beveridge, W.T. Scrugs, A.J. O'Toole, D. Bolme, K.W. Bowyer, B.A. Draper, G.H. Givens, Y.M. Lui, H. Sahibzada, Joseph A. Scallan III, S. Weimer, Overview of the multiple biometrics grand challenge, in *Proceedings of the 3rd IAPR/IEEE International Conference on Biometrics* (2009), pp. 705–714
9. P.N Belhumeur, D.W. Jacobs, D Kriegman, N. Kumar, Localizing parts of faces using a consensus of exemplars, in *Computer Vision and Pattern Recognition (CVPR)* (2011), pp. 545–552
10. Y.Y. Boykov, M.P. Jolly, Interactive graph cuts for optimal boundary and region segmentation of objects in N-D images, in *Proceedings of the IEEE International Conference on Computer Vision (ICCV)* (2001), pp. 105–112
11. C. Rother, V. Kolmogorov, A. Blake, "GrabCut": interactive foreground extraction using iterated graph cuts, in *Proceedings of ACM Transactions on Graphics (SIGGRAPH)* (2004), pp. 309–314
12. L. Grady, Random walks for image segmentation. IEEE Trans. Pattern Anal. Mach. Intell. (TPAMI) **28**(11), 1768–1783 (2006)
13. P. Viola, M. Jones, Robust real-time face detection. Int. J. Comput. Vis. **57**, 137–154 (2004)
14. C. Zhang, Z. Zhang, A survey of recent advances in face detection. Tech. Rep. MSR-TR-2010-66, June 2010
15. J. Li, Y. Zhang, Learning surf cascade for fast and accurate object detection, in *2013 IEEE Conference on Computer Vision and Pattern Recognition (CVPR)* (2013), pp. 3468–3475
16. X. Zhu, D. Ramanan, Face detection, pose estimation, and landmark localization in the wild, in *2012 IEEE Conference on Computer Vision and Pattern Recognition (CVPR)* (2012), pp. 2879–2886
17. P.F. Felzenszwalb, R.B. Girshick, D. McAllester, D. Ramanan, Object detection with discriminatively trained part-based models. IEEE Trans. Pattern Anal. Mach. Intell. **32**(9), 1627–1645 (2010)

18. X. Yu, J. Huang, S. Zhang, W. Yan, D.N. Metaxas, Pose-free facial landmark fitting via optimized part mixtures and cascaded deformable shape model, in *2013 IEEE International Conference on Computer Vision (ICCV)* (2013), pp. 1944–1951
19. D. Chen, S. Ren, Y. Wei, X. Cao, J. Sun, Joint cascade face detection and alignment, in *Computer Vision ECCV 2014*, eds. by D. Fleet, T. Pajdla, B. Schiele, T. Tuytelaars. Lecture Notes in Computer Science, vol. 8694 (Springer International, New York, 2014), pp. 109–122
20. G. Ghiasi, C. Fowlkes, Occlusion coherence: localizing occluded faces with a hierarchical deformable part model, in *CVPR* (2014)
21. M. Mathias, R. Benenson, M. Pedersoli, L. Van Gool, Face detection without bells and whistles, in *Computer Vision ECCV 2014*, eds. by D. Fleet, T. Pajdla, B. Schiele, T. Tuytelaars. Lecture Notes in Computer Science, vol. 8692 (Springer International, New York, 2014), pp. 720–735
22. C. Garcia, M. Delakis, Convolutional face finder: a neural architecture for fast and robust face detection. IEEE Trans. Pattern Anal. Mach. Intell. **26**(11), 1408–1423 (2004)
23. H. Jin, Q. Lin, H. Lu, X. Tong, Face detection using one-class SVM in color images, in *Proceedings of the International Conference on Signal Processing* (2004)
24. Y. LeCun, L. Bottou, Y. Bengio, P. Haffner, Gradient-based learning applied to document recognition. Proc. IEEE **86**(11), 2278–2324 (1998)
25. T. Serre, L. Wolf, S. Bileschi, M. Riesenhuber, T. Poggio, Robust object recognition with cortex-like mechanisms. IEEE Trans. Pattern Anal. Mach. Intell. **29**(3), 411–426 (2007)
26. A. Krizhevsky, I. Sutskever, G.E. Hinton, Imagenet classification with deep convolutional neural networks, in *NIPS* (2012), pp. 1097–1105
27. X. Xiong, F. de la Torre, Supervised descent method and its applications to face alignment, in *2013 IEEE Conference on Computer Vision and Pattern Recognition (CVPR)*, June 2013, pp. 532–539
28. R. Gross, I. Matthews, J.F. Cohn, T. Kanade, S. Baker, Multi-PIE, in *Proceedings of the 8th IEEE International Conference on Automatic Face and Gesture Recognition (FG 2008)* (2008), pp. 1–8

Face Detection Coupling Texture, Color and Depth Data

Loris Nanni, Alessandra Lumini, Ludovico Minto, and Pietro Zanuttigh

Abstract In this chapter, we propose an ensemble of face detectors for maximizing the number of true positives found by the system. Unfortunately, combining different face detectors increases both the number of true positives and false positives. To overcome this difficulty, several methods for reducing false positives are tested and proposed. The different filtering steps are based on the characteristics of the depth map related to the subwindows of the whole image that contain the candidate faces. The most simple and easiest criteria to use, for instance, is to filter the candidate face region by considering its size in metric units.

The experimental section demonstrates that the proposed set of filtering steps greatly reduces the number of false positives without decreasing the detection rate. The proposed approach has been validated on a dataset of 549 images (each including both 2D and depth data) representing 614 upright frontal faces. The images were acquired both outdoors and indoors, with both first and second generation Kinect sensors. This was done in order to simulate a real application scenario. Moreover, for further validation and comparison with the state-of-the-art, our ensemble of face detectors is tested on the widely used BioID dataset where it obtains 100 % detection rate with an acceptable number of false positives.

A MATLAB version of the filtering steps and the dataset used in this paper will be freely available from http://www.dei.unipd.it/node/2357.

1 Introduction

The goal of face detection is to determine the location of faces in an image. It is one of the most studied problems in computer vision, due partly to the large number of applications requiring the detection and recognition of human beings and the availability of low-cost hardware. Face detection has also attracted a lot of

L. Nanni (✉) • L. Minto • P. Zanuttigh
DEI, University of Padova, Via Gradenigo 6, 35131 Padova, Italy
e-mail: nanni@dei.unipd.it; mintolud@dei.unipd.it; zanuttigh@dei.unipd.it

A. Lumini
DISI, Università di Bologna, Via Sacchi 3, 47521 Cesena, Italy
e-mail: alessandra.lumini@unibo.it

© Springer International Publishing Switzerland 2016
M. Kawulok et al. (eds.), *Advances in Face Detection and Facial Image Analysis*,
DOI 10.1007/978-3-319-25958-1_2

attention in the research community because it is a very hard problem, certainly more challenging than face localization in which a single face is assumed to be located inside an image [1]. Although human faces have generally the same appearance, several personal variations (like gender, race, individual distinctions, and facial expression) and environment conditions (like pose, illumination, and complex background) can dramatically alter the appearance of human faces. A robust face detection system must overcome all these variations and be able to perform detection in almost any lighting condition. Moreover, it must manage to do all this in real-time.

Over the past 25 years, many different face detection techniques have been proposed [2], motivated by the increasing number of real world applications requiring recognition of human beings. Indeed, face detection is a crucial first step for several applications ranging from surveillance and security systems to human–computer interface interaction, face tagging, behavioral analysis, as well as many other applications [3].

The majority of existing techniques address face detection from a monocular image or a video-centric perspective. Most algorithms are designed to detect faces using one or more camera images, without additional sensor information or context. The problem is often formulated as a two-class pattern recognition problem aimed at classifying each subwindow of the input image as either containing or not containing a face [4].

The most famous approach for frontal 2D detection is the Viola–Jones algorithm [5], which introduced the idea of performing an exhaustive search of an image using Haar-like rectangle features and then of using Adaboost and Cascade algorithm for classification. The importance of this detector, which provides high speed, can be measured by the number of approaches it has inspired, such as [6–9]. Amongst them, SURF cascades, a framework recently introduced by Intel labs [10] that adopts multi-dimensional SURF features instead of single dimensional Haar features to describe local patches, is one of the top performers. Another recent work [11] that compared many commercial face detectors (Google Picasa, Face.com acquired by Facebook, Intel Olaworks, and the Chinese start-up Face++) showed that a simple vanilla deformable part model (a general purpose object detection approach which combines the estimation of latent variables for alignment and clustering at training time and the use of multiple components and deformable parts to handle intra-class variance) was able to outperform all the other methods in face detection.

The Viola–Jones algorithm and its variants are capable of detecting faces in images in real-time, but these algorithms are definitely affected by changeable factors such as pose, illumination, facial expression, glasses, makeup, and factors related to age. In order to overcome problems related to these factors, 3D face detection methods have been proposed. These new methods take advantage of the fact that the 3D structure of the human face provides highly discriminatory information and is more insensitive to environmental conditions.

The recent introduction of several consumer depth cameras has made 3D acquisition available to the mass market. Among the various consumer depth cameras, Microsoft Kinect is the first and the most successful device. It is a depth sensing device that couples the 2D RGB image with a depth map (RGB-D) computed using the structured light principle which can be used to determine the depth of every object in the scene. A second generation of the sensor exploiting the Time-of-flight principle has been recently introduced. The depth information is not enough precise to differentiate among different individuals, but can be useful to improve the robustness of a face detector.

Because each pixel in Kinect's depth map indicates the distance of that pixel from the sensor, this information can be used both to differentiate among different individuals at different distances and to reduce the sensitivities of the face detector to illumination, occlusions, changes in facial expression, and pose. Several recent approaches have used depth maps or other 3D information for face detection and several have been tested on the first benchmark datasets collected by Kinetic devices for 3D face recognition [12] and detection [13]. Most exiting 3D approaches use depth images combined with gray-level images to improve detection rates. In [14] Haar wavelets on 2D images are first used to detect the human face, and then face position is refined by structured light analysis. In [15] depth-comparison features are defined as pixel pairs in depth images to quickly and accurately classify body joints and parts from single depth images. In [16] a similar method for robust and accurate face detection based on square regions comparison is coupled with Viola Jones face detector. In [1, 17] depth information is used to reduce the number of false positive and improve the percentage of correct detection. In [18] biologically inspired integrated representation of texture and stereo disparity information are used to reduce the number of locations to be evaluated during the face search. In [19] the additional information obtained by the depth map improves face recognition rates. This latter method involves texture descriptors extracted both from color and depth information and classification based on random forest. Recently, 3D information is used by DeepFace [20] to perform a 3D model-based alignment that is coupled with large capacity feedforward models for effectively obtaining a high detection rate.

An improved face detection approach based on information of the 2D image and the depth obtained by Microsoft Kinect 1 and Kinect 2 is proposed in this paper.

The proposed method in this chapter is based on an ensemble of face detectors. One advantage of using an ensemble to detect faces is that it maximizes the number of true positives; a major disadvantage, however, is that it increases both the number of false positives and the computation time. The main aim of this work, an update of our previous paper [1], is to propose a set of filtering step for reducing the number of false positives while preserving the true positive rate. To achieve this goal, the following approaches:

- SIZE: the size of the candidate face region is calculated according to the depth data, removing faces that are the too small or too large.
- STD: images of flat objects (e.g. candidate face found in a wall) or uneven objects (e.g. candidate face found in the leaves of a tree) are removed using the depth map and a segmentation approach based on the depth map.

- SEG: a segmentation step based on the depth map is used to segment a candidate image into homogenous regions; images whose main region is smaller than a threshold (with respect to the candidate image dimension) are then filtered out.
- ELL: using the segmented depth-image, an ellipse fitting approach is employed to evaluate whether the larger region can be modeled as an ellipse. The fitting cost is evaluated to decide whether or not to remove the candidate face.
- EYE: an eye detection step is used to find eyes in the candidate image and reject regions having a very low eye detection score.
- SEC: a neighborhood of the candidate face region is considered in the depth map. Neighbor pixels whose depth value is close to the face mean depth are assigned to a number of radial sectors. The lower sectors should contain a higher number of pixels.

The proposed approach has been validated on a dataset composed by 549 samples (containing 614 upright frontal faces) that include both 2D and depth images. The experimental results prove that the proposed filtering steps greatly reduce the number of false positives without decreasing the detection rate. For a further validation, the ensemble of face detectors is also tested on the widely used BioID dataset [21], where it obtains 100 % detection rate with an acceptable number of false positives. We want to stress that our approach outperforms the approach proposed in [22], which works better than such powerful commercial face detectors as Google Picasa, Face.com—Facebook, Intel Olaworks, and the Chinese start-up Face++.

The arrangement of this chapter is as follows. In Sect. 2 the whole detection approach is described, starting from the base detectors and moving on to explain all the filtering steps. In Sect. 3 the experiments on the above mentioned benchmark datasets are presented, including a description of the self-collected datasets, the definition of the testing protocol, and a discussion of the experimental results. Finally, in Sect. 4 the chapter is concluded and some future research directions are presented. The MATLAB code developed for this chapter and the datasets will be freely available.

2 The Proposed Approach

The proposed method is based on an ensemble of several well-known face detectors. As a first step we perform face detection on the color images using a low acceptance threshold in order to have high recall. This results in low precision since many false positive occur in the search. As a second step the depth map is aligned to the color image, and both are used to filter out false positives by means of several criteria designed to remove non-face images from the final list. In order to better handle non-upright faces, the input color images are also rotated {20°, −20°} before detection. In the experiments the use of rotated images for adding poses is denoted by a *.

We perform experiments on the fusion of four face detectors:

- ViolaJones(VJ) [5], which is probably the most diffused face detector due to its simplicity and very fast classification time. VJ uses simple image descriptors, based on Haar wavelets extracted in a low computational time from the integral image. Classification is performed using an ensemble of AdaBoost classifiers for selecting a small number of relevant descriptors, along with a cascade combination of weak learners for classification. This approach requires a long training time but it is very fast for testing. The precision of VJ strictly relies on the threshold σ used to classify a face an input subwindow.

- SN [23] is a face detector[1] based on local descriptors and Successive Mean Quantization Transform (SMQT) features that is applied to a Split up sparse Network of Winnows (SN) classifier. The face detector extracts SMQT features by a moving a patch of 32×32 pixels that is repeatedly downscaled and resized in order to find faces of different sizes. SMQT is a transform for automatic enhancement of gray-level images that reveals the structure of the data and removes properties such as gain and bias. As a result SMQT features overcome most of the illumination and noise problems. The detection task is performed by a Split up sparse Network of Winnows as the classifier, which is a sparse network of linear units over a feature space that can be used to create lookup-tables. SN precision in face detection can be adjusted by a sensitivity parameter σ that can be tuned to obtain low to high sensitivity values. In the original implementation $\sigma_{min} = 1$ and $\sigma_{max} = 10$.

- FL [22] is a method that combines an approach for face detection that is a modification of the standard Viola–Jones detection framework with a module for the localization of salient facial landmark points. The basic idea of this approach is to scan the image with a cascade of binary classifiers (a multi-scale sequence of regression tree-based estimators) at all reasonable positions and scales. An image region is classified as containing a face if it successfully passes all the classifiers. Then a similar ensemble is used to infer the position of each facial landmark point within a given face region. Each binary classifier consists of an ensemble of decision trees with pixel intensity comparisons as binary tests in their internal nodes. The learning process consists of a greedy regression tree construction procedure and a boosting algorithm. The reported results show performance improvement with respect to several recent commercial approaches.

- RF [24] is a face detector based on face fitting, which is the problem of modeling a face shape by inferring a set of parameters that control a facial deformable model. The method, named Discriminative Response Map Fitting (DRMF), is a novel discriminative regression approach for the Constrained Local Models (CLMs) framework which shows impressive performance in the generic face

---

[1] http://www.mathworks.com/matlabcentral/fileexchange/loadFile.do?
objectId = 13701&objectType = FILE.

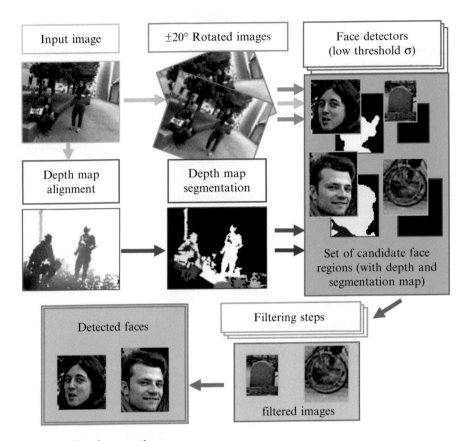

Fig. 1 Outline of our complete system

fitting scenario. RF precision can be adjusted by a sensitivity parameter σ, which can be tuned to obtain a lower or a higher sensitivity value.

In Fig. 1 a schema of our complete system is outlined. In the first step one or more face detectors (the final configuration is a result of the experimental section) are employed for an "imprecise" detection using a low acceptance threshold, then in the second step all the candidate face regions are filtered out according to several criteria (detailed below in the following subsections) that take advantage of the presence of the depth map.

The second step exploits the information contained in the depth data to improve face detection. First calibration between color and depth data is computed according to the method proposed in [25]: the positions of the depth samples in the 3D space are first computed using the intrinsic parameters of the depth camera and then reprojected in the 2D space using both the color camera intrinsic parameters and the extrinsic ones between the two cameras. Then a color and a depth value are associated with each sample (to speed-up the approach, this operation can be

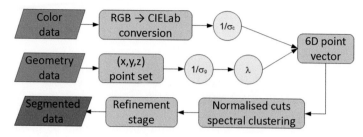

Fig. 2 Architecture of the proposed segmentation scheme

performed only for the regions containing a candidate face). Finally, filtering is applied in order to remove false positives from the set of candidate faces. In the next subsections, the depth map alignment and segmentation approach and all the filtering rules used in this work are detailed. In Fig. 4 some candidate images properly filtered out by the different filtering rules are shown.

2.1 Depth Map Alignment and Segmentation

The color image and the depth map are jointly segmented using the approach shown in Fig. 2. The employed approach associates to each sample a multi-dimensional vector and then clusters the set of vectors associated to the image using the Mean Shift algorithm [26] following an approach similar to [27].

As shown in Fig. 2 the procedure has two main stages: first a six-dimensional representation of the points in the scene is built from the geometry and color data and then second the obtained point set is segmented using Mean Shift clustering.

Each sample in the acquired depth map correspond to a 3D point of the scene p_i, $i = 1, \ldots, N$. After the joint calibration of the depth and color cameras, it is possible to reproject the depth samples over the corresponding pixels in the color image and to associate to each point the spatial coordinates x, y, and z of p_i and its R, G, and B color components. Notice that these two representations lie in completely different spaces and cannot be directly compared.

In order to obtain multi-dimensional vectors suited for the clustering algorithm, the various components need to be comparable. All color values are converted to the CIELAB perceptually uniform space. This provides a perceptual significance to the Euclidean distance between the color vectors that will be used in the clustering algorithm. We can denote the color information of each scene point in the CIELAB space with the 3-D vector:

$$ p_i^c = \begin{bmatrix} L(p_i) \\ a(p_i) \\ b(p_i) \end{bmatrix}, \ i = 1, \ldots, N $$

The geometry is instead simply represented by the 3-D coordinates of each point, i.e., by:

$$p_i^g = \begin{bmatrix} x(p_i) \\ y(p_i) \\ z(p_i) \end{bmatrix}, \; i = 1, \ldots, N$$

As previously noted the scene segmentation algorithm should be insensitive to the relative scaling of the point-cloud geometry and should bring geometry and color distances into a consistent framework. Therefore, all components of p_i^g are normalized with respect to the average of the standard deviations of the point coordinates in the three dimensions $\sigma_g = (\sigma_x + \sigma_y + \sigma_z)/3$, thus obtaining the vector:

$$\begin{bmatrix} \bar{x}(p_i) \\ \bar{y}(p_i) \\ \bar{z}(p_i) \end{bmatrix} = \frac{3}{\sigma_x + \sigma_y + \sigma_z} \begin{bmatrix} x(p_i) \\ y(p_i) \\ z(p_i) \end{bmatrix} = \frac{1}{\sigma_g} \begin{bmatrix} x(p_i) \\ y(p_i) \\ z(p_i) \end{bmatrix}$$

In order to balance the relevance of color and geometry in the merging process, the color information vectors are also normalized by the average of the standard deviations of the L, a, and b components. The final color representation, therefore, is:

$$\begin{bmatrix} \bar{L}(p_i) \\ \bar{a}(p_i) \\ \bar{b}(p_i) \end{bmatrix} = \frac{3}{\sigma_L + \sigma_a + \sigma_b} \begin{bmatrix} L(p_i) \\ a(p_i) \\ b(p_i) \end{bmatrix} = \frac{1}{\sigma_c} \begin{bmatrix} L(p_i) \\ a(p_i) \\ b(p_i) \end{bmatrix}$$

From the above normalized geometry and color information vectors, each point is finally represented as

$$p_i^f = \begin{bmatrix} \bar{L}(p_i) \\ \bar{a}(p_i) \\ \bar{b}(p_i) \\ \lambda \bar{x}(p_i) \\ \lambda \bar{y}(p_i) \\ \lambda \bar{z}(p_i) \end{bmatrix}$$

The parameter λ controls the contribution of color and geometry to the final segmentation. High values of λ increase the relevance of geometry, while low values of λ increase the relevance of color information. Notice that at the two extrema the algorithm can be reduced to a color-based segmentation ($\lambda = 0$) or to a geometry (depth) only segmentation $\left(\lambda \to \infty\right)$. A complete discussion on the effect of this parameter and a method to automatically tune it to the optimal value is presented in [27].

Fig. 3 Color image, depth map, and segmentation map

The computed vectors p_i^f are then clustered in order to segment the acquired scene. Mean shift clustering [26] has been used since it obtains an excellent trade-off between the segmentation accuracy and the computation and memory resources. A final refinement stage is also applied in order to remove regions smaller than a predefined threshold typically due to noise. In Fig. 3 an example of segmented image is reported.

2.2 Image Size Filter

The image size filter (SIZE) rejects candidate images according to their size. The size of a candidate face region is extracted from the depth map. Assuming that the face detection algorithm returns the 2D position and dimension in pixels (w_{2D}, h_{2D}) of a candidate face region, its 3D physical dimension in mm (w_{3D}, h_{3D}) can be estimated as:

$$w_{3D} = w_{2D} \frac{\bar{d}}{f_x}$$
$$h_{3D} = h_{2D} \frac{\bar{d}}{f_y}$$

where f_x and f_y are the Kinect camera focal lengths computed by the calibration algorithm in [25] and $\bar{d}$ is the average depth of the samples within the face candidate bounding box. Note how $\bar{d}$ is defined as the median of the depth samples; this is done in order to reduce the impact of noisy samples in the average computation.

The candidate regions out of a fixed range [0.075, 0.35] centimeters are rejected.

2.3 Flatness\Unevenness Filter

Another source of significant information that can be obtained from the depth map is the flatness\unevenness of the candidate face regions. For this filter a segmentation procedure is applied. Then from each face candidate region the standard deviation of the pixels of the depth map that belong to the larger segment is calculated. Regions having a standard deviation (STD) out of a fixed range [0.01, 2] are removed.

This method is slightly different to the one proposed in our previous work [1] (called STD° in the experimental section), where the flatness\unevenness was calculated using the whole candidate window.

2.4 Segmentation Based Filtering

From the segmented version of the depth image, it is possible to evaluate some characteristics of the candidate face based on its dimension with respect to its bounding box and its shape (which should be elliptical). According to these considerations, we define two simple filtering rules. The first evaluates the relative dimension of the larger region with respect to the whole candidate image (SEG). We have rejected the candidate regions where the area of the larger region is less than 40 % of the whole area.

The second considers the fitting score of an ellipse fitting approach[2] to evaluate whether the larger region can be modeled as an ellipse (ELL). The candidate regions with a cost higher than 100 are rejected.

2.5 Eye Based Filtering

The presence of two eyes is another good indicator of a face. In this work two efficient eye detectors are applied to candidate face regions [28]. The score of the eye detector is used to filter out regions having a very low probability of containing two eyes (EYE).

The first approach is a variant of the PS model proposed in [28]. PS is a computationally efficient framework for representing a face in terms of an undirected graph $G = (V, E)$, where the vertices V correspond to its parts (i.e., two eyes, one nose, and one mouth) and the edge set E characterizes the local pairwise spatial relationship between the different parts. The PS approach proposed in [28] enhances the traditional PS model to handle the complicated appearance and structural changes of eyes under uncontrolled conditions.

The latter approach is proposed in [29] where the color information is used to build an eye map for emphasizing the iris area. Then, a radial symmetry transform is applied both to the eye map and the original image. Finally, the cumulative result of the transforms indicates the positions of the eye.

---

[2]http://it.mathworks.com/matlabcentral/fileexchange/3215-fit-ellipse.

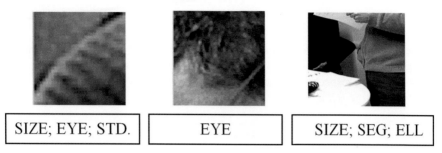

| SIZE; EYE; STD. | EYE | SIZE; SEG; ELL |

Fig. 4 Some samples of images filtered out by the different filtering rules

Fig. 5 Partitioning of a neighborhood of the candidate face region into 8 sectors (*gray area*). Lower sectors 4, 5 are depicted in *dark gray*

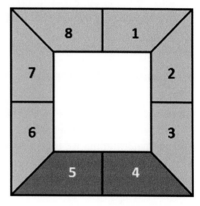

Only the face candidates where a pair of eyes is found with a score higher than a prefixed threshold (1 for the first approach and 750 for the latter) are retained (Fig. 4).

2.6 Filtering Based on the Analysis of the Depth Values

Excluding some critical cases like people lying on the floor, it is reasonable to assume the person body is present in the lower volume right under the face, while the remaining surrounding volume is likely to be empty. We exploit this observation in order to reject candidate faces whose neighborhood shows a different pattern from the expected one.

In particular, we enlarge the rectangular region associated to the candidate face in order to analyze a neighborhood of the face in the depth map, i.e., all the pixels which belong to the extended region but are not part of the smaller face box. The region is then partitioned into a number of radial sectors centered to the center of the candidate face. We used eight sectors in our experiments, see Fig. 5. For each sector S_i we count the number of pixels n_i whose depth value d_p is close to the average depth value of the face $\bar{d}$, i.e., .

$$n_i = \left| \left\{ p \; : \; \left| d_p - \overline{d} \right| < t_d \wedge p \in S_i \right\} \right|$$

where we used $t_d = 50cm$. Finally, the number of pixels per sector is averaged on the two lower sectors (S_4 and S_5) and on the remaining ones, obtaining the two values n_u and n_l respectively. The ratio between n_l and n_u is then computed:

$$\frac{n_l}{n_u} = \frac{\frac{1}{2}(n_4 + n_5)}{\frac{1}{6}(n_1 + n_2 + n_3 + n_6 + n_7 + n_8)}$$

If the ratio drops below a certain threshold t_r, then the candidate face is removed. We set $t_r = 0.8$ in our experiments.

This approach is named SEC in the experimental section.

3 Experimental Results

For our experiments we use four datasets of faces acquired in an unconstrained setup ("in the wild") for purposes other than face detection. For preliminary experiments and parameter tuning, we used a separate set of images appositely collected for this aim. The images will be publicly available as a part of the Padua FaceDec dataset. All four datasets contain upright frontal images possibly with a limited degree of rotation, and all are composed of color images and their corresponding depth map:

- Microsoft hand gesture [30] is a dataset collected for gesture recognition composed of images of ten different people performing gestures; most of the images in the datasets are quite similar to each other, and each image contains only one face. Only a subset of the whole dataset (42 images) has been selected for our experiments and manually labeled with the face position.
- Padua Hand gesture [31] is a dataset similar to the previous one that was collected for gesture recognition purposes and composed of images from ten different people; each image contains only one face. A subset of 59 images has been manually labeled and selected for our experiments.
- Padua FaceDec [1], is a dataset collected and labeled for the purpose of face detection. It contains 132 images acquired both outdoors and indoors with the Kinect sensor at the University campus in Padua. Some images contain more than one face and some contain no faces. The images capture one or more people performing various daily activities (e.g., working, studying, walking, and chatting) in an unconstrained setup. Images have been acquired different hours of the day in order to account for varying lighting conditions, and some faces are partially occluded by objects and other people. For these reasons, this dataset is more challenging than previous datasets.
- Padua FaceDec2, is a new dataset acquired for the purpose of this chapter by a second generation Kinect sensor. For each scene a 512×424 depth map and a 1920×1080 color image have been acquired. The dataset includes 316 images of

Table 1 Datasets characteristics

Dataset	No. images	Color resolution	Depth resolution	No. faces	Difficulty
Microsoft hand gesture	42	640×480	640×480	42	Low
Padua hand gesture	59	1280×1024	640×480	59	Low
Padua FaceDec	132	1280×1024	640×480	150	High
Padua FaceDec2	316	1920×1080	512×424	363	High
MERGED	549	–	–	614	High

both indoor and outdoor scenes with people in various positions and challenging situations, e.g., with the head tilted with respect to the camera or very close to other objects. Some images contain more than one face and some contain no faces. Note that, even if the second generation Kinect, differently from the first version, is able to work outdoor, the outdoor depth data is noise, thus making the recognition problem more challenging. The depth data has been retro-projected over the color frame and interpolated to the same resolution thus obtaining two aligned depth and color fields.

A summary of the datasets characteristics is reported in Table 1.

The four datasets have been merged to form a single larger dataset consisting of 549 images containing 614 faces (only upright frontal faces with a maximum rotation of $\pm 30°$ have been considered). The parameter optimization of the face detectors has been performed manually and, despite the different origin and characteristics of the images included in the final set, the selected parameter optimizations have been fixed for all the images. The MERGED dataset is not an easy dataset to classify, as illustrated in Fig. 6, which presents some samples the face detectors could not accurately detect (even when executed with a very low recognition threshold). The collected dataset contains images with various lighting conditions, see Fig. 7.

Moreover, for comparing the face detectors and the proposed ensemble with other approaches proposed in the literature, we perform comparisons on the well-known BioID dataset, the foremost benchmark for upright frontal face detection. It is composed by 1521 images of 23 different people acquired during several identification sessions. The amount of rotation in the facial images is small. All the images are gray-scale and do not have a depth map. As a result, most of the filters proposed in this work are not applicable to the BioID dataset. Nonetheless, this dataset provides a means of comparing the approach proposed in this chapter with other state-of-the-art methods and is one way of showing the effectiveness of our ensembles.

The performance of the proposed approach is evaluated according the following well-known performance indicators:

Fig. 6 Some samples where faces are not correctly detected

- **Detection rate (DR)**: the detection rate, also known as *recall* and *sensitivity*, is the fraction of relevant instances that are retrieved, i.e. the ratio between the number of faces correctly detected and the total number of faces (manually labelled) in the dataset. Let d_l (d_r) be the Euclidean distance between the manually extracted C_l (C_r) and the detected C'_l (C'_r) left (right) eye position,[3] the relative error of detection is defined as $ED = \max(d_l, d_r)/d_{lr}$ where the normalization factor d_{lr} is the Euclidean distance of the expected eye centers used to make the measure independent of the scale of the face in the image and of image size.
- **False positives (FP)**: it is the number of candidate faces not containing a face.

---

[3]The face detectors FL and RF give the positions of the eye centers as the output, while for VJ and SN the detected eye position is assumed to be a fixed position inside the face bounding box.

Fig. 7 Some samples with various lighting conditions

- **False negatives (FN)**: it is the number of faces not retrieved, i.e. the candidate faces erroneously excluded by the system. This value is correlated to the detection rate, since it can be obtained as (1-Detection Rate) × N°faces.

Table 2 Performance of the four face detectors and some ensembles (last five rows) on the MERGED dataset (* denotes the use of adding poses)

Face detector(σ)/ensemble	+Poses	DR	FP	FN
VJ(2)	No	55.37	2528	274
RF(−1)	No	47.39	4682	323
RF(−0.8)	No	47.07	3249	325
RF(−0.65)	No	46.42	1146	329
SN(1)	No	66.61	508	205
SN(10)	No	46.74	31	327
FL	No	78.18	344	134
VJ(2)*	Yes	65.31	6287	213
RF(−1)*	Yes	49.67	19475	309
RF(−0.8)*	Yes	49.67	14121	309
RF(−0.65)*	Yes	49.02	5895	313
SN(1)*	Yes	74.59	1635	156
SN(10)*	Yes	50.16	48	306
FL*	Yes	83.39	891	102
FL+ RF(−0.65)	No	83.06	1490	104
FL+ RF(−0.65) + SN(1)	No	86.16	1998	85
FL+ RF(−0.65) + SN(1)*	Mixed	88.44	3125	71
FL* + SN(1)*	Yes	87.79	2526	75
FL* + RF(−0.65) + SN(1)*	Mixed	90.39	3672	59

In this dataset (due to the low quality of several images) we considered a face detected in an image if ED < 0.35.

The first experiment is aimed at comparing the performance of the four face detectors and their combination by varying the sensitivity factor σ (when applicable) and the detection procedure (i.e., when using or not using additional poses with 20°/−20° rotation).

For each detector in Table 2, the value fixed for the sensitivity threshold is shown in parentheses. We also compare in Table 2 different ensembles of face detectors. To reduce the number of false positives, all the output images having a distance $md \leq 30$ pixels are merged together.

From the result in Table 2, it is clear that the adding poses is not so useful for the RF face detector. This probably is due to the fact that RF has already been trained on images containing rotated faces. Moreover, using added poses strongly increases the number of false positives, as might be expected. For the ensembles, we report only the most interesting results. As can be seen in Table 2, combining more high performing approaches clearly boosts the detection rate performance. Unfortunately, the ensembles based on the three face detectors have too many false negatives.

Another interesting result is that of the 3125 false positives of "FL+ RF(−0.65) + SN(1)*" 2282 are found where the depth map has valid values, the

Table 3 Performance of the same four face detectors and ensembles used above on the BioID dataset

Face detector(σ)/ensemble	+Poses	DR (ED<0.15)	DR (ED<0.25)	DR (ED<0.35)	FP
VJ(2)	No	13.08	86.46	99.15	517
RF(−1)	No	87.84	98.82	99.08	80
RF(−0.8)	No	87.84	98.82	99.08	32
RF(−0.65)	No	87.84	98.82	99.08	21
SN(1)	No	71.27	96.38	97.76	12
SN(10)	No	72.06	98.16	99.74	172
FL	No	92.57	94.61	94.67	67
VJ(2)*	Yes	13.08	86.46	99.15	1745
RF(−1)*	Yes	90.53	99.15	99.41	1316
RF(−0.8)*	Yes	90.53	99.15	99.41	589
RF(−0.65)*	Yes	90.53	99.15	99.41	331
SN(1)*	Yes	71.33	96.52	97.90	193
SN(10)*	Yes	72.12	98.36	99.87	1361
FL*	Yes	92.57	94.61	94.67	1210
FL+ RF(−0.65)	No	98.42	99.74	99.74	88
FL+ RF(−0.65) + SN(10)	No	99.15	99.93	99.93	100
FL+ RF(−0.65) + SN(1)*	Mixed	99.15	100	100	281
FL* + SN(1)*	Yes	98.03	99.87	99.93	260
FL* + RF(−0.65) + SN(1)*	Mixed	99.15	100	100	1424

others are found where the values of depth map is 0 (i.e., the Kinect has not been able to compute the depth value due to occlusion, low reflectivity, too high distance or other issues), while all the true positives are found where exists the depth map.

In Table 3 we report the performance of the same face detectors on the BioID dataset. It is interesting to note that the creation of adding poses is not mandatory; if the acquisition is in a constrained environment, the performance is almost the same with or without the addition of artificial poses. Using artificial poses strongly increases the number of false positives. It is clear that different face detectors exhibit different behaviors as is evident by the fact that each is able to detect different faces. As a result of this diversity, the ensemble is able to improve the best stand-alone approaches. Another interesting result is the different behaviors exhibited by the same face detectors on the two different datasets. For instance, RF works very well on the BioID dataset but rather poorly on our dataset that contains several low quality faces. Regardless, in both datasets the same ensemble outperforms the other approaches.

The next experiment is aimed at evaluating the filtering steps detailed in Sect. 2. Since the first experiments proved that the best configuration (i.e., trade-off between performance and false positives) is the ensemble composed by FL+ RF(−0.65) + SN(1)*, the filtering steps are performed on the results of this detector. The performance after each filter or combination of filters is reported in Tables 4 and 5. The computation time reported in seconds is evaluated on a Xeon E5-1620 v2 −

Table 4 Performance of different filtering steps on the MERGED dataset

Filter	DR	FP	FN	Time
SIZE	88.44	1247	71	0.000339
STD	88.44	2207	71	0.010865
SEG	88.44	2144	71	0.008088
ELL	88.44	1984	71	0.010248
EYE	88.44	1580	71	19.143445
SEC	88.11	1954	73	0.015302
STD°	87.79	2265	75	0.008280

Table 5 Performance obtained combining different filtering steps on the MERGED dataset

Filter combination	DR	FP	FN
SIZE	88.44	1247	71
SIZE + STD	88.44	1219	71
SIZE + STD + SEG	88.27	1193	72
SIZE + STD + SEG + ELL	88.11	1153	73
SIZE + STD + SEG + ELL + EYE	88.11	1050	73
SIZE + STD + SEG + ELL + SEC + EYE	86.97	752	80

Table 6 Performance obtained maximizing the reduction of the FP

Filter	DR	FP	FN
SIZE	87.79	944	75
SIZE + STD	87.79	908	75
SIZE + STD + SEG	87.79	877	75
SIZE + STD + SEG + ELL	87.30	852	78
SIZE + STD + SEG + ELL + EYE	86.16	560	85
SIZE + STD + SEG + ELL + SEC + EYE	84.85	431	93

3.7 GHz – 64 GB Ram using Matlab R2014a on a candidate region of 78×78 pixels without parallelizing the code (note, however, that the different filters can be run in parallel).

It is clear that SIZE is the most useful criterion for removing the false positive candidates found by the ensemble of face detectors. The second best approach is the eye detector EYE. Although it works quite well, it has a high computational time. As a result, it cannot be used in all applications. The other approaches are less useful if taken individually; however, since they do not require a high computational cost, they can be useful in sequence to decrease the number of false positives. In applications where real-time detection is not mandatory (e.g. in face tagging), EYE filtering can be used in the ensemble to further reduce the number of false positives without decreasing the number of true positives.

As a final experiment, we report in Table 6 the performance of the different filters and combinations of filters by partially relaxing the thresholds, which greatly decreases the number of false positives even though some true positives are lost.

From the results of Table 6 it is evident that the proposed approach works even better than FL (which is known in the literature as one of the best performing face detectors). Even though these results have been obtained on a small dataset, we are confident that they are realistic and would perform comparatively in real-world conditions since the images contained in MERGE are very different from each other and include both images with a single frontal face and images acquired "in the wild" with multiple faces.

4 Conclusion

In this work a smart false positive-reduction method for face detection is proposed. An ensemble of state-of-the-art face detectors is combined with a set of filtering steps calculated both from the depth map and the color image. The main goal of this approach is to obtain accurate face detection with few false positives. This goal is accomplished using a set of filtering steps: the size of candidate face regions; the flatness or unevenness of the candidate face regions; the size of the larger cluster of the depth map of the candidate face regions; ellipse fitting to evaluate whether the region can be modeled as an ellipsis; and an eye detection step. The experimental section demonstrates that the proposed set of filtering steps reduces the number of false positives with little effect on the detection rate.

In conclusion, we show that the novel facial detection method proposed in this work is capable of taking advantage of a depth map to obtain increased effectiveness even under difficult environmental illumination conditions. Our experiments, which were performed on a merged dataset containing images with complex backgrounds acquired in an unconstrained setup, demonstrate the feasibility of the proposed system.

We are aware that the dimensions of the datasets used in our experiments are lower than most benchmark datasets containing only 2D data. It is in our intention to continue acquiring images to build a larger dataset with depth maps. In any case, the reported results obtained on this small dataset have statistical significance. As a result, we can confirm that the depth map provides criteria that can result in a significant reduction of the number of false positives.

References

1. L. Nanni, A. Lumini, F. Dominio, P. Zanuttigh, Effective and precise face detection based on color and depth data. Appl. Comput. Inform. **10**(1), 1–13 (2014)
2. C. Zhang, Z. Zhang, A survey of recent advances in face detection. Microsoft Research Technical Report, MSR-TR-2010-66, June 2010
3. Z. Zeng, M. Pantic, G.I. Roisman, T.S. Huang, A survey of affect recognition methods: audio, visual, and spontaneous expressions. IEEE Trans. Pattern Anal. Mach. Intell. **31**(1), 39–58 (2009)

4. H.L. Jin, Q.S. Liu, H.Q. Lu, Face detection using one-class based support vectors, in *Proceedings 6th IEEE International Conference Automatic Face Gesture Recognition*, Hoboken, NJ 07030 USA, (2004), pp. 457–462
5. P. Viola, M.J. Jones, Rapid object detection using a boosted cascade of simple features, *Proceedings of the 2001 IEEE Computer Society conference on Computer Vision and Pattern Recognition*, **1**, 511–518 (2001)
6. C. Küblbeck, A. Ernst, Face detection and track in video sequences using the modified census transformation. Image Vis. Comput. **24**(6), 564–572 (2006)
7. C. Huang, H. Ai, Y. Li, S. Lao, High-performance rotation invariant multiview face detection. IEEE Trans. Pattern Anal. Mach. Intell. **29**(4), 671–686 (2007)
8. J. Wu, S. Charles Brubaker, M.D. Mullin, J.M. Rehg, Fast asymmetric learning for cascade face detection. IEEE Trans. Pattern Anal. Mach. Intell. **30**, 369–382 (2008)
9. M. Anisetti, Fast and robust face detection, in *Multimedia Techniques for Device and Ambient Intelligence*, Chapter 3 (Springer, US, 2009). ISBN: 978-0-387-88776-0
10. J. Li, Y. Zhang, Learning surf cascade for fast and accurate object detection, in *IEEE Conference on Computer Vision and Pattern Recognition (CVPR)* (IEEE, 2013) Hoboken, NJ 07030 USA
11. M. Mathias, et al. Face detection without bells and whistles, in *Computer Vision–ECCV 2014* (Springer International Publishing, 2014), Zurich, Switzerland, pp. 720–735
12. F. Tsalakanidou, D. Tzovaras, M.G. Strintzis, Use of depth and colour eigenfaces for face recognition. Pattern Recogn. Lett. **24**(910), 1427–1435 (2003)
13. R.I. Hg, P. Jasek, C. Rofidal, K. Nasrollahi, T.B. Moeslund, G. Tranchet, An RGB-D database using Microsoft's Kinect for windows for face detection, in *2012 Eighth International Conference on Signal Image Technology and Internet Based Systems (SITIS)* (IEEE, 2012), Hoboken, NJ 07030 USA, pp. 42–46
14. M.-Y. Shieh, T.-M. Hsieh, Fast facial detection by depth map analysis. Math. Problems Eng. (2013), Article ID 694321, pp. 10, doi: 10.1155/2013/694321
15. J. Shotton, A. Fitzgibbon, M. Cook, T. Sharp, M. Finocchio, R. Moore, A. Kipman, A. Blake, Real-time human pose recognition in parts from single depth images. CVPR **2**, 3 (2011)
16. R. Mattheij, E. Postma, Y. Van den Hurk, P. Spronck, Depth-based detection using Haarlike features, in *Proceedings of the BNAIC 2012 Conference* (Maastricht University, The Netherlands, 2012), pp. 162–169
17. M. Anisetti, V. Bellandi, E. Damiani, L. Arnone, B. Rat, A3FD: Accurate 3D face detection, in *Signal Processing for Image Enhancement and Multimedia Processing* (Springer, US, 2008), pp. 155–165
18. F. Jiang, M. Fischer, H.K. Ekenel, B.E. Shi, Combining texture and stereo disparity cues for real-time face detection. Signal Process. Image Commun. **28**(9), 1100–1113 (2013)
19. G. Goswami, S. Bharadwaj, M. Vatsa, R. Singh, On RGB-D face recognition using Kinect, in *2013 IEEE Sixth International Conference on Biometrics: Theory, Applications and Systems (BTAS)* (IEEE, 2013), Hoboken, NJ 07030 USA, pp. 1–6
20. Y. Taigman, M. Yang, M.A., Ranzato, L. Wolf, Deepface: Closing the gap to human-level performance in face verification, in *2014 IEEE Conference on Computer Vision and Pattern Recognition (CVPR)* (IEEE, 2014), Hoboken, NJ 07030 USA, pp. 1701–1708
21. O. Jesorsky, K. Kirchberg, R. Frischholz, in *Face Detection Using the Hausdorff Distance*, eds. by J. Bigun, F. Smeraldi. Audio and Video based Person Authentication—AVBPA, (Springer, 2001), Berlin, Germany, pp. 90–95
22. N. Markuš, M. Frljak, I.S. Pandžić, J. Ahlberg, R. Forchheimer, Fast localization of facial landmark points, in *Proceedings of the Croatian Computer Vision Workshop*, Zagreb, Croatia, (2014)
23. M. Nilsson, J. Nordberg, I. Claesson, Face detection using local SMQT features and split up SNOW classifier. IEEE Int. Conf. Acoust. Speech Signal Process. (ICASSP), **2**, Hoboken, NJ 07030 USA, 589–592 (2007)
24. A. Asthana, S. Zafeiriou, S. Cheng, M. Pantic, Robust discriminative response map fitting with constrained local models, in *2013 IEEE Conference on Computer Vision and Pattern Recognition (CVPR)* (IEEE, 2013), Hoboken, NJ 07030 USA, pp. 3444–3451

25. D. Herrera, J. Kannala, J. Heikkilä, Joint depth and color camera calibration with distortion correction. IEEE Trans. Pattern Anal. Mach. Intell. **34**, 2058–782 (2012)
26. D. Comaniciu, P. Meer, Mean shift: a robust approach toward feature space analysis. IEEE Trans. Pattern Anal. Mach. Intell. **24**(5), 603–619 (2002)
27. C. Dal Mutto, P. Zanuttigh, G.M. Cortelazzo, Fusion of geometry and color information for scene segmentation. IEEE J. Sel. Top. Signal Process. **6**(5), 505–521 (2012)
28. X. Tan, F. Song, Z.-H. Zhou, S. Chen, Enhanced pictorial structures for precise eye localization under incontrolled conditions, in *IEEE Conference on Computer Vision and Pattern Recognition 2009 (CVPR 2009)*, Hoboken, NJ 07030 USA, 20–25 June 2009, pp. 1621, 1628
29. E. Skodras, N. Fakotakis, Precise localization of eye centers in low resolution color images. Image Vision Comput. (2015). doi:10.1016/j.imavis.2015.01.006
30. Z. Ren, J. Meng, J. Yuan. Depth camera based hand gesture recognition and its applications in human-computer-interaction, in *Proceedings of ICICS* (2011), Hoboken, NJ 07030 USA, pp. 1–5
31. F. Dominio, M. Donadeo, P. Zanuttigh, Combining multiple depth-based descriptors for hand gesture recognition. Pattern Recogn. Lett. **50**, 101–111 (2014)

Lighting Estimation and Adjustment for Facial Images

Xiaoyue Jiang, Xiaoyi Feng, Jun Wu, and Jinye Peng

Abstract For robust face detection and recognition, the problem of lighting variation is considered as one of the greatest challenges. Lighting estimation and adjustment is a useful way to remove the influence of illumination for images. Due to the different prior knowledge provided by a single image and image sequences, algorithms dealing with lighting problems are always different for these two conditions. In this chapter we will present a lighting estimation algorithm for a single facial image and a lighting adjustment algorithm for image sequences. To estimate the lighting condition of a single facial image, a statistical model is proposed to reconstruct the lighting subspace where only one image of each subject is required. For lighting adjustment of image sequences, an entropy-based optimization algorithm is proposed to minimize the difference between consequent images. The effectiveness of those proposed algorithms are illustrated on face recognition, detection and tracking tasks.

1 Introduction

Face detection, recognition and tracking are difficult due to variations caused by pose, expression, occlusion and lighting (or illumination), which make the distribution of face object highly nonlinear. Lighting is regarded as one of the most critical factors for robust face recognition. Current attempt to handle lighting variation is either to find the invariant features and representations or to model the variation. The gradient based algorithm [1, 2] attempted to find illumination invariant features. While the other kind of algorithms tried to extract the reflectance

X. Jiang (✉)
School of Electronics and Information, Northwestern Polytechnical University,
Xi'an 710072, China
e-mail: xjiang@nwpu.edu.cn

X. Feng • J. Wu
Northwestern Polytechnical University, Xi'an 710072, China

J. Peng
Northwestern University, Xi'an 710069, China

© Springer International Publishing Switzerland 2016
M. Kawulok et al. (eds.), *Advances in Face Detection and Facial Image Analysis*,
DOI 10.1007/978-3-319-25958-1_3

35

information from the observations in different lighting conditions, including the algorithms based on Quotient Images [3–8] and the algorithms based on Retinex theory [9–12]. But these methods cannot extract sufficient features for accurate recognition.

Early work on modeling lighting variation [13, 14] showed that a 3D linear subspace can represent the variation of a Lambertian object under a fixed pose when there is no shadow. With the same Lambertian assumption, Belhumeur and Kriegman [15] showed that images illuminated by an arbitrary number of point light sources formed a convex polyhedral cone, i.e. the illumination cone. In theory, the dimensionality of the cone is finite. They also pointed out that the illumination cone can be approximated by a few properly chosen images. Good recognition results of the illumination cone in [16] demonstrated its representation for lighting variation.

Recent research is mainly focused on the application of low-dimensional subspace to lighting variation modeling. With the assumption of Lambertian surface and non-concavity, Zhou et al. [17] found a set of basis images from training images based on photometric methods, while Ramamoorith and Hanrahan [18] and Basri and Jacobs [19] independently introduced the spherical harmonic (SH) subspace to approximate the illumination cone. Chen et al. [20] decomposed the original image into lighting and reflectance maps and build up lighting subspace from lighting maps. Sparse representation-based algorithm also showed its effectiveness in dealing with illumination problem [21–24], while the lack of training images always limited its application. In order to construct the lighting subspace, a lot of algorithms also applied the 3D model of faces to handling lighting variations [25–30]. However, recovering the 3D information from images is still an open problem in computer vision.

Lee et al. [31] built up a subspace that is nearest to the SH subspace and has the largest intersection with the illumination cone, called the nine points of light (9PL) subspace. It has a universal configuration for different subjects, i.e. the subspace is spanned by images under the same lighting conditions for different subjects. In addition, the basis images of 9PL subspace can be duplicated in real environments, while those of the SH subspace cannot because its the basis images contain negative values. Therefore the 9PL subspace can overcome the inherent limitation of SH subspace. Since the human face is neither completely Lambertain nor entirely convex, SH subspace can hardly represent the specularities or cast shadows (not to mention inter-reflection). The basis images of 9PL subspace are taken from real environment, they already contain all the complicated reflections of the objects. Therefore the 9PL subspace can give a more detailed and accurate description of lighting variation.

In practice, the requirement of these nine real images cannot always be fulfilled. Usually there are fewer gallery images (e.g. one gallery image) per subject, which can be taken under arbitrary lighting conditions. In this chapter, we propose a statistical model for recovering the 9 basis images of the 9PL subspace from only one gallery image. Based on the estimation of lighting coefficient for a facial image, its basis images can be constructed with the prior knowledge about the distribution of the basis images. An overview of the proposed basis construction algorithm

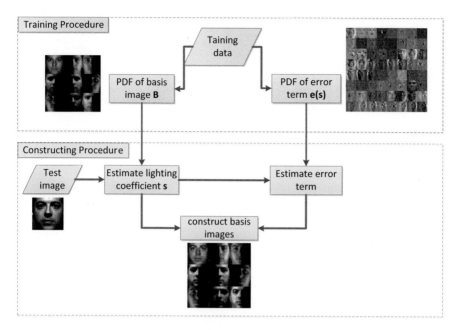

Fig. 1 Flowchart of constructing basis images for lighting subspace

is shown in Fig. 1. Zhang and Samaras [25] presented a statistical method for recovering the basis images of SH subspace instead. In their training procedure, geometric and albedo information is still required for synthesizing the harmonic images. In contrast, the proposed method requires only some real images that can be easily obtained in real environment. Since the recovered basis images of the 9PL subspace contain all the reflections caused by the shape of faces, such as cast shadows, specularities, and inter-reflections, better recognition results are obtained, even under extreme lighting conditions. Compared with other algorithms based on 3D model [25, 28, 29], the proposed algorithm is entirely a 2D algorithm, which has much lower computational complexity. For more details of the algorithm is presented in paper [32].

Lighting variations will also bring challenges for the processing of image sequences. Even though there are a lot of algorithms to deal with lighting problem for images, there are a few investigations for image sequences particularly. In fact most lighting algorithms are too complicated for image sequences, and do not consider the special characteristic of sequences. Also, the scene can be more complex in real applications. Thus it is impossible to build lighting models for different kind of subjects and adjust the lighting condition for each of them independently. However, the lighting adjustment can be applied to the whole scene according to an optimal lighting condition. In this chapter, we also propose a two-step entropy-based lighting adjustment algorithm for image sequences. An overview of the adjustment algorithm is shown in Fig. 2. Paper [33] presents more details about the lighting adjustment algorithm.

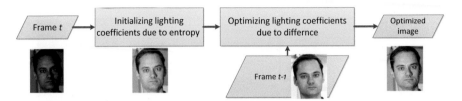

Fig. 2 Flowchart of lighting adjustment algorithm for image sequences

In the lighting adjustment algorithm, the difference between successive frames is used as a cost function for optimizing a lighting model of an image sequence. For each image frame, to ensure convergence, the proposed lighting adjustment algorithm works in two steps. First initial values of the lighting model parameters are estimated using the entropy of the current image as measure. These values are then used as initial guesses for a constrained least squares optimization problem, considering two successive frames. It is worth pointing out that the proposed algorithm did not only try to do shadow removal, which is only one aspect of lighting problem because there is no shadow but highlight or un-uniform lighting distribution in an image sometimes. Also, we adjusted the global lighting conditions to be more uniform and enhance the local features of the image as well.

This chapter is organized as follows. In Sect. 2, we briefly summarize the methods of low-dimensional linear approximation of the illumination cone, including the SH subspace and the 9PL subspace. The training of our statistical model and the application of the model for recovering basis images from only one gallery image are described in Sects. 3 and 4 respectively. Section 5 is dedicated to the experimental results. Then perception based lighting model is introduced in Sect. 6. In Sect. 7 we give an overview of the proposed lighting adjustment algorithm for image sequence. Section 8 discusses qualitative and quantitative results of the lighting adjustment in the case of facial features detection and tracking. Finally, conclusions are drawn in Sect. 9.

2 Approximation of the Illumination Cone

Belhumeur and Kriegman [15] proved that the set of n-pixel images of a convex object that had a Lambertian surface illuminated by an arbitrary number of point light sources at infinity formed a convex polyhedral cone, called the illumination cone $\mathscr{C}$ in $\mathscr{R}^n$. Each point in the cone is an image of the object under a particular lighting condition, and the entire cone is the set of images of the object under all possible lighting conditions. Any images in the illumination cone $\mathscr{C}$ (including the boundary) can be determined as a convex combination of extreme rays (images) given by

$$I_{ij} = max(\tilde{B}\tilde{s}_{ij}, 0) \tag{1}$$

where $\tilde{s}_{ij} = \tilde{b}_i \times \tilde{b}_j$ and $\tilde{B} \in \mathfrak{R}^{n \times 3}$. Every row of $\tilde{B}$, $\tilde{b}_i$, is a three element row vector determined by the product of the albedo with the inward pointing unit normal vector of a point on the surface. There are at most $q(q-1)$ extreme rays for $q \leq n$ distinct surface normal vectors. Therefore the cone can be constructed with finite extreme rays and the dimensionality of the lighting subspace is finite. However, building the full illumination cone is tedious, and the low dimensional approximation of the illumination cone is applied in practice.

From the view of signal processing, the reflection equation can be considered as the rotational convolution of incident lighting with the albedo of the surface [18]. The spherical harmonic functions $Y_{lm}(\theta, \phi)$ are a set of orthogonal basis functions defined in the unit sphere, given as follows,

$$Y_{lm}(\theta, \phi) = N_{lm} P_l^m(\cos \theta) \exp^{im\phi} \tag{2}$$

where $N_{lm} = \sqrt{\frac{2l+1}{4\pi} \frac{(l-m)!}{(l+m)!}}$, (θ, ϕ) is the spherical coordinate (θ is the elevation angle, which is the angle between the polar axis and the z-axis with range $0 \leq \theta \leq 180°$, and ϕ is the azimuth angle with the range $-180° \leq \phi \leq 180°$). P_l^m is the associated Legendre function, and the two indices meet the conditions $l \geq 0$ and $l \geq m \geq -l$. Then functions in the sphere, such as the reflection equation, can be expanded by the spherical harmonic functions, which are basis functions on the sphere. Images can be represented as a linear combination of spherical harmonic functions. The first three order ($l \leq 3$) basis can account for 99 % energy of the function. Therefore the first three order basis functions (altogether 9) can span a subspace for representing the variability of lighting. This subspace is called the spherical harmonic (SH) subspace.

Good recognition results reported in [19] indicates that the SH subspace $\mathscr{H}$ is a good approximation to the illumination cone $\mathscr{C}$. Given the geometric information of a face, its spherical harmonic functions can be calculated with Eq. (2). These spherical harmonic functions are synthesized images, also called harmonic images. Except the first harmonic image, all the others have negative values, which cannot be obtained in reality. To avoid the requirement of geometric information, Lee et al. [31] found a set of real images which can also serve as a low dimensional approximation to illumination cone based on linear algebra theory.

Since the SH subspace $\mathscr{H}$ is good for face recognition, it is reasonable to assume that a subspace $\mathscr{R}$ close to $\mathscr{H}$ would be likewise good for recognition. $\mathscr{R}$ should also intersect with the illumination cone $\mathscr{C}$ as much as possible. Hence a linear subspace $\mathscr{R}$ which is meant to provide a basis for good face recognition will also be a low dimensional linear approximation to the illumination cone $\mathscr{C}$. Thus subspace should satisfy the following two conditions [31]:

1. The distance between $\mathscr{R}$ and $\mathscr{H}$ should be minimized.
2. The unit volume ($vol(\mathscr{C} \cap \mathscr{R})$) of $\mathscr{C} \cap \mathscr{R}$ should be maximized (the unit volume is defined as the volume of the intersection of $\mathscr{C} \cap \mathscr{R}$ with the unit ball).

Note that $\mathscr{C} \cap \mathscr{R}$ is always a subcone of $\mathscr{C}$; therefore maximizing its unit volume is equivalent to maximize the solid angle subtended by the subcone $\mathscr{C} \cap \mathscr{R}$. If $\{\tilde{I}_1, \tilde{I}_2, \cdots, \tilde{I}_k\}$ are the basis images of $\mathscr{R}$. The cone $\mathscr{R}_c \subset \mathscr{R}$ is defined by $\tilde{I}_k$,

$$\mathscr{R}_c = \{I | I \in \mathscr{R}, I = \sum_{k=1}^{M} \alpha_k \tilde{I}_k, \alpha_k \geq 0\} \tag{3}$$

is always a subset of $\mathscr{C} \cap \mathscr{R}$. In practice the subcone $\mathscr{C} \cap \mathscr{R}$ is taken as $\mathscr{R}_c$ and the subtended angle of $\mathscr{R}_c$ is maximized. $\mathscr{R}$ is computed as a sequence of nested linear subspace $\mathscr{R}_0 \subseteq \mathscr{R}_1 \subseteq \cdots \subseteq \mathscr{R}_i \subseteq \cdots \subseteq \mathscr{R}_9 = \mathscr{R}$, with $\mathscr{R}_k (k > 0)$ a linear subspace of dimension i and $\mathscr{R}_0 = \emptyset$. First, EC denotes the set of (normalized) extreme rays in the illumination cone $\mathscr{C}$; and EC_k denotes the set obtained by deleting k extreme rays from EC, where $EC_0 = EC$. With $\mathscr{R}_{k-1}$ and EC_{k-1}, the sets EC_k and R_k can be defined iteratively as follows:

$$\tilde{I}_k = \arg \max_{I \in EC_{k-1}} \frac{dist(I, \mathscr{R}_{k-1})}{dist(I, \mathscr{H})} \tag{4}$$

where $\tilde{I}_k$ denotes the element in EC_{k-1}. $\mathscr{R}_k$ is defined as the space spanned by $\mathscr{R}_{k-1}$ and $\tilde{I}_k$. $EC_k = EC_{k-1} \backslash \tilde{I}_k$. The algorithm stops when $\mathscr{R}_9 \equiv \mathscr{R}$ is reached. The result of Eq. (4) is a set of nine extreme rays that span $\mathscr{R}$ and there are nine directions corresponding to these nine extreme rays. For different subjects, the nine lighting directions are qualitatively very similar. By averaging Eq. (4) of different subjects and maximizing this function as follows:

$$\tilde{I}_k = \arg \max_{I \in EC_{k-1}} \sum_{p=1}^{N} \frac{dist(I^p, \mathscr{R}_{k-1}^p)}{dist(I^p, \mathscr{H}^p)} \tag{5}$$

where I^p denotes the image of subject p taken under a single light source. H^p is the SH subspace of subject p. $\mathscr{R}_{k-1}^p$ denotes the linear subspace spanned by images $\{\tilde{I}_1^p, \cdots, \tilde{I}_k^p\}$ of subject p. The universal configuration of nine light source direction is obtained. They are $(0, 0)$, $(68, -90)$, $(74, 108)$, $(80, 52)$, $(85, -42)$, $(85, -137)$, $(85, 146)$, $(85, -4)$, $(51, 67)$. The directions are expressed in spherical coordinates as pairs of (ϕ, θ), Fig. 3a illustrates the nine basis images of a person from the Yale face database B [16].

3   Statistical Model of Basis Images

According to the universal configuration of lighting directions, we can apply nine images taken under controlled environment to spanning the 9PL linear subspace. However, even these nine images may not be available in some situations. Thus, we propose a statistical method for estimating the basis images from one gallery image.

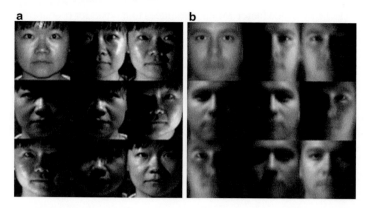

Fig. 3 The basis images of 9PL subspace. (**a**) images taken under certain lighting conditions can serve as the basis images of the object. (**b**) The mean images of the basis images estimated from the bootstrap data set

To build the statistical model, we must find the probability density function (PDF) of basis images and the PDF of the error term. Due to the limited amount of the training data, we use the bootstrap method to estimate the statistics of basis images. The recovering step is to estimate the corresponding basis images from one single image of a novel subject under arbitrary lighting conditions. For a given image, we first estimate its lighting coefficient. Then according to the maximum a posteriori (MAP) estimation, we obtain an estimation of the basis images. Finally, we apply the recovered subspace to face recognition. The probe image is identified as the face whose lighting subspace is closest in distance to the image.

Given nine basis images, we can construct images under arbitrary lighting conditions as follows,

$$I = B\mathbf{s} + e(\mathbf{s}) \qquad (6)$$

where $I \subset \Re^{d \times 1}$ is the image vector. $B \subset \Re^{d \times 9}$ is the matrix of nine basis images, every column of which is the vector of the basis image. $\mathbf{s} \subset \Re^{d \times 1}$ is the vector of lighting coefficients which denotes the lighting conditions of the image. Error term $e(\mathbf{s}) \subset \Re^{d \times 1}$ is related to the pixels' position and lighting conditions.

For a novel image, we estimate its basis images through the maximum a posterior (MAP) estimation. That is

$$B_{MAP} = \arg \max_{B} P(B|I) \qquad (7)$$

According to the Bayes rule

$$P(B|I) = \frac{P(I|B)P(B)}{P(I)} \qquad (8)$$

where $P(I)$ is the evidence factor which guarantees that posterior probabilities would sum to one. Then Eq. (7) can become

$$B_{MAP} = \arg \max_B (P(I|B)P(B)) \tag{9}$$

In order to recover the set of basis images from an image with Eq. (9), one should know the PDF of the basis images, i.e. $P(B)$, and the PDF of the likelihood, i.e. $P(I|B)$. Assuming the error term of Eq. (6) is normally distributed with mean $\mu_e(x, \mathbf{s})$ and variance $\sigma_e^2(x, \mathbf{s})$, we can deduce that the PDF of the likelihood $P(I|B)$, is also Gaussian with mean $B\mathbf{s} + \mu_e(\mathbf{s})$ and variance $\sigma_e^2(\mathbf{s})$ according to Eq. (6).

We assume that the PDF of the basis images B are Gaussians of unknown means μ_B and covariances C_B as in [25, 34]. The probability $P(B)$ can be estimated from the basis images of the training set. In our experiments, the basis images of 20 different subjects from the extended Yale face database B [16] are introduced to the bootstrap set. Note that, the basis images of every subject are real images which were taken under certain lighting conditions. The lighting conditions are determined by the universal configurations of the 9PL subspace. The sample mean μ_B and sample covariance matrix C_B are computed. Figure 3b shows the mean basis images, i.e. μ_B.

The error term $e(\mathbf{s}) = I - B\mathbf{s}$ models the divergence between the real image and the estimated image which is reconstructed by the low dimensional subspace. The error term is related to the lighting coefficients. Hence, we need to know the lighting coefficients of the training images before estimating the error term. For a training image, its lighting coefficient can be estimated by solving the linear equation $I = B\mathbf{s}$. For every subject in the extended Yale face database B, there are 64 images under different lighting conditions. We use the images of the same 20 subjects whose basis images are used for training the mean value of basis, i.e. μ_B, for computing the statistics of the error. Under a certain lighting condition, we estimate the lighting coefficients of every subject's image, i.e. $\mathbf{s}_k^p$ (the lighting coefficients of the pth subject's image under the lighting condition $\mathbf{s}_k$. The mean value of different subjects' lighting coefficients can be the estimated coefficients $(\bar{\mathbf{s}}_k)$ for that lighting condition, i.e. $\bar{\mathbf{s}}_k = \sum_{p=1}^N \mathbf{s}_k^p / N$. Then, under a certain lighting condition, the error term the of the pth subject's image is

$$e_p(\bar{\mathbf{s}}_k) = I_k^p - B_p \bar{\mathbf{s}}_k \tag{10}$$

where I_k^p is the training image of the pth subject under lighting condition $\mathbf{s}_k$ and B_p is the basis images of the pth subject. Following the above assumption, we can estimate the mean $\mu_e(\bar{\mathbf{s}}_k)$ and variance $\sigma_e^2(\bar{\mathbf{s}}_k)$ of the error term.

4   Estimating the Basis Images

As described in the previous section, the basis images of a novel image can be recovered by using the MAP estimation. But before applying the statistical model to estimating the basis images, one needs to first estimate the lighting coefficient **s** of the image. Then, the error term under those lighting conditions can be estimated.

4.1   Estimating Lighting Coefficients

Lighting influences greatly the appearance of an image. Under similar illumination, images of different subjects will appear almost the same. The difference between the images of the same subject under different illuminations is always larger than that between the images of different subjects under the same illumination [35]. Therefore we can estimate the lighting coefficients of a novel image with an interpolation method. The kernel regression is a smooth interpolation method [36]. It is applied to estimating the lighting coefficients. For every training image, we have their corresponding lighting coefficients. For a novel image I_n, its lighting coefficient is given by

$$\mathbf{s} = \frac{\sum_{k=1}^{M} w_k \mathbf{s}_k^p}{\sum_{k=1}^{M} w_k} \tag{11}$$

$$w_k = \exp(-\frac{[D(I_n, I_k^p)]^2}{2(\sigma_{I_k^p})^2}) \tag{12}$$

where $D(I_n, I_k^p) = \|I_n - I_k^p\|_2$ is the L_2 norm of the image distance. $\sigma_{I_k^p}$ determines the weight of test image I_k^p in the interpolation. M is the number of images that have similar lighting condition. In the training set, every subject has 64 different images and there are altogether 20 different subjects. Thus, for a novel image, there will be $M = 20$ images with similar illumination. In our experiment, we assign the farthest distance of these 20 images from the probe image to $\sigma_{I_k^p}$. $\mathbf{s}_k^p$ is the lighting coefficient of image I_k^p.

4.2   Estimating the Error Term

The error term denotes the difference between the reconstructed image and the real image. This divergence is caused by the fact that the 9PL subspace is the low-dimensional approximation to the lighting subspace, and it only accounts for the low frequency parts of the lighting variance. Given the statistics of the error term under known illumination, i.e. $\mu_e(\bar{s}_k)$, $\sigma_e^2(\bar{s}_k)$, those under a new lighting condition can be estimated, also via the kernel regression method [34].

$$\mu_e(\mathbf{s}) = \frac{\sum_{k=1}^{M} w_k \mu_e(\bar{s}_k)}{\sum_{k=1}^{M} w_k} \tag{13}$$

$$\sigma_e^2(\mathbf{s}) = \frac{\sum_{k=1}^{M} w_k \sigma_e^2(\bar{s}_k)}{\sum_{k=1}^{M} w_k} \tag{14}$$

$$w_k = \exp(-\frac{[D(\mathbf{s}, \bar{s}_k)]^2}{2[\sigma_{\bar{s}_k}]^2}) \tag{15}$$

where $D(\mathbf{s}, \bar{s}_k) = \|\mathbf{s} - \bar{s}_2\|_2$ is the L_2 norm of the lighting coefficient distance. Like $\sigma_{I_k^p}$, $\sigma_{\bar{s}_k}$ determines the weight of the error term related to the lighting coefficients $\bar{s}_k$. Also, we assign the farthest lighting coefficient distance of these 20 images from the probe image to $\sigma_{\bar{s}_k}$.

4.3 Recovering the Basis Images

Given the estimated lighting coefficients $\mathbf{s}$ and the corresponding error term $\mu_e(\mathbf{s})$, $\sigma_e^2(\mathbf{s})$, we can recover the basis images via the MAP estimation. If we apply the log probability, omit the constant term, and drop $\mathbf{s}$ for compactness, Eq. (9) can become

$$\arg\max_{B} \left(-\frac{1}{2}(\frac{I - B\mathbf{s} - \mu_e}{\sigma_e})^2 - \frac{1}{2}(B - \mu_B)C_B^{-1}(B - \mu_B)^T \right) \tag{16}$$

To solve Eq. (16), we estimate the derivatives,

$$-\frac{2}{\sigma_e^2}(I - B\mathbf{s} - \mu_e)\mathbf{s}^T + 2(B - \mu_B)C_B^{-1} = 0 \tag{17}$$

Then we rewrite Eq. (17) as a linear equation,

$$AB = b \tag{18}$$

where $A = \frac{\mathbf{s}\mathbf{s}^T}{\sigma_e^2} + C_B^{-1}$ and $b = \frac{I - \mu_e}{\sigma_e^2}\mathbf{s} + C_B^{-1}\mu_B$. The solution of the linear equation is $B = A^{-1}b$. Using the Woodbury's identity [37], we can obtain an explicit solution

$$B_{MAP} = A^{-1}b$$

$$= \left(C_B - \frac{C_B \mathbf{ss}^T C_B}{\sigma_e^2 + \mathbf{s}^T C_B \mathbf{s}} \right) \left(\frac{I - \mu_e}{\sigma_e^2} \mathbf{s} + C_B^{-1} \mu_B \right)$$

$$= \left(\frac{I - \mu_B \mathbf{s} - \mu_e}{\sigma_e^2 + \mathbf{s}^T C_B \mathbf{s}} \right) C_B \mathbf{s} + \mu_B \tag{19}$$

From Eq. (19), the estimated basis image is composed of the term of character-istics, $\left(\frac{I - \mu_B \mathbf{s} - \mu_e}{\sigma_e^2 + \mathbf{s}^T C_B \mathbf{s}} C_B \mathbf{s} \right)$, and the term of mean, μ_B. In the term of characteristics, $(I - \mu_B \mathbf{s} - \mu_e)$ is the difference between the probe image and the images recon-structed via the mean basis images.

4.4 Recognition

The most direct way to perform recognition is to measure the distance between probe images and the subspace spanned by the recovered basis images. Every column of B is one basis image. However, the basis images are not orthonormal vectors. Thus we perform the QR decomposition on B to obtain a set of orthonormal basis, i.e. the matrix Q. Then the projection of probe image I to the subspace spanned by B is $QQ^T I$, and the distance between the probe image I and the subspace spanned by B can be computed as $\|QQ^T I - I\|_2$. In the recognition procedure, the probe image is identified as the subspace with minimum distance from it.

5 Experiments on Lighting Estimation

The statistical model is trained by images from the extended Yale face database B [16]. With the trained statistical model, we can reconstruct the lighting subspace from only one gallery image, which should be insensitive to lighting variation. Thus, recognition can be achieved across illumination conditions.

5.1 Recovered Basis Images

To recover the basis images from a single image, the lighting coefficients of the image should be estimated first. Then we estimate the error terms of the image. Finally, the basis images of the image can be obtained with Eq. (19).

Although the images of the same object are under different lighting conditions, the recovered basis images should be similar. The probe images are from the Yale face database B. There are 10 subjects and 45 probe images per subject. According to the lighting conditions of the probe images, they can be grouped into 4 subsets as

Table 1 The subsets of Yale face database B

	subset1	subset2	subset3	subset4
Illumination	0–12	13–25	26–50	50–77
Number of images	70	120	120	140

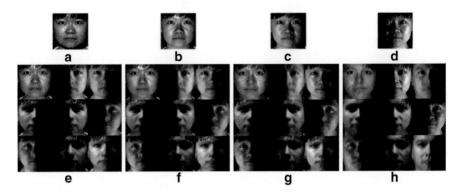

Fig. 4 Recovered basis images. (**a**)–(**d**) are images in subset 1–4 of Yale face database B respectively. (**e**)–(**h**) are recovered basis images from image (**a**)–(**d**) respectively

in [16]. The details can be found in Table 1. From subset1 to subset4, the lighting conditions become extreme. For every subject, we recover its basis images from only one of its probe images each time. Then we can obtain 45 sets of basis images for every subject. Figure 4 shows the basis images recovered from an image of each subset. $\bar{\sigma}_{basis}$ (the mean standard deviation of the 45 sets of basis images of 10 subjects) is 7.76 intensity levels per pixel, while $\bar{\sigma}_{image}$ (the mean standard deviation of the original 45 probe images of 10 subjects) is 44.12 intensity levels per pixel. From the results, we can see that the recovered basis images are insensitive to the variability of lighting. Thus we can recover the basis images of a subject from its images under arbitrary lighting conditions.

5.2 Recognition

Recognition is performed on the Yale face database B [16] first. We take the frontal images (pose 0) as the probe set, which is composed of 450 images (10 subjects, 45 images per subject). For every subject, one image is used for recovering its lighting subspace and the 44 remaining images are used for recognition. The comparison of our algorithm with the reported results is shown in Table 2.

Our algorithm reconstructed the 9PL subspace for every subject. The recovered basis images also contained complicated reflections on faces, such as cast shadows, specularities, and inter-reflection. Therefore the recovered 9PL subspace can give a more detailed and accurate description for images under different lighting conditions. As a result, we can get good recognition results on images with different

Table 2 The recognition error rate of different recognition algorithms on Yale face database B

Algorithms	subset1 & 2	subset3	subset4
Correlation [16]	0.0	23.3	73.6
Eigenfaces [16]	0.0	25.8	75.7
LTV [9]	0.2	21.4	24.5
S&L(LOG-DCT) [5]	0.0	14.0	15.7
S&L(NPL-QI) [5]	0.0	3.2	13
Linear subspace [16]	0.0	0.0	15.0
Cones-attached [16]	0.0	0.0	8.6
Cones-cast [16]	0.0	0.0	0.0
harmonic images-cast [16]	0.0	0.0	2.7
3D based SH model [25]	0.0	0.3	3.1
SH model extreme [25]	7.75	7.5	6.8
BIM (30 bases) [28]	0.0	0.0	0.7
Wang et al. [29]	0.0	0.0	0.1
9PL-real [31]	0.0	0.0	0.0
Intrinsic lighting subspace [20]	0.0	0.0	5.71
Our algorithm	0.0	0.0	0.72

lighting conditions. Also, the reported results of 'cone-cast', 'harmonic images-cast' and '9PL-real' showed that better results can be obtained when cast shadows were considered. Although paper [28, 29] also use one image to adjust lighting conditions, they need to recover the 3D model of the face first. Our algorithm is a completely 2D-based approach. Computationally, it is much less expensive compared with those 3D based methods. The basis images of a subject can be directly computed with Eq. (19) while the recognition results are comparable to those from the 3D-based methods. Compared with the results of Quotient-based methods [5, 9], the results of lighting-subspace-based methods [20, 25, 28, 29, 31] are much better. That is due to the loss of appearance information in Quotient-based methods. Actually, these information contained a lot of low-frequency information which is important for recognition.

5.3 Multiple Lighting Sources

An image taken under multiple lighting sources can be considered as images taken under a single lighting source being superimposed. Through interpolation, the lighting coefficients of images taken under single lighting are linearly combined to approximate those of the image taken under multiple-lighting. Here we also apply the statistical model trained on the extended Yale Database B to basis images estimation. Similarly the lighting coefficients of images are estimated through

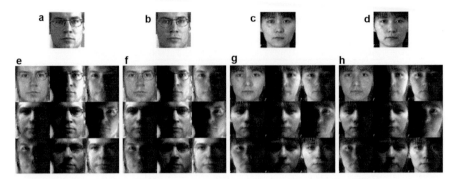

Fig. 5 Recovered basis images. (**a**) and (**b**) are images in PIE database, (**e**) and (**f**) are estimated basis images from image (**a**) and (**b**), respectively. (**c**) and (**d**) are images in AR database, (**g**) and (**h**) are estimated basis images from image (**c**) and (**d**), respectively

Table 3 Recognition rate on different databases

Face database	PIE	AR
$\bar{\sigma}_{basis}$	11.01	11.34
$\bar{\sigma}_{image}$	285	38.59
Recognition rate (%)	98.21	97.75

interpolation. Then the error term can be estimated according to the lighting coefficients. Finally, the basis images are recovered.

In the PIE face database [38], there are 23 images per subject taken under multiple lighting sources, and altogether 69 subjects. We recover 23 sets of the basis images from the 23 images of every subject respectively. With these estimated basis images, we perform recognition on the 1587 images (23 images per person) 23 times. We also estimate basis for images in the AR database [39]. We select randomly 4 images under different illumination per subject (image 1, 5, 6, 7) and recover the respective basis images from those images. Recognition is performed on 504 images (126 subjects and 4 images per subject) 4 times. Samples of the recovered basis images from images in the PIE and ARdatabases are shown in Fig. 5. The average recognition rates, the mean standard deviation of the recovered basis images ($\bar{\sigma}_{basis}$) and the mean standard deviation of the gallery images ($\bar{\sigma}_{images}$) are presented in Table 3. The results show that the statistical model trained by images taken under single lighting source can also be generalized to images taken under multiple lighting sources. The recognition results on PIE dataset is also compared with some lighting pre-processing methods reported in paper [40], as shown in Table 4. From the results, we can see that the proposed algorithm performed better than those reflectance estimation methods [9–12]. That is due to the accurate description of appearance in different lighting conditions with estimated basis images.

Table 4 Comparison with other lighting preprocessing methods on the dataset of PIE

Algorithm	LTV [9]	TT [10]	LDCT [11]	LN [12]	Our algorithm
Recognition rate	89.1	94.4	92.1	97.1	98.21

6 Perception-Based Lighting Model

The Human Vision System (HVS) can adapt very well under enormously changed lighting conditions. People can see well at daytime and also at night. That is due to the accurate adaptation ability of the HVS. However, image capturing devices seldom have this adaptation ability. For an image taken under extreme lighting conditions, such as the images shown in first row of Fig. 7b, a proper lighting adjustment algorithm should not only adjust the brightness of the images, but also enhance the features of the image, especially for the dark regions. To reach this goal, we propose to reduce the light variations by an adaptive adjustment of the image. Here, we employ a model of photoreceptor adaptation in Human Vision System [41] in which three parameters (α, f, m) control the lighting adjustment. The adjusted image Y is modeled as a function of these lighting parameters and the input image X as:

$$Y(\alpha, m, f; X) = \frac{X}{X + \sigma(X_a)} V_{max} \qquad (20)$$

where σ, referred to as *semi-saturation constant*, X_a the *adaptation level*, and V_{max} determines the maximum range of the output value (we use $V_{max} = 255$ to have grey image output in the range of $[0, 255]$). The semi-saturation constant σ describes the image intensity and its contrast through the parameters f and m, respectively [41]:

$$\sigma(X_a) = (fX_a)^m \qquad (21)$$

Adaptation Level I_a If we choose the average intensity of the image as the adaptation level I_a, the adjustment is global. It does not perform any specific processing to the darker or brighter region and some details in those regions may be lost. To compensate the details, the local conditions of every point should be considered. We can use the bi-linear interpolation to combine the global adaptation I_a^{global} and local adaptation $I_a^{local}(x, y)$ as,

$$I_a(x, y) = \alpha I_a^{local}(x, y) + (1 - \alpha)I_a^{global} \qquad (22)$$

$$I_a^{local}(x, y) = K(I(x, y)) \qquad (23)$$

$$I_a^{global} = mean(I) \qquad (24)$$

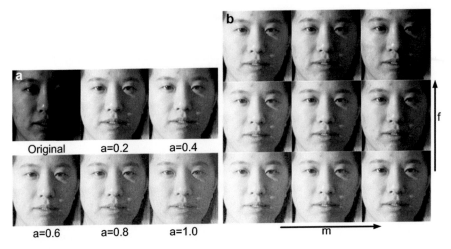

Fig. 6 (a) Adaptation level v.s. α parameter, (b) lighting adjustment v.s. m and f parameters

Different kernel $K(\bullet)$ can be applied to extract the local information. Gauss kernel is the most commonly used one. The interpolation of the global and local information will adjust the details. In Fig. 6a, with the increasing of the parameter α, the details become notable gradually. When $\alpha = 1$, i.e. $I_a = I_a^{local}$, all the details are expressed out including the noise.

Parameter f and m The other two parameters f and m control the intensity and contrast, respectively. Parameter f is the multiplier in the adaptation function, i.e. to every point's adaptation level $I_a(x, y)$, f magnifies them with the same scale. The brightness of the whole image will be enhanced or suppressed accordingly.

The alternation of brightness can be shown only when changes on f is large enough. In [41], the parameter f is suggested to be rewritten in the following form

$$f = exp(-f')\tag{25}$$

With a comparative smaller changing range of f', f can alter the brightness of the image.

Parameter m is an exponent in the adaptation function. Different from the parameter f, m magnifies every $I_a(x, y)$ with a different scale based on its adaptation value. Therefore, parameter m can emphasize the difference between every point, i.e. the contrast. In Fig. 6b, the parameter α is fixed. With the increment of m, the contrast of the image is enhanced in every row. And in every column, the brightness of the image is enhanced with the increase of f.

7   Image Sequence Lighting Adjustment

In capturing an image sequence the influence of the scene lighting may not be neglected. Often the variations of the lighting conditions cannot be avoided while recording, and therefore lighting adjustment methods must be used before further processing. In this paper, we propose a tow-steps lighting adjustment approach. First, the initial optimal parameters, α_k^0, f_k^0, m_k^0 of each frame $X_k; k = 1, \cdots N$ are calculated using entropy as an objective function. These values are then used as initial guesses for a constrained least squares optimization problem for further refinement of those parameter. In this step, the objective function is the difference between the adjusted previous frame Y_{k-1} and the current frame X_k. The two steps are detailed in the following sections, and experimental results are presented in Sect. 8.

7.1   Single Image Enhancement

It is well known that an image with large entropy value indicates that the distribution of its intensity values is more uniform, i.e. each intensity value has almost the same probability to appear in the image. Hence, the image cannot be locally too bright or too dark. Entropy $H(x)$, defined as:

$$H(X) = -\sum_{i=0}^{255} p(i) log_2(p(i)) \tag{26}$$

where $p(i)$ is the probability of the intensity values i in the whole image, can be employed to evaluate image lighting quality. When all the intensity values have the same probability in the image, the entropy can reach its maximum value 8. However, not all the images can reach the entropy $H(X) = 8$ when they are in their best situation. The optimal entropy value, H_o, is image content dependent. In this paper, we set $H_o = 7$ as the expected optimal entropy for all the images. Therefore the objective function for the lighting adjustment of every single image is

$$J_1(\alpha, m, f) = \underset{\substack{a\in[0,1]; m\in[0.3,1) \\ f\in[\exp(-8),\exp(8)]}}{\arg\min} |H(Y(\alpha, m, f; X)) - H_o| \tag{27}$$

The lighting parameter α controls the adaptation level of the images, as in Eq. (23). It can adjust the image much more than the other two parameters (f, m). Therefore an alternate optimization strategy is used [42]. First, the parameter α is optimized with fixed m and f. Then the parameter m and f are optimized with fixed α. These two optimizations are repeated until convergence. To initialize, we estimate $\hat{\alpha}$ with fixed m and f which are selected according to the luminance situation of

a   b

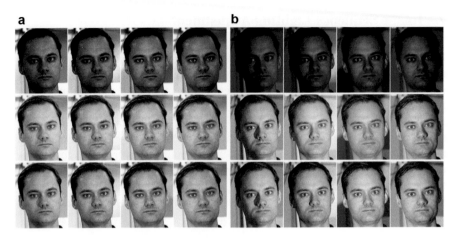

Fig. 7 Lighting adjustment results of frame 1–4 in L1 and L2. (**a**) and (**b**) are results of L1 and L2: *from top to bottom* are original images, entropy-based optimization, and 2-step optimization results, respectively

the image. The contrast-control parameter m can be determined by the key k of the image [41], as

$$m = 0.3 + 0.7k^{1.4} \qquad (28)$$

The key of the image evaluates the luminance range of the image and is defined as

$$k = \frac{L_{max} - L_{av}}{L_{max} - L_{min}} \qquad (29)$$

where L_{av}, L_{min}, L_{max} are the log average, log minimum and log maximum of the luminance respectively. For color images, we use the luminance image computed as $L = 0.2125I_r + 0.7154I_g + 0.0721I_b$, where I_r, I_g, I_b are the red, green, blue channels. The brightness-control parameter f is set to 1. Then the simplex search algorithm [43] is applied for determining the optimal $\hat{\alpha}$. Fixing the value $\hat{\alpha}$ in J_1, the simplex search algorithm is then used to search for optimal $\hat{m}$ and $\hat{f}$. The alternate optimization will stop when the objective function J_1 is smaller than a given threshold.

This approach can adjust an image to have suitable brightness and contrast. Also, it can enhance the local gradient features of the image due to the adjustment of the parameter α. However, entropy does not relate to intensity directly. Different images can have the same entropy value while their brightness is different. For example, the images in the second row of Fig. 7a, b, being the lighting adjusted results of the images of the first row, have the same entropy values, but their lighting conditions are not similar. Consequently, for a sequence of images, we still need to adjust the brightness and contrast of successive frames to be similar and therefore enhance their features.

7.2   Lighting Adjustment of Successive Images

In video sequences, the difference between successive frames is due to object and/or camera motions and lighting changes. Whereas the former differences are exploited in object tracking and camera motion estimation, the latter, i.e. lighting differences, are such that the required brightness constancy assumption for tracking gets violated. In this paper, we show that for tracking of slow movement in a sequence captured by a fixed camera, the lighting problem can be reduced by applying a lighting adjustment method. Indeed, the lighting of the overall sequence could be made more uniform (in a sequential manner) by considering the changes between successive frames. We propose to use the difference between successive frames as an objective function to estimate the optimal lighting parameters of the current frame X_j, provided that the previous frame X_{j-1} has been adjusted, i.e. given Y_{j-1}:

$$J_2(\alpha, m, f) = \underset{\substack{\alpha \in [0,1]; m \in [0.3,1) \\ f \in [\exp(-8), \exp(8)]}}{\arg \min} \sum_x \sum_y \left(Y(\alpha, m, f; X_j(x, y)) - Y_{j-1}\right)^2 \quad (30)$$

With Eq. (20), the difference $e(\alpha, m, f) = Y(\alpha, m, f; X_j) - Y_{j-1}$ between frames can be written as (for simplicity we drop the pixel index (x, y)):

$$e = \frac{X_j}{X_j - (f_j X_{a_j})^{m_j}} - \frac{X_{j-1}}{X_{j-1} - (f_{j-1} X_{a_{j-1}})^{m_{j-1}}} \quad (31)$$

When searching for the optimal parameters for the objective function J_2, the derivatives over different parameters need to calculate.

If the two images concerned are the same, the difference between these images is minimum, at the same time, the difference between the inverse of images will also reach to its minimum value. Therefore, we calculate the difference between the inverse of adjusted images to simplify the computation of derivatives, as

$$\begin{aligned}
\tilde{e} &= \frac{X_j - (f_j X_{a_j})^{m_j}}{X_j} - \frac{X_{j-1} - (f_{j-1} X_{a_{j-1}})^{m_{j-1}}}{X_{j-1}} \\
&= \frac{(f_j X_{a_j})^{m_j}}{X_j} - \frac{(f_{j-1} X_{a_{j-1}})^{m_{j-1}}}{X_{j-1}}
\end{aligned} \quad (32)$$

Let $\hat{Y}_{j-1} = (f_{j-1} X_{a_{j-1}})^{m_{j-1}}/X_{j-1}$ and apply *log* to both side of Eq. (32), we can simplify the difference between frames further as

$$\begin{aligned}
\hat{e} &= \log \frac{(f_j X_{a_j})^{m_j}}{X_j} - \log \hat{Y}_{j-1} \\
&= m_j \log f_j + m_j \log X_{a_j} - \log X_j - \log \hat{Y}_{j-1}
\end{aligned} \quad (33)$$

Then the objective function J_2 can be rewritten as

$$\hat{J}_2(\alpha_j, m_j, f_j) = \underset{\substack{\alpha \in [0,1]; m \in [0.3,1) \\ f \in [\exp(-8), \exp(8)]}}{\arg\min} \sum_x \sum_y \left(m_j \log f_j + m_j \log X_{aj} - \log X_j - \log \hat{Y}_{j-1} \right)^2 \quad (34)$$

This formulation allows easily estimating the partial derivatives, and we apply the interior-point algorithm [44] to solve the optimization problem $\hat{J}_2$, with initial values of the lighting parameters α_j^0, f_j^0 and m_j^0 obtained by minimizing Eq. (27).

8 Experiments on Lighting Adjustment

The proposed lighting adjustment algorithms of the previous section have been tested on the PIE facial database [38], from which we selected images under different lighting conditions to compose 3 test sequences, here referred to as L1, L2 and L3. We intend to take these sequences as typical examples to demonstrate the performance of the algorithm in slight lighting variations (L1), overall dark sequences (L2) and suddenly changing light variations (L3). To show the benefits of the proposed image sequence lighting adjustment approach, we compare it to state-of-art lighting adjustment methods for single images, namely, the quotient image (QI) algorithm [4], and the well known histogram equalization (HE) approach.

The lighting conditions of the test sequences can be described as follows. Sequence L1 and L2 are composed of 19 frames taken from the same person. The first row of Fig. 7 shows the first 4 frames of L1 and L2. The images in L1 are taken with ambient lighting and 19 different point light sources. The positions of these light points, are 10, 07, 08, 09, 13, 14, 12, 11, 06, 05, 18, 19, 20, 21, and 22, respectively. The images in L2 are taken under the same light point source but without ambient lighting, so they appear to be more dark. Sequence L3 is composed of 39 images which come from L1 and L2 alternately. Thus the lighting condition of the images in L3 is ambient lighting on and off alternately. The first row of Fig. 9 shows the frames 9–14 of L3.

To evaluate the lighting quality of the adjusted images, the *key value* (Eq. (29)) and entropy are depicted in Fig. 8. The *key value* of an image evaluates the luminance range of the image. The entropy, being the mean entropy of the 3 color channels, relates to the distribution of the intensity values in each channel.

The key value of all adjusted frames and the original sequence of L3 are shown in Fig. 8d. The key value zigzags due to the alternate brightness of the original sequence L3. For a sequence with reduced lighting variation the key value should stay constant throughout the sequence. Therefore, we show the variance of the key value in Fig. 8b. For all the 3 test sequences, the variance of the key value of the results of the proposed 2-step optimization algorithm is smaller than that of the other algorithms except HE algorithm. However, HE algorithm costs the entropy value of images, whose results are even worse than the original images (Fig. 8a). The reason

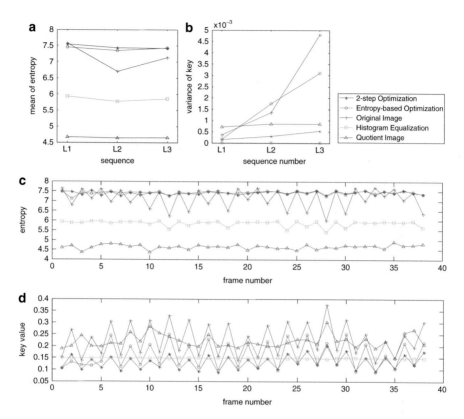

Fig. 8 Entropy and image key curves. (**a**) and (**b**) are the mean entropy and the variance of key of all the frames in the original sequences and adjusted results of the sequences, respectively. (**c**) and (**d**) are the entropy value and key value of every frame in L3 and different adjusted results of L3, respectively

is that HE algorithm can make the intensity distribution uniform only by skipping values in the intensity range [0,255] of the adjusted images, thereby leaving many gaps in the histogram of the adjusted images. The entropy value of the QI results are the smallest because of the loss of the low frequency information in the images. The proposed algorithm is the largest in the mean of entropy, Fig. 8a, and we can also see from Fig. 9a that these resemble most the intensity value distribution of the original images. Our goal is indeed not to change the image appearance dramatically (as compared to QI) but only to obtain a good lighting quality. Therefore, it is quite normal that we couldn't improve L1 sequence so much, which is already captured at a reasonable lighting quality with the ambient light. However, we were still able to adjust its brightness to be more uniform while keeping its high image quality, as shown in Fig. 7a. On the other hand, our 2-step algorithm enhanced the image lighting quality significantly for the sequences L2 and L3 containing images taken under extreme lighting conditions.

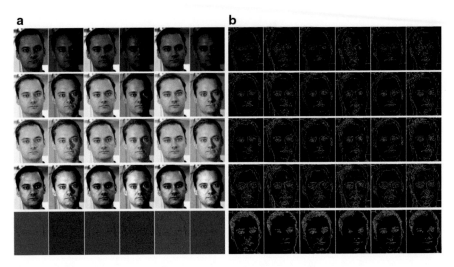

Fig. 9 Lighting adjustment results of frame 9–14 in sequence L3. (**a**) *From top to bottom* are the original images, entropy-based optimization, 2-step optimization, histogram equalization (HE), and quotient image (QI) results, respectively. (**b**) The edge of corresponding images in (**a**)

Next, we examine the effect of the lighting adjustment methods on the object's edges of Fig. 9b to determine if the methods are appropriate as pre-processing for feature detection methods. Considering the edges in the adjusted images, our proposed algorithm enhances the feature of images. This is especially the case for those images taken in a dark environment. Also, highlight are compensated and the influence of shadows on the edges are reduced. The HE algorithm was able to enhance the contrast of the image but at the same time it enhanced noise as well. As we already mentioned, the QI algorithm removed most low frequency information of the image thereby included some important features of the image.

The advantage of the image difference-based optimization step is illustrated for facial feature tracking (on the sequences L1–L3). We demonstrate that the difficulty of tracking a modified object appearance due to lighting changes can be overcome by employing our proposed algorithm as pre-processing. In this paper, we focus on the results of a template-based eye and mouth corner tracker. That tracker is part of a previously developed approach to automatically locate frontal facial feature points under large scene variations (illumination, pose and facial expressions) [45]. This approach consisted of three steps: (1) we use a kernel-based tracker to detect and track the facial region; (2) we constrain a detection and tracking of eye and mouth facial features by the estimated face pose of (1) by introducing the parameterized feature point motion model into a Lukas-Kanade tracker; (3) we detect and track 83 semantic facial points, gathered in a shape model, by constraining the shapes rigid motion and deformation parameters by the estimated face pose of (1) and by the eyes and mouth corner features location of (2).

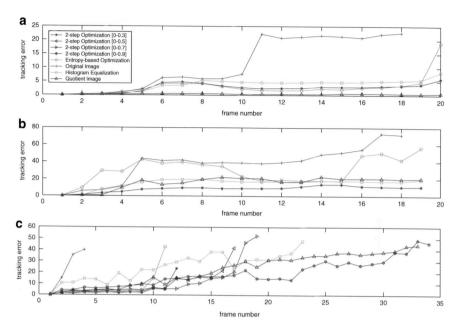

Fig. 10 Feature points tracking error. (**a**)–(**c**) are tracking errors for sequence L1 to L3, respectively

The performance of the tracking of the eyes and mouth corners (6 feature points) on the original and adjusted image sequences L1–L3 is displayed in Fig. 10. The tracking error per frame is calculated as the average distance between the real positions (manually identified) and the tracked positions of the 6 feature points in the image. When tracking was lost, the graph is truncated. Figure 10a shows that all adjustments of the sequence L1 allow to track until the end of that sequence. The QI shows the smallest tracking error because it enhances the gradient features in the image, but at the cost of obtaining visually unpleasant images (see last row of Fig. 9a). Compared to the HE results, our two-step optimization does reach a better tracking performance. Because the initial lighting variations in sequence L1 are not that big, the entropy-step alone may already improve the tracking. The benefit of the image difference-based optimization step becomes obvious via the tracking error graphs of the dark sequence L2 in Fig. 10b. Here, the tracking errors on the 2-step optimization are the smallest. This shows that local features are enhanced very well, but also that taking care of correspondences between images is indeed important. QI and HE adjustments perform worse in tracking. For QI, the reason is that it may enhance the local features (gradients) only when the noise level is not high, i.e. images taken in good lighting conditions such as in L1. On the alternating dark and light sequence L3 the tracking of the original and entropy-optimized sequence is very quickly lost, as shown in Fig. 10c. It is thus crucial to take into account the sequence aspects in lighting adjustment. It is worth noting that the tracking for our proposed algorithm results was lost only when a part of the image were

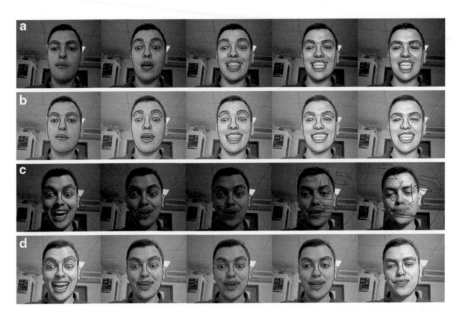

Fig. 11 Shape model results. (**a**) results on original sequence M1 with 10 frames (there are frame 1, 3, 5, 7, 10 *from left to right*), (**b**) corresponding results on adjusted sequence M1; (**c**) results on original sequence M2 with 14 frames (there are frame 1, 4, 6, 9, 14 *from left to right*), (**d**) corresponding results on adjusted sequence M2

in deep shadow (such as frame 12, 17 and 19). Although no adjustment method can track until the end of the sequence, we see that a larger enhancement of the local features may allow to track longer (reduced entropy). That was done by enlarging the alpha range from $[0, 0.3]$ to $[0, 0.9]$ in the 2-step optimization (Eqs. (27) and (30)). When comparing the errors before tracking was lost, we see that reducing frame differences, especially with small alpha range, increases the accuracy of the tracking. This shows that enhancing image sequence quality can also help to track.

Then we tested the constrained shape model tracking (step (3) of [45]) on a sequence [46] adjusted by the 2-step lighting optimization. Before adjustment, shown in Fig. 11a, c, some tracked features could not be well delineated due to the illumination changes in the image sequence. The intensity and texture of the face image were improved by our lighting adjustment and therefore all shape points were tracked more efficiently as shown in Fig. 11b, d.

9   Conclusion

Lighting is always a crucial problem for image based pattern recognition due to the reason that lighting is the main factor that makes the image. Therefore when lighting changes, images of objects will also change such that difficulties arise in detecting, recognizing and tracking them throughout images.

To deal with the practical requirement of few training images, we built a statistical model based on the 9PL theory. With the MAP estimation, we can recover the basis images from one gallery image under arbitrary lighting conditions, which could be single lighting source or multiple lighting sources. The experimental results based on the recovered subspace are comparable to those from other algorithms that require lots of gallery images or the geometric information of the subjects. Even in extreme lighting conditions, the recovered subspace can still appropriately represent lighting variation. The recovered subspace retains the main characteristics of the 9PL subspace. Based on our statistical model, we can build the lighting subspace of a subject from only one gallery image. It avoids the limitation of requiring tedious training or complex training data, such as many gallery images or the geometric information of the subject. After the model has been trained well, the computation for recovering the basis images is quite simple and without the need of 3D models. Besides faces, the proposed algorithm is able to be generalized to model lighting variation for images of other objects in a fixed pose.

Currently, most of the algorithms that deal with the lighting problems are only aimed at adjusting the lighting condition of one image. Furthermore, it is not practical to build lighting models for each object in a complicated scene, especially when the object is unknown beforehand. For the application of image sequences processing, we proposed a 2-step lighting adjustment algorithm to reduce the variations of lighting in an image sequence. First, an entropy-based algorithm is applied to calculate initial lighting parameters of a perceptual lighting model. Then the difference between current and previous frames is employed as an objective function for the further optimization of those lighting parameters. Using this criteria, successive frames are adjusted to have similar brightness and contrast. Image lighting quality, measured by entropy and key value, but also local features are enhanced. The proposed two-step lighting adjustment algorithm can be applied to any image sequences besides facial sequences. We did demonstrate the effectiveness of the proposed algorithm for subsequent image processing, such as detection and tracking.

Acknowledgements This material is based upon work supported by the PhD Programs Foundation of Ministry of Education of China (Grant No. 20136102120041, 20116102120031), National High-tech Research and Development Program of China(863 Program) (No. 2014AA015201), National Natural Science Foundation of China (No. 61103062, No. 61502388), and the Fundamental Research Funds for the Central Universities (No. 3102015BJ(II)ZS016).

References

1. Y. Gao, M.K. Leung, Face recognition using line edge map. IEEE Trans. Pattern Anal. Mach. Intell. **24**(6), 764–779 (2002)
2. H. Zhou, A.H. Sadka, Combining perceptual features with diffusion distance for face recognition. IEEE Trans. Syst. Man Cybern. Part C **41**(5), 577–588 (2011)
3. A. Shashua, T. Riklin-Raviv, The quotient image: class-based re-rendering and recognition with varying illuminations. IEEE Trans. Pattern Anal. Mach. Intell. **23**(2), 129–139 (2001)

4. H. Wang, S.Z. Li, Y. Wang, Generalized quotient image, in *IEEE Conference on Computer Vision and Pattern Recognition (CVPR)*, vol. 2 (IEEE, New York, 2004), pp. 498–505

5. X. Xie, W.S. Zheng, J. Lai, P.C. Yuen, C.Y. Suen, Normalization of face illumination based on large-and small-scale features. IEEE Trans. Image Process. **20**(7), 1807–1821 (2011)

6. X. Zhao, S.K. Shah, I.A. Kakadiaris, Illumination normalization using self-lighting ratios for 3d2d face recognition, in *European Conference on Computer Vision (ECCV) Workshops and Demonstrations* (Springer, New York, 2012), pp. 220–229

7. Y. Fu, N. Zheng, An appearance-based photorealistic model for multiple facial attributes rendering. IEEE Trans. Circuits Syst. Video Technol. **16**(7), 830–842 (2006)

8. M. De Marsico, M. Nappi, D. Riccio, H. Wechsler, Robust face recognition for uncontrolled pose and illumination changes. IEEE Trans. Syst. Man Cybern. Syst. **43**(1), 149–163 (2013)

9. T. Chen, W. Yin, X.S. Zhou, D. Comaniciu, T.S. Huang, Total variation models for variable lighting face recognition. IEEE Trans. Pattern Anal. Mach. Intell. **28**(9), 1519–1524 (2006)

10. X. Tan, B. Triggs, Enhanced local texture feature sets for face recognition under difficult lighting conditions. IEEE Trans. Image Process. **19**(6), 1635–1650 (2010)

11. W. Chen, M.J. Er, S. Wu, Illumination compensation and normalization for robust face recognition using discrete cosine transform in logarithm domain. IEEE Trans. Syst. Man Cybern. B Cybern. **36**(2), 458–466 (2006)

12. X. Xie, K.M. Lam, An efficient illumination normalization method for face recognition. Pattern Recogn. Lett. **27**(6), 609–617 (2006)

13. P.W. Hallinan, A low-dimensional representation of human faces for arbitrary lighting conditions, in *IEEE Conference on Computer Vision and Pattern Recognition (CVPR)* (IEEE, New York, 1994), pp. 995–999

14. S.K. Nayar, H. Murase, Dimensionality of illumination in appearance matching, in *IEEE International Conference on Robotics and Automation*, vol. 2 (IEEE, New York, 1996), pp. 1326–1332

15. P.N. Belhumeur, D.J. Kriegman, What is the set of images of an object under all possible illumination conditions? Int. J. Comput. Vis. **28**(3), 245–260 (1998)

16. A.S. Georghiades, P.N. Belhumeur, D. Kriegman, From few to many: illumination cone models for face recognition under variable lighting and pose. IEEE Trans. Pattern Anal. Mach. Intell. **23**(6), 643–660 (2001)

17. S.K. Zhou, G. Aggarwal, R. Chellappa, D.W. Jacobs, Appearance characterization of linear lambertian objects, generalized photometric stereo, and illumination-invariant face recognition. IEEE Trans. Pattern Anal. Mach. Intell. **29**(2), 230–245 (2007)

18. R. Ramamoorthi, P. Hanrahan, On the relationship between radiance and irradiance: determining the illumination from images of a convex lambertian object. J. Opt. Soc. Am. A **18**(10), 2448–2459 (2001)

19. R. Basri, D.W. Jacobs, Lambertian reflectance and linear subspaces. IEEE Trans. Pattern Anal. Mach. Intell. **25**(2), 218–233 (2003)

20. C.P. Chen, C.S. Chen, Intrinsic illumination subspace for lighting insensitive face recognition. IEEE Trans. Syst. Man Cybern. B Cybern. **42**(2), 422–433 (2012)

21. A. Wagner, J. Wright, A. Ganesh, Z. Zhou, H. Mobahi, Y. Ma, Toward a practical face recognition system: robust alignment and illumination by sparse representation. IEEE Trans. Pattern Anal. Mach. Intell. **34**(2), 372–386 (2012)

22. L. Zhuang, A.Y. Yang, Z. Zhou, S.S. Sastry, Y. Ma, Single-sample face recognition with image corruption and misalignment via sparse illumination transfer, in *IEEE Conference on Computer Vision and Pattern Recognition (CVPR)* (IEEE, New York, 2013), pp. 3546–3553

23. L. Li, S. Li, Y. Fu, Discriminative dictionary learning with low-rank regularization for face recognition, in *10th IEEE International Conference Automatic Face and Gesture Recognition* (2013)

24. Y. Zhang, M. Shao, E. Wong, Y. Fu, Random faces guided sparse many-to-one encoder for pose-invariant face recognition, in *International Conference on Computer Vision (ICCV)* (2013)

25. L. Zhang, D. Samaras, Face recognition under variable lighting using harmonic image exemplars, in *IEEE Conference on Computer Vision and Pattern Recognition (CVPR)*, vol. 1 (IEEE, New York, 2003), I–19
26. Z. Wen, Z. Liu, T.S. Huang, Face relighting with radiance environment maps, in *IEEE Conference on Computer Vision and Pattern Recognition (CVPR)*, vol. 2 (IEEE, New York, 2003), II–158
27. L. Zhang, S. Wang, D. Samaras, Face synthesis and recognition from a single image under arbitrary unknown lighting using a spherical harmonic basis morphable model, in *IEEE Conference on Computer Vision and Pattern Recognition (CVPR)*, vol. 2 (IEEE, New York, 2005), pp. 209–216
28. J. Lee, J. Moghaddam, H. Pfister, R. Machiraju, A bilinear illumination model for robust face recognition, in *International Conference on Computer Vision (ICCV)*, vol. 2 (IEEE, New York, 2005), pp. 1177–1184
29. Y. Wang, Z. Liu, G. Hua, Z. Wen, Z. Zhang, D. Samaras, Face re-lighting from a single image under harsh lighting conditions, in *IEEE Conference on Computer Vision and Pattern Recognition (CVPR)* (IEEE, New York, 2007), pp. 1–8
30. X. Zhao, G. Evangelopoulos, D. Chu, S. Shah, I.A. Kakadiaris, Minimizing illumination differences for 3d to 2d face recognition using lighting maps. IEEE Trans. Cybern. **44**(5), 725–736 (2014)
31. K.C. Lee, J. Ho, D. Kriegman, Acquiring linear subspaces for face recognition under variable lighting. IEEE Trans. Pattern Anal. Mach. Intell. **27**(5), 684–698 (2005)
32. X. Jiang, Y.O. Kong, J. Huang, R. Zhao, Y. Zhang, Learning from real images to model lighting variations for face images, in *European Conference on Computer Vision (ECCV)* (2008), pp. 284–297
33. X. Jiang, P. Fan, I. Ravyse, H. Sahli, J. Huang, R. Zhao, Y. Zhang, Perception-based lighting adjustment of image sequences, in *Asian Conference on Computer Vision (ACCV)* (2009), pp. 118–129
34. T. Sim, T. Kanade, Combining models and exemplars for face recognition: an illuminating example, in *Proceedings of the CVPR 2001 Workshop on Models versus Exemplars in Computer Vision*, vol. 1 (2001)
35. Y. Adini, Y. Moses, S. Ullman, Face recognition: the problem of compensating for changes in illumination direction. IEEE Trans. Pattern Anal. Mach. Intell. **19**(7), 721–732 (1997)
36. C.G. Atkeson, A.W. Moore, S. Schaal, Locally weighted learning for control, in *Lazy Learning* (Springer, New York, 1997), pp. 75–113
37. L.L. Scharf, *Statistical Signal Processing*, vol. 98 (Addison-Wesley, Reading, 1991)
38. T. Sim, S. Baker, M. Bsat, The cmu pose, illumination, and expression (pie) database, in *Fifth IEEE International Conference on Automatic Face and Gesture Recognition* (IEEE, New York, 2002), pp. 46–51
39. A.M. Martinez, The ar face database. CVC Technical Report, vol. 24 (1998)
40. H. Han, S. Shan, X. Chen, W. Gao, A comparative study on illumination preprocessing in face recognition. Pattern Recogn. **46**(6), 1691–1699 (2013)
41. E. Reinhard, K. Devlin, Dynamic range reduction inspired by photoreceptor physiology. IEEE Trans. Vis. Comput. Graph. **11**(1), 13–24 (2005)
42. X. Jiang, P. Sun, R. Xiao, R. Zhao, Perception based lighting balance for face detection, in *Asia Conference on Computer Vision, ACCV06* (2006), pp. 531–540
43. J.A. Nelder, R. Mead, A simplex method for function minimization. Comput. J. **7**, 308–313 (1965)
44. R.A. Waltz, J.L. Morales, J. Nocedal, D. Orban, An interior algorithm for nonlinear optimization that combines line search and trust region steps. Math. Program. **107**(3), 391–408 (2006)
45. Y. Hou, H. Sahli, I. Ravyse, Y. Zhang, R. Zhao, Robust shape-based head tracking, in *Proceedings of the Advanced Concepts for Intelligent Vision Systems (ACIVS 2007)*, eds. by J. Blanc-Talon, W. Philips, D. Popescu, P. Scheunders. Springer Lecture Notes in Computer Science, vol. 4678 (2007), pp. 340–351
46. F. Dornaika, F. Davoine, Simultaneous facial action tracking and expression recognition in the presence of head motion. Int. J. Comput. Vis. **76**(3), 257–281 (2008)

Advances, Challenges, and Opportunities in Automatic Facial Expression Recognition

Brais Martinez and Michel F. Valstar

Abstract In this chapter we consider the problem of automatic facial expression analysis. Our take on this is that the field has reached a point where it needs to move away from considering experiments and applications under in-the-lab conditions, and move towards so-called in-the-wild scenarios. We assume throughout this chapter that the aim is to develop technology that can be deployed in practical applications under unconstrained conditions. While some first efforts in this direction have been reported very recently, it is still unclear what the right path to achieving accurate, informative, robust, and real-time facial expression analysis will be. To illuminate the journey ahead, we first provide in Sect. 1 an overview of the existing theories and specific problem formulations considered within the computer vision community. Then we describe in Sect. 2 the standard algorithmic pipeline which is common to most facial expression analysis algorithms. We include suggestions as to which of the current algorithms and approaches are most suited to the scenario considered. In Sect. 3 we describe our view of the remaining challenges, and the current opportunities within the field. This chapter is thus not intended as a review of different approaches, but rather a selection of what we believe are the most suitable state-of-the-art algorithms, and a selection of exemplars chosen to characterise a specific approach. We review in Sect. 4 some of the exciting opportunities for the application of automatic facial expression analysis to everyday practical problems and current commercial applications being exploited. Section 5 ends the chapter by summarising the major conclusions drawn.

1 Facial Expression Theory: Three Models

How humans perceive and interpret facial expressions, be it in terms of mental models of emotions and affective states [102], social signals [141], or indicators of health [131], has been widely studied from the perspective of human psychology. These studies have given rise to several theories of how to encode, represent and interpret facial expressions. When the Computer Vision community first tried to

B. Martinez (✉) • M.F. Valstar
School of Computer Science, Jubilee Campus, Wollaton Road, Nottingham NG8 1BB, UK
e-mail: brais.martinez@nottingham.ac.uk; michel.valstar@nottingham.ac.uk

© Springer International Publishing Switzerland 2016
M. Kawulok et al. (eds.), *Advances in Face Detection and Facial Image Analysis*,
DOI 10.1007/978-3-319-25958-1_4

define the problem of the machine analysis of facial expressions, it was only natural to resort to the psychology theories and adopt some of their theories, conventions and coding systems.

In the absence of a unique comprehensive and widely accepted theory multiple Computer Vision approaches to modelling expressive facial behaviour emerged. We describe the following three (non-exhaustive) problem definitions: recognition of prototypical facial expressions, analysis of facial muscle actions, and dimensional affect recognition. Social Signal Processing, which aims to interpret facial displays as social cues [141], and Behaviomedics, which aims to detect medical conditions based on abnormal expressive behaviour [131], can be framed as higher level behaviour interpretation approaches, which can make predictions based on one or more of the three definitions listed above.

One important clarification is the distinction between facial expressions and emotions/affect. The former refers to the *signal* used to convey a message, in this case the facial movements and appearance changes associated to the expression. The latter relates instead to the *message*, i.e., to what the subject wants to convey through the facial expression [24]. For example, a smile (the physical stretching of the lip corners) is a signal, while the message can be e.g. happiness, embarrassment or amusement, depending on context [2]. Some authors prefer to use the term *facial display* to refer to the signal, but the term facial expression is more commonly used and we suffice here with clarifying the distinction between single and message.

Categorical Approach Darwin was the first to theorise that humans have a universal, evolutionary developed and thus in-born way of expressing and understanding a set of so-called basic or prototypical emotions [30]. Further proof of this early work was presented by Ekman [40], who extended the set of basic emotions to six: anger, disgust, fear, happiness, sadness and surprise. Recently, contempt has been added as a seventh basic emotion [81]. As a consequence, for these expressions there is a direct link between the signal (which is what can be *observed* and is therefore the subject of computer vision), and the message. It is because of this very attractive property that this is still the most common perspective to facial expression analysis in Computer Vision.

There are however some other very important shortcomings to the categorical approach, which are becoming more prominent with the advancement of the state of the art. The most relevant of those is the fact that humans make use of a much wider range of facial expressions for everyday communication than the six basic expressions, with some expressions even conveying combinations of the six basic emotions [38]. It is often said that there are approximately 7000 different expressions that people frequently use in everyday life. Furthermore, some of the expressions can have multiple interpretations depending on the context in which they are shown. For example, smiles are often displayed while a person feels embarrassed or is in pain instead of happy. It is thus reasonable to separate the analysis of the signal and, subsequently, analysing the message associated.

Facial Action Coding System (FACS) [42] FACS is a taxonomy of human facial expressions, and is the most commonly used system to objectively describe the

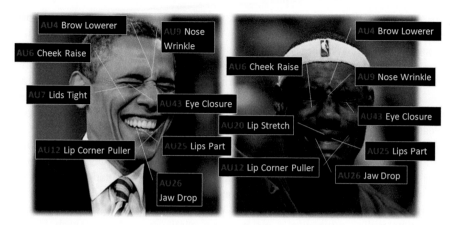

Fig. 1 Two expressive images and the list of active AU (together with their physical meaning) that objectively describe the facial expression

facial expression signal by human observers. FACS was originally developed by Ekman and Friesen [41], while a revised version was presented in [42]. It currently specifies 32 atomic facial muscle actions, named Action Units (AU), and 14 other additional Action Descriptors (ADs) (e.g. bite). FACS has five different intensity levels (not counting neutral), and provides the basis for encoding the temporal segments onset, apex and offset.[1] Being designed for human observers, AUs are described in terms of visible appearance changes. They therefore appear to be a prime candidate for Computer Vision-based detection. Two example images with their associated AU annotations are show in Fig. 1.

Any facial action can be unequivocally encoded in terms of FACS, no message interpretation is required. This allows a two-stage approach to expression analysis, where an expression is first automatically detected in terms of FACS AUs, and interpretation of the meaning of the message being delayed to a second analysis stage. The Computer Vision community has defined a set of problems related to the automatic analysis of AU, such as AU detection [135], AU intensity estimation [63], and the automatic detection of the AU temporal segments [58].

While the system is defined using objective parameters and inter-rater reliability is relatively high, annotation is taxing and very time consuming, and in addition annotators require expert training to be able to produce consistent annotations. Furthermore, the annotation of AU intensities is particularly challenging given the small variation between consecutive levels. Commonly inter-rater reliability for AU intensity coding is lower than that for AU occurrence coding [137].

---

[1]While FACS does not explicitly define temporal phases, there's a large amount of consensus on how to code them. See e.g. [132].

Dimensional Approach While the Categorical approach is only concerned with a small and discrete set of emotions, it is obvious that the complexity of the emotions and affect exhibited by humans has a much wider range and subtleties. The dimensional approach represents affect continuously and multi-dimensionally [48]. The circumplex of affect [105] is the most common dimensional approach model. It represent the affective state of a subject through two continuous-valued variables indicating arousal (ranging from relaxed to aroused) and valence (from pleasant to unpleasant). It is conjectured that each basic emotion corresponds to specific (ranges of) values within the circumplex of affect, while other emotions can be equally mapped into this representation. Dimensions other than valence and arousal can also be considered, augmenting the representational power of the model. Some of the extra dimension most commonly used include power, dominance and expectation. Computer Vision approaches within this problem definition aim at automatically estimating a continuous value for each of the dimensions considered, most commonly on a frame-by-frame basis. The predictions are thus both continuous in time and in value [92].

2   The Standard Algorithmic Pipeline

In the following we will describe the standard algorithmic pipeline for facial expression recognition. While the target of inference depends on the adopted facial expression theory, the considerations regarding the algorithmic pipeline are typically common to each of them, with only the inference layer being specific for the problem of choice. We divide the algorithmic process into three major components: pre-processing (which includes face alignment and face registration), feature extraction, and machine learning. We briefly summarise the major aims and challenges of each of these steps, and we include suggestions of best practice and recommend existing state-of-the-art algorithms that constitute good choices when attempting to build an automatic facial expression recognition system.

Pre-Processing The pre-processing step aims to align and normalise the visual information contained in the face, so that the features extracted capture as much semantic meaning as possible. Features are typically computed at image locations defined in terms of the face bounding box or the face shape (i.e. with respect to facial landmark locations). The alignment step thus consists of registering the coordinate systems in which features are computed, so that they have the same semantic meaning between images. This step is aimed fundamentally at eliminating irrelevant variability in the input signal coming from misalignment, alleviating the effects of head pose variation and identity. The whole process is depicted in Fig. 2.

Face Detection Any face analysis process starts with the detection of the face. The vast majority of the standard datasets contain mostly near-frontal head poses, have good image quality and resolution, and present very few partial occlusions (e.g. no sunglasses, hand gestures covering the mouth, etc). It is thus unsurprising that the

Fig. 2 A standard pre-processing algorithmic pipeline: given an input image, face detection is performed, and facial landmark detection follows. The face shape is used to compute a transformation bringing the face image to an upright position and resize it to a pre-defined scale

Viola and Jones (V&J) face detection algorithm [142] has been deemed sufficiently robust and accurate for most works.

However, moving to in-the-wild imagery requires the use of better face detection algorithms. For example, practical applications cannot guarantee frontal view of the face, and the V&J algorithm simply fails in such cases. Furthermore, the precision of the face detection is an important factor determining the quality of the subsequent facial landmarking step, in particular when tackling challenging scenarios. Some face detection algorithms resulting in state-of-the-art performance include [95, 154], who applied the Deformable Parts Model [44] framework for face detection, and Mathias et al. [80], who use a variant of the V&J model based on the use of multi-channel images. An interesting resource is the Face Detection Database benchmark [53], which offers a comparison under pre-defined conditions and metrics of many of the top-performing face detection algorithms. While the dataset used typically contains images with lower quality than those necessary for facial expression analysis, it is still a good resource, as it typically contains up-to-date benchmarking results for state-of-the-art face detection models.

Facial Landmarking While face detection is the only mandatory step enabling feature extraction, it is advisable to also perform facial landmarking. The localisation of fine-grained inner-facial structures, such as the corner of the eyes and the tip of the nose, allows for a much better registration of the face. It also allows for the direct extraction of geometric features, as well as more powerful local features (see the feature extraction section below).

The appearance of discriminative regression-based facial landmarking approaches [19, 79] has transformed the practical performance of face alignment, avoiding the need for semi-automatic approaches. For 2D imagery, the most prominent facial landmark detector nowadays is the Supervised Descent Method (SDM) [151]. Its success is due to a very high accuracy on images taken from realistic scenarios, an extremely simple implementation and a very low computational cost. The authors provide a publicly available implementation as

Fig. 3 Facial landmarks
automatically detected by the
author's implementation of
[151]. The *green dots*
represent landmarks that are
stable through flexible face
motions such as expressions,
while *blue dots* are landmarks
that can be displaced in the
presence of expressions

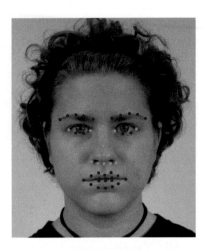

part of the publicly available IntraFace software.[2] This includes a pre-trained model
offering excellent performance. Figure 3 depicts the set of 49 landmarks detected
by the software as provided by the authors.[3] It is however possible to further
fine-tune the detection results, for example by applying a generative model to the
output of the SDM, e.g. [130]. Generative models for facial landmarking tend
to be less robust but can be more precise than discriminative ones. While these
methodologies are precise for a classical "webcam scenario", where the subject
is typically looking towards the camera and keeping a roughly frontal pose, more
unconstrained scenarios can result in poor performance. It is then necessary to
resort to more robust methods [78, 129, 152]. A complementary resource is that
of [161], which is designed to achieve a globally optimal of the face fitting loss
function. While the fitting is not as precise as other methods, its robustness can
serve as an excellent starting point to some of the algorithms mentioned before in
more complicated situations.

It is very common to apply a facial landmark detection algorithm to every frame
in a sequence independently. However, the use of a tracking algorithm can result in
further performance improvement and robustness to several factors such as partial
occlusions. A few works have addressed the problem of facial landmark tracking.
For example, the implementation of the SDM [151] also provides tracking-specific
models. These models differ from the detection models in that the initial shape is
assumed to be much more accurate. This is due to the use of the previous landmark
location to initialise the search rather than using the initialisation provided by the

---

[2]http://www.humansensing.cs.cmu.edu/intraface/.

[3]It is interesting to note that different methods define the set of landmarks to be detected differently.
The widely most common nowadays are the 49 landmarks depicted in Fig. 3, The 66 landmarks
that result from adding 17 landmarks laid on the face contour, and the set of 68 landmarks that
result from adding 2 extra landmarks on the inner lip mouth corners.

face detector bounding box. This method however still disregards important aspects that can be exploited by tracking algorithms: the appearance consistency and the temporal consistency. Appearance consistency was incorporated to this model by Asthana et al. [5], who presented an incremental version of the SDM algorithm. Thus, the models used for inference are adapted in an online manner to the specific characteristics of the test sequence at hand. This results in more stable and precise tracking, in particular for long sequences. However, none of these works impose temporal smoothness on the estimated landmark locations. The consequence is detection jitter, that can hinder the use of geometric features (see below).

Although some works tackle 3D-specific landmark localisation, it is less clear which works can be regarded as the state-of-the-art. Notable examples include efforts making use of regression forest [29]. In this work, the features are relative depth differences between two specific locations. It is however possible to directly extend one of the works for 2D appearance to this case by re-training the appearance models. The appearance model will in this case be trained on the 3D appearance projected on the image plane. An example of a direct adaptation of a 2D facial landmarking method to the 3D case is that of [8] (source code is publicly available), in which the authors adapt the Constrained Local Models method of [108]. Some methods do however apply some feature descriptors that are specific to 3D imagery, such as [97] (only 8 landmarks are detected, although profile faces are considered too), or [20].

Registration Once the facial landmarks have been localised, they can be used to register the faces. For 2D imagery, this involves computing a transformation aligning the detected facial landmarks with a predefined reference shape. The face appearance can then be registered using the transformation computed to register the face shape [61]. Face appearance registration is only necessary when features are used that intent to encode this appearance. A Procrustes transformation is the most widely used registration transform. It involves translation, in-plane rotation, and isotropic scaling parameters (totalling 4 parameters). The difference with respect to an affine transformation is the isotropic scaling (shearing), which reduces the degrees of freedom from 6 to 4. Using a subset of the landmarks, specifically the stable points under expressions (see Fig. 3), to compute the registration transformation can yield some benefit when computing holistic representations (see the text regarding feature extraction bellow).

One of the alternative registration strategies worth mentioning is frontalisation. While this approach has not been properly validated for facial expression recognition, it is an important topic of research within the wider face analysis community. We can distinguish two cases: the frontalisation of the face shape, and the frontalisation of the face appearance. The former task is a significantly easier. One approach to this was proposed by Rudovic and Pantic [104] who used coupled Gaussian Process regression to learn the projection of points in a mesh from non-frontal to frontal view. It is worth noticing that fitting a 3D shape model to a set of 2D landmarks is possible [108]. This is done by finding the 3D shape parameters so that the average point-to-point Euclidean distance between the original 2D landmarks and the projection of the 3D shape is minimised. This is an interesting approach as the 3D shape can then be rotated into a frontal view without distorting expressive information. The 3D face shape is however just an estimation, and therefore even with highly precise 2D landmark detection the computed appearance transformation is likely to corrupt the visual data and result in poor performance. The precision lost and impact caused by this in practice is not yet clear.

Frontalising the face appearance is a very challenging problem. It has very recently received attention [49], partially due to its applicability to face recognition. How to frontalise the face appearance without distorting the expressive information is a very complex problem, and the best way to do so is not yet clear. Some early works in this direction have opted for transforming the face to a frontal neutral face by using a piecewise affine transformation of the face. While this transformation eliminates the configural information from the face, some of the appearance information relating to the wrinkles and bulges produced by facial muscle activations is still kept [4]. This is however an obviously sub-optimal way of frontalisation when it comes to facial expression recognition due to the elimination of important information from the face appearance.

These considerations are not equally relevant for 3D imagery, as head pose rotation is handled in a natural way. Registration in this case is thus reduced to translation, scaling, and dealing with self-occluded parts of the face.

Feature Extraction The choice of face representation is regarded as one of the key aspects of facial expression analysis, and many of the existing works focus on improving this step. The main challenge is that nuisance factors such as subject identity, head pose variation, illumination conditions, or even alignment errors have a larger impact on the appearance than expressive behaviour [109]. Thus, the challenge of feature extraction is to produce features robust to the nuisance factors and yet preserving the expressive information.

It is possible to divide the different feature extraction approaches into geometric, appearance, motion and hybrid features. Geometric features encode information based only on the facial shape locations [132], appearance features encode pixel intensity information instead [61], and motion features are constructed based on a dense registration of appearances between (consecutive) frames [66]. Hybrid features combine at least two of these types of features. We however will not dwell on motion features in this chapter due to their practical shortcomings. When using 3D imagery, it is possible to construct 3D-specific features by incorporating depth information, such as for example the curvature of the face surface at a given point.

Appearance Features We distinguish the following aspects characterising the feature extraction approach: the *feature type* used to represent an image region, and the *representation strategy*, which defines the face regions used to represent it. That is to say, the feature types are *how* appearance is encoded, while the representation strategy defines *what* is encoded.

When referring to the representation strategy, it is common to distinguish between holistic and part-based representations. Holistic representations use global face coordinates to extract the features, while part-based representations apply

Fig. 4 Different representation strategies for facial expression recognition

the feature descriptor to patches defined in terms of the facial landmark or facial component locations. Examples of these strategies are shown in Fig. 4. Both these strategies have different properties: part-based methods offer a very good registration. Since the patches represented are defined in terms of the facial landmarks, the features represent the same part of the face for every example. It is also easier to construct features robust to head pose rotations and illumination variations: head pose rotations can be approximated locally by affine transformations, and illumination variations are approximately locally homogeneous. As a drawback, they have some in-built robustness to the displacement of facial landmarks, which is typical of expressive behaviour. Furthermore, they might not capture the full face appearance (the cheeks for example do not contain landmarks). Holistic representations instead represent the full face and is sensitive to landmark displacements and represent the full face appearance, but lack some of the positive properties of local representations.

Each feature type has different properties and levels of robustness against the different nuisance factors. This also defines the representation strategy they are most suitable for. For example, LBP features [94] are common, and most often used with a holistic representation [114]. This is due to their robustness to illumination changes and to poor registrations. They also tend to encode local information rather than the face structure. However, since they are histogram-based, they are used in combination with a strategy called tiling. Tiling divides the full face bounding box in a grid manner. Then, features are extracted on each sub-patch, and all the resulting feature vectors are concatenated into a single one [60]. Instead, local features are commonly used with HOG features [28]. This is due to HOG features being very robust to affine transformations (as those related to head pose rotation in part-based representations) and to uniform illumination changes. Instead, if they were applied holistically, local fine-grained information would be shadowed by coarser face structure information. Instead, Gabor features can be applied both in a holistic and local representations, although only the Gabor magnitude is used to increase the robustness to misalignment. This is due to the capability of Gabor features to capture local structures and, specifically, bulges and wrinkles typical of facial muscle activations. Historically, Gabor wavelets were one of the first features used for facial expression recognition [76]. Their very large feature dimensionality and the challenge posed by finding the right parametrisation of the Gabor features are the major drawbacks.

Geometric Features They encode relations between the face shape locations, by for example computing the location of a landmark respect to the mean (neutral) face, the distance between two landmarks, or the angle formed by the segments joining three landmarks [132]. These features are very attractive due to their intuitive interpretation. They are easy to implement, and run very fast (once the landmarks are detected). Geometric features are invariant to illumination conditions, and non-frontal head poses can be dealt with by registering the shapes to a frontal head pose. They can easily be applied to 3D too [123]. In fact, geometric features might be even more interesting in the 3D case since distances on 3D are more meaningful than distances on the image plane.

Learned Features Another way of dividing types of features is into the categories hand-crafted and learned. In this ontology, all the features mentioned above are termed 'hand-crafted', as they are the result of mathematical descriptors designed with certain properties in mind, for example the illumination invariance of LBP or the scale-invariance of SIFT. While many of these have proven to be very effective, they are basically the result of expert knowledge of the domain. Another approach to creating features is to learn them from data. This has become very popular with the advent of Deep Learning, in particular Deep Convolutional Neural Networks and (stacked) auto-encoders.

The beauty of learned features is that they can be learned in an unsupervised manner. All that is needed is a very large amount of relevant data. For example, auto-encoders are Artificial Neural Networks that take an image as input, have one hidden-layer with fewer neurons than input nodes, and the output layer is again the same dimensionality as the input layer. The learning task is then to reconstruct the input image at the input, where feature learning happens by the fact that the lower number of hidden neurons effectively forces a dimensionality reduction. Because this is an unsupervised approach, one does not need a large dataset of annotated facial expressions, one merely needs a very large set of images with faces. The latter is very easy to obtain by making use of the Internet.

3D Features The appearance of 3D imagery has resulted in the proposition of a wide range of feature descriptors. Some of them are extensions to 3D of an existing 2D feature type, such as the adaptation of LBP features to 3D [106]. Instead, some 3D features are specific of this modality [77], and can encode aspects like the curvature of the surface [113]. 3D features are invariant to illumination conditions and to head pose variations, making them very interesting in practise. Since 3D feature design is a relatively recent and understudied problem compared to 2D features, this is an interesting area of research with new features being proposed on a regular basis. It is thus only natural that there is not currently a standard and widely accepted 3D feature descriptor for facial expression recognition.

Machine Learning While the previous steps of the algorithmic pipeline are shared among different affect sub-problems, the Machine Learning techniques used are generally specific for each problem. Inference (i.e. prediction) of expressions can be targeted at frame-level labelling or sequence-level labelling. To be specific, frame-level inference assigns a separate label prediction to every frame, whereas sequence-level prediction assigns one (possibly multi-dimensional) label to a number of frames that make up the sequence.

Sequence labelling was often considered in early works and datasets due to its simplicity. It is however a restrictive scenario, as it requires a mechanism for segmenting the input data into segments. Almost every work, past and present, assumes the availability of pre-segmented sequences, which is generally unrealistic in practical scenarios. However, provided such a segmentation is available, it is then possible to directly apply sequence-based classifiers such as HMM [23], or to classify the sequence based on the majority vote of a frame-level classifier [133]. Other techniques such as multiple-instance labelling have been proposed as well [125].

Frame-level inference can be performed with a number of methodologies. For example, Support Vector Machines, Boosting/Ensemble Learning techniques or logistic regression are all reasonable choices for classification problems. It is interesting to note that facial expression recognition can rely on a multi-class classifier [70, 114] (only one out of the k classes is assigned) or on multiple binary classifiers [60] (multiple classes can be active simultaneously). Regression techniques, such as valence and arousal prediction or AU intensity estimation, are better tackled with regression techniques such as Support Vector or Relevance Vector Regression.

The performances attained by different ML techniques in regards to frame-level predictions are comparable, so in practice is little gain to be attained by trying different frame-based classifiers in terms of classification accuracy. However, considering output correlations is a much more attractive aspect. Again, the correlations to be consider can vary depending on the problem considered. All frame-level approaches have strong correlations on the temporal dimension, so it is possible to exploit the fact that the labelling of consecutive frames has to be consistent and smooth. For example, a positive label in-between negative labels can be frequent if no temporal information is used. However, this labelling pattern is impossible in practice, as an expressive event cannot span only one frame. When co-occurrence of multiple labels for a single frame is possible, then the correlations (such as co-occurrences) between the different labels at a specific temporal point can also be exploited. This is both the case for the automatic analysis of AU [107, 126] and for the analysis of continuous affect dimensions [92].

Another interesting aspect, often ignored, is that of feature fusion. Feature fusion happens when more than one combination of feature type and representation strategy are considered. The underlying idea is that instead of studying which is the best-performing feature, they should be considered instead as complementary. The problem is then defined as finding the *best combination* of different feature types and representations [112]. This interpretation can even be extended to the problem of finding the optimal fusion strategy of 2D and 3D information [127]. While feature fusion can be attained by simply concatenating the features together, feature fusion can also be seen as a learning problem [54].

Finally, other standard aspect of ML refers to the use of unsupervised (e.g. PCA) or supervised feature selection. While this is advisable in general due to the typically large dimensionality of feature vector representing the face, these are however standard techniques. We will thus not discuss them further in this chapter.

3 Challenges

Below we will address what the authors consider to be the most pressing challenges in automatic facial expression recognition. This includes obtaining task-representative data, issues around obtaining ground truth, dealing with occlusions, and modelling dynamics, among others.

Long-Term Challenges The first major long-term challenge of facial expression recognition is attaining fast and reliable in-the-wild performance. Nowadays works are designed and tested using imagery recorded under controlled lab conditions, a bias caused by the dependence on available standard datasets containing this kind of imagery. In-the-lab imagery displays subjects who maintain a frontal or near-frontal head pose, images are acquired under controlled illumination conditions (typically frontal with respect to the subject to avoid cast shadows), self occlusions are not considered (e.g. subjects are instructed not to cover their face or data with self-occlusions is removed), and the image quality is typically high.

Expressive behaviour is often elicited using video clip stimuli, or involve human-computer interaction tasks. Both scenarios reduce the complexity of the data significantly. Instead, in-the-wild conditions do not constrain any of these characteristics. There is an obvious association between in-the-wild data and the sought-after automatic face analysis technologies in real-world applications given that most real-world applications cannot constrain the data acquisition conditions.

The second main challenge concerns the integration of the analysis of human facial expression analysis in a high-level framework modelling human behaviour. Human behaviour is currently analysed from different perspectives, of which facial expressive behaviour is just one aspect. If we are to understand humans, then we should aim for a joint view on human behaviour. For example, cues from audio and verbal content should be included, and facial expressions and head-pose should be jointly analysed, rather than separately as is currently the case. Besides obtaining a *big picture*, i.e., having a fully multi-modal view of communicative intent, taking multiple cues into account can naturally help disambiguate the message as well as improve performance of each of the specific sub-problems, including that of automatic facial expression analysis.

Data Any learning problem is primarily determined by the data available. The dependence of performance on the quality and quantity of data can hardly be overestimated. An inferior method trained with more abundant or higher quality data will most often result in better performance than a superior method trained with lower-quality or less abundant data. This is particularly dramatic in the case of facial expression recognition.

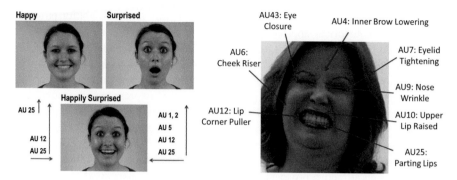

Fig. 5 Examples of non-prototypical facial expressions. *Left*: Happy, surprised and happily surprised with their associated AU as defined by Du et al.[38]. *Right*: facial expression of pain with AU annotations [74]

The first main factor is the wide range of facial expressions humans are able to display and interpret. For example, it has been shown that humans exhibit up to 7000 AU combinations in everyday life [110]. Due to the way AUs are defined, this means that each of these combinations results in a distinctly different visual input (although many of these combinations could result in the same high-level interpretation). While facial expression problems within the categorical approach consider only around six facial expressions, these categories cover only a small portion of our expressive behaviour. Examples of the many other non-prototypical expressions include the automatic analysis of facial expressions of pain [74], or the work of Du et al. [38], where it was argued that some facial expressions are the result of combining more than one basic emotion. For example a facial expression can convey both happiness and surprise simultaneously, resulting in what the authors call compound facial expressions (see Fig. 5).

The second main aspect that highlights the needs of more data is the large impact on the face appearance of factors of variation other than facial expressive behaviour. These include subject identity, illumination conditions, head pose variations, errors in the face registration, or factors such as the camera resolution, lens distortion, and acquisition noise. All of these factors can be considered nuisance factors, and result in an increase of the intra-class variability. Learning using standard ML models in the presence of such high intra-class variance results in the requirement of large and varied sets of data. In such cases training data needs to be abundant and varied enough to cover all of the different factors of variation. While this was already a challenging aspect of facial expression analysis for in-the-lab conditions, the aim towards in-the-wild facial expression analysis magnifies these considerations and thus scales the need for data, or alternatively the need for methods that can reduce the intra-class variability such as illumination independent descriptors.

In summary, facial expressions result in an extremely wide range of visual signals. Their expressive richness is much more varied than that considered within the classical options provided by the eight classes (including contempt and neutral) allowed within the Ekmanian categorical theory. If this large variability is to be considered, a categorical approach would imply recording task-specific datasets. This quickly becomes inefficient as the number of classes considered grows. One alternative that scales well in the number of facial expression classes considered is to learn models for facial AU analysis. Since they are a low-dimensional set of atomic units encoding the physical properties of facial expressions, the same AU models can be applied to the analysis of any other expression, from the compound expressions of [38] to expressions of pain (see for example Fig. 5). For example, the abundant data recorded for the analysis of prototypical facial expressions (augmented with AU labels) could be used to analyse facial expressions of pain. While this is an attractive prospect, this approach also has two caveats: Firstly, the cost of manual annotation of AU labels is much higher than the labelling of facial expressions in a categorical manner. Secondly, the signal to be detected is typically more subtle, thus magnifying the challenges posed by the nuisance factors.

It is however clear that the field is veering away from prototypical expressions, and the construction of publicly-available datasets capturing a growing range of real-world variability would be of great help. It is still unclear whether a categorical approach or the creation of datasets with AU annotations is the best way forward. Either way, because of the inability to handle unseen categories (categorical approaches), or because of the large intra-class vs. inter-class variation (AU analysis), the conclusion is that large amounts of well-annotated data of increased variability and complexity is probably still the most beneficial contribution to automatic facial expression analysis.

In particular, few datasets currently consider a naturalistic scenario. Some of the rare examples include Affectiva-MIT facial expressions in the wild [84], which contains a large dataset of subjects watching eliciting material in front of a computer screen. This database is annotated in terms of facial action units. Another in-the-wild dataset is that used for the Acted Facial Expressions in the Wild challenge [36], which contains clips from films labelled in terms of the six basic facial expressions. Other datasets not focusing on prototypical expressions include pain estimation-related datasets such as the UNBC-McMaster shoulder pain dataset [75] and the EmoPain dataset [6], the compound facial expressions of emotions database [38], the MAHNOB-Laughter dataset [99], the AVEC 2013 audio-visual depression corpus [136], and the SEMAINE corpus of dyadic interactions [88], which has been partially AU annotated for the FERA 2015 challenge [137].

Semi-Automatic Annotation Due to the above considerations, it seems clear that a purely manual annotation of a sufficiently large amount of data for facial expression recognition is extremely challenging or even impossible. A reasonable approach to this task is to use a tool capable of providing an approximate but fully automatic labelling, which would then be refined by the expert annotator [36]. One important aspect of this approach is the capability of identifying the limitations of current software, and thus avoiding annotating "redundant" information, i.e., information already successfully encoded by the ML model.

It is however surprising that despite the large amount of effort put in both manual annotation and on the construction of tools facilitating annotation (e.g. [17, 64]), these two areas have been treated as isolated steps, with the annotation effort simply preceding the learning stage. Annotation tools for facial expressions are not necessarily targeted to the creation of ML models. In fact, one of the justifications of the research on the automatic modelling of facial expressions is their potential use to ease the annotation burden on researchers studying human behaviour from a psychology or sociology perspective. Given the extreme technical challenge posed by the creation of a functional fully automatic annotation tool, it is only reasonable that a middle ground solution (i.e., a semi-automatic labelling tool) is created first. Such a tool would thus be beneficial in two senses: it could assist researchers of other fields by easing the annotation process required for their studies, and could have an important impact on the amount and quality of the annotated data at the disposal of researchers studying the creation of automatic facial expression analysis models.

Label Subjectivity The subjective criterion of the manual annotator often plays an important role on which labels are assigned to a specific data point. When these subjective effects are large, then the number of manual annotators required to obtain a consistent labelling grows, and with that the resources required to perform the manual annotation grow accordingly. Measures of inter-rater reliability can be used to assess how subjective a specific annotation task is [115]. Important factors affecting the inter-rater reliability include both the expertise of the manual annotator for that specific annotation task, and the nature of the annotation task.

Manual annotation for the categorical theory results in very high inter-rater reliability, in particular if the problem is restricted to the six prototypical expressions. Considering more classes can result in more challenging annotation processes. It is then possible to consider two annotation scenarios, that of free labelling (where any label the annotator can think of is allowed) or that of forced-choice annotation, where the annotator are given a set of options to which they are restricted. While free-choice annotation is sometimes used in psychology and sociology, it is in general not considered in computer vision problems.

The manual annotation of facial AUs is far more challenging than for those relying on the categorical theory. Since the aim is to annotate the data into classes with which humans are less familiar with, expert training is required, which is time consuming and expensive, and makes finding human annotators a hard task. Annotation is very laborious, and this can result in errors on the side of the annotator. Furthermore, some AU sub-problems, such as AU intensity estimation, are very challenging even for expert annotators since the differences between the different intensity levels are very subtle. It is thus common that AU annotations are not carried out by only one manual annotator, but rather several annotators. It is however noteworthy that, despite this increased difficulty on the annotation process, facial AU are defined in terms of objective factors, which is a great advantage when annotators are properly trained. Furthermore, the constraint that annotators must be

FACS certified improves the quality of the annotators. These two reasons alleviate to a large degree the problems introduced by the challenging task, and generally only minor discrepancies are observed in AU labelling.

Annotating continuous affect dimensions poses an even greater challenge than for any other facial expression analysis approach. Annotators are asked to code what they think the person they observe is feeling, and ratings are thus inherently subjective. The low inter-rater reliability of dimensional affect annotation is probably the most pressing problem of computational models that aim to automatically predict a dimensional affect approach. Unlike facial AUs, there is no manual to follow and no objective instructions of how to annotate continuous dimensions exist. No training program is available, and thus it is unclear how to improve the level of expertise of a novel annotator. This problem is accentuated by the subjective nature of the task at hand. Categorical approaches focusing on the six basic emotions have the advantage of a simple labelling space and a theory linking the emotional state to the (observable) facial expression display. Facial AU are designed for objectivity and omit making reference to the emotional state of a subject. None of these two aspects are true for the continuous affect dimensions: the link between facial expressions and emotional state is inherently ambiguous and subject to interpretation.

Another challenging aspect is the trend to seek continuous annotation also in the temporal sense. That is to say, the task of the annotator is to assign a continuous-valued label for every moment in time within the sequence. Temporally continuous annotation of affective dimensions is currently hindered by two major drawbacks: the need to provide a (subjective) label to each frame even when no signal is observed, and the current annotation strategy employed.

The first drawback is a reflection of the practical differences between psychology and computer science. It makes sense from a psychology point of view to ask a manual annotator to provide his (subjective) opinion of the state of a subject in a temporally continuous manner, as we as humans are able to (approximately) infer the emotional state of a subject based on a set of sparse signs spread throughout the video. For example, nodding while listening indicates agreement throughout, but even under strong agreement nodding events are sparse with respect to time. Instead, current ML methods are tasked with inference based on the signal (i.e., the observable facial expression) at a specific time frame. The label might however not correspond to the signal: expressions are not always there, while instead the subject's emotional state does not cease to exist. Another problem is the reaction time of annotators, which varies both between raters and within a single rater, for example when an annotator grows tired or becomes distracted. All these issues result in the need to either modify the labelling strategy, or re-think the way we apply ML to this problem.

The second drawback relates to the specific annotation strategy followed. Standard manual annotation strategies are inapplicable in this case, as frame-by-frame precise labelling would be too time consuming. Furthermore, manual annotators need the dynamic information provided by the video to provide a good judgement of the emotional state of a subject. As a solution to these problems, continuous affect annotations are typically carried out online (as the annotator watches a video in real time) by using a joystick [26]. The manual annotator shifts the joystick up and down to indicate the label value, and the annotation program records the exact position of the joystick at the time each frame of the sequence is displayed on a screen. While this solves the aforementioned problems, it introduces other important challenges. Firstly, human reactions are not immediate. Secondly, judgement is poorer than if the annotator was provided with the possibility to play the video back and forth. Thus, the inter-rater reliability is typically very low, needing up to 50 manual annotators to get sufficiently consistent labels [87]. Furthermore, the labels will be shifted in time because of the reaction delay, and each annotator has a different reaction time. While some efforts have been made towards sorting these limitations (e.g. [93]), practical drawbacks of this annotation strategy are still very high.

A final consideration regarding labelling consistency refers to how the labelling strategy is defined on new problems. A very good exemplar of this is that of automatic pain intensity estimation. The work by Lucey et al. [75] was the first large and systematic dataset containing pain estimation annotations within the computer science community. The authors opted for creating the ground truth labels for the pain intensity based on manually annotated facial AUs. Since AUs can be annotated in an objective manner, encoding the intensity of pain expressions as a function of the AU intensities is a reasonable option. Alternatively, Aung et al. [6] opted for using the same joystick-based annotation tool used for continuous affect annotation. This better reflects the extra judgement humans are able to produce: following the AU-based pain annotation strategy, a smile would be encoded as a painful event, while instead humans are able to immediately understand that in many situations this is not the case. Which annotation strategy is the best for new problems such as automatic pain estimation is largely an open (and fundamental) research problem.

Avoiding Dataset Biases Researchers typically validate their algorithms on standard publicly available benchmark datasets for the problem at hand. This means that there is a risk that the main aim of research becomes maximising performance on specific datasets, which are very likely to contain biases. While exploiting specific biases can boost performance on a specific dataset, this practice is unacceptable if the aim is to solve the general problem. This is a classic problem of overfitting versus generalisation. Facial expressions are no different in this sense. Many works have exploited unrealistic assumptions resulting from dataset biases.

The most common bias when dealing with categorical approaches is the assumption of pre-segmented sequences. This is still an interesting scenario worth exploring, both because the technologies developed could potentially be extended to the general (unsegmented) scenario, and because some practical applications can actually constrain the scenario to the pre-segmented case (e.g., when analysing the reaction of a user to an ad). It is however important to bear in mind that this is a specific scenario with an intrinsic bias that does not generalise. Other biases are more damaging, as the bias exploited cannot be assumed in any practical setting. One such bias is the assumption that sequences start on a neutral frame, or at least that a frame displaying a neutral face of every subject is available (e.g., [13]).

These assumptions are good examples of the care that a reader should take to judge these methods. The assumption of a neutral frame at the start of the sequence cannot be extended beyond the scope of datasets where this bias is present. Instead, the assumption of the availability of a neutral frame could be extended to the automatic estimation/identification of neutral frames within the sequence [10]. This however will result in a lower performance due to errors in the automatic estimation.

Other specific biases that can be exploited are the absence of complex lighting conditions, as many datasets are recorded under controlled illumination conditions. If care is not taken, the field can end up putting efforts in attaining an "algorithmic local maximum". Such a warning was for example included in the work on [21], as they presented a study with a quantitative performance evaluation for different feature representations which included raw pixel intensities. Using raw pixel intensities directly and without the employment of illumination invariant features will only work in such artificial datasets.

Finding a Better Representation As previously mentioned, a major challenge regarding facial expression recognition is how to handle modes of face appearance variation other than facial expressions. While some ML methods can be employed to deal with this issue, using adequate features has a dramatic effect in terms of performance. It thus comes as no surprise that many of the recent works on facial expression recognition have focused on employing a variety of features, or even proposing new face representations [61, 109].

Many works have focused on studying the relative merit of individual feature types within an arbitrarily-defined set of features. There has not been any wide-spread agreement on which features perform best, with different studies yielding different relative feature rankings [21, 133]. This is likely due to several reasons: no work has performed a really exhaustive characterisation of performance in terms of the features used, and the feature configurations used have been at times sub-optimal. One such a study, performed rigorously and as exhaustively as possible, would be beneficial to understand what are the strong and weak aspects of each of the possible feature representations. Other factors include the specific dataset used for evaluation, and problem tackled. For example, facial AU datasets include annotations for different subsets of AUs, which means that average performances are hard to compare.

It is likely that combining multiple feature types is the best way to proceed. The most common ways to fuse features are to perform the so-called feature-level fusion and decision-level fusion [98]. Feature-level fusion simply concatenates all features into a single vector, while decision-level fusion instead trains a separate model for each of the feature vectors, and then learns how to combine them into a final solution, typically by computing a weighted average of the feature-specific predictions [62]. Multiple Kernel Learning (MKL) for combining multiple features has lately been regarded as a better way of fusing features [112], as the multiple kernels can model the different underlying data distributions. Performance using MKL is often superior to performances attained using feature-level because badly-performing features do not degrade the overall performance. Similarly, it typically offers superior performance to decision-level fusion because of its ability to feed information of the final fusion scores back into the individual models. This framework has however been relatively under-explored and focuses mostly on the fusion of geometric and (one single type of) appearance features. Further exploring this potential seems like a reasonable way forward.

Another promising line of research is the use of dynamic appearance descriptors, such as those belonging to the TOP family of features [1, 60, 159]. The use of spatio-temporal appearance information is consistent with the nature of the task, as after all the problem is essentially identifying and analysing *actions*, which are inherently events with a temporal nature. Dynamic appearance features are constructed by extracting 2D features from each of the three orthogonal planes (TOP) and concatenating them into a single vector. By extracting 2D features from three orthogonal planes, the dimensionality of the final vector is kept to a reasonably low level (typically three times that of static features) compared to full-fledged volume-based descriptors (which have an exponential increase in the number of features when moving from 2D to 3D). In this spirit, TOP extensions have been made of the LBP features [159], LPQ features [57], or the successful LGBP features [1]. It is however possible to define spatio-temporal appearance in different ways, either following a more classical feature extraction strategy [155] or a bag-of-words type of representation [124].

Some research has also started analysing expressive behaviour at the level of *events* rather than on a frame-by-frame basis. This means that the aim of inference is to find the start and end frames of a specific event, and thus the frame-level labelling is obtained as a by-product of this inference strategy. This problem definition is usually referred to facial AU analysis, as it is the only problem which systematically considers a test scenario with unsegmented events. How to effectively exploit this paradigm is however unclear right now. There has been some prior work that uses bag-of-words representations to this end. It is then possible to combine the bag-of-words representation with the structured output framework described by Blaschko and Lampert [15] for the case of facial expression analysis [22, 116]. However bag-of-words representations are somewhat poor in terms of the information they encode. An alternative approach was proposed in [37], where frame-level inference and event-level inference were combined. It is interesting to bear in mind that event-level representations are interesting when dealing with unsegmented data, which is often not the case when studying categorical problems. However, lessons learnt on classifying categorical data are likely to be transferable to an unsegmented scenario by following an event-based approach.

All of the existing feature performance considerations will need to be confirmed or revised for in-the-wild imagery. Most of the studies have been carried out for in-the-lab data, and asking whether the acquired knowledge will extend to this more general case is thus a valid question. For example, the relative merits of features robust to head pose variation and to different illumination conditions (e.g., local HOG features) are likely to gain in importance.

One important aspect that might gain relevance under in-the-wild imagery is the use of mid-level representations. The use of mid-level representations for human action recognition, a problem typically boasting larger variability than for human faces, has proven very effective in the past. This approach has been spearheaded within the action recognition literature by models such as poselets [16]. The same idea has been extended to other problems, among them facial expressions [72]. This is however a first attempt at this kind of mid-level representations for facial expressions. The increasing variability of in-the-wild imagery might result in a surge of such approaches.

An unavoidable question given the massive success of deep learning techniques in a wide variety of computer vision problems is whether they can also provide a significant boost of the state of the art for facial expression recognition. Surprisingly, there are extremely few works tackling facial expression recognition from a deep learning perspective, and even less published at high impact conferences. One example to highlight is that of [73], where Deep Belief Networks were used. However, the very popular Convolutional Neural Networks has so far yielded performances below state of the art [47]. Given the popularity of deep learning techniques, this is very likely due to the so-called *positive publishing bias*, for which negative results are unlikely to be published, rather than lack of attempts from the research community. Whether this is due to the inherent mismatch between deep learning and the problem at hand, or due to the specific forms of deep learning techniques used remains to be seen. After all, current techniques have been typically developed for general object recognition rather than for fine-grained categorisation. The development of some deep learning feature extractor, much in the spirit of the popular AlexNet [68], that could be used for effective facial expression recognition, would be a massive addition to the field.

A final consideration concerns the typically large dimensionality of face representations, and how this can be reduced. While it is possible to directly apply variance-based dimensionality reduction techniques such as PCA, it is likely that some facial expression information will be eliminated from the data. Face appearance changes less due to facial expressions than because of differences in identity, head pose or even illumination. It is thus reasonable to hypothesise that expressive behaviour will partially be encoded in low-energy PCA dimensions. An alternative approach in the literature has consisted in the encoding of only part of the face appearance, justified by the spatially localised nature of facial expressions. For example, it is possible to enforce L_1 sparsity constraints on the regions of the face used [160], or to learn part-based models in combination with a decision-level fusion [59]. Exploiting the spatially localised nature of expressions and integrating this knowledge within the inference methods is a line of research with high potential. This is much more the case when dealing with AUs, as they present a particularly strong spatial localisation.

Occlusions The first challenge towards dealing with occlusions is face alignment. Current methods for face alignment, such as [151], show some in-built robustness to partial occlusions due to the use of HOG features for face appearance modelling, and the use of the full face appearance to perform inference. Furthermore, some works have proposed extensions specifically targeted to dealing with partial occlusions [18, 156], including both the robustness to partial occlusions and the identification of which landmarks are occluded.

The next natural step, and one widely missing in the literature, is how to incorporate this knowledge within the facial expression recognition step. Learning with partial occlusions could be seen as a special case of learning with corrupted features. Some works have proposed learning algorithms robust to these cases (e.g., [56, 139]), but these have not been applied to facial expression recognition problems. In any case, it seems suboptimal not to exploit the information regarding which features or face parts are occluded, which could be automatically estimated from the face alignment step. How to integrate this information into these kind of algorithms, or envisioning some other new way of tackling learning under partial occlusions, is an important and understudied challenge within facial expression recognition literature.

Dynamics Facial expressions are actions by nature. While this has been exploited in terms of enforcing temporal consistency of the labelling, and through the use of spatio-temporal features, the dynamics of facial expressions on a global sense are not yet fully understood nor exploited. In fact, it is unclear even how to attempt to model dynamics. It is possible to distinguish between intra-class and inter-class dynamics, as well as distinguishing between short and long term dynamics. Another useful distinction is between pairwise dynamic relations and higher-order dynamic relations.

Intra-class or intrinsic dynamics encode the temporal relations within a single labelling problem, while inter-class or extrinsic dynamics refer to the temporal relations between different or even heterogeneous problems. Intra-class dynamics encode for example the fact that the frame-level labelling for that specific AU should be temporally smooth. If a frame is labelled as neutral between positive frames, it is most likely a false negative. Similar mechanisms can be used for any temporally-structured output problem, in our case any facial AU or dimensional affect problem. Examples of these mechanisms are the use of HMM (in practise the frame-level relation between the frame data and the frame label is often encoded using a discriminative method with a confidence output [133]), and discriminative graphs such as a CRF [138]. These methods typically capture short-term dynamics, i.e., they relate one frame to the next (see below for a more detailed explanation). How to encode intra-class dynamics with temporal range of more than approximately 5 s is still unclear.

Inter-class or extrinsic dynamics instead capture temporal correlations of co-occurrences between different classes. Some examples include the use of Dynamic Bayesian Networks used in [126] for AU detection, while [92] used structured output regression for the combined estimation of Valence and Arousal. However, all of these cases use temporal correlations among labels relating to the same problem. It is possible to use heterogeneous problems that temporally correlate to the problem at hand. One such mechanism was for example used for combining multi-model information for interaction modelling [89], exploiting temporal correlations across modalities. It is possible to see the relevance of this problem when considering a scenario of dynamic interaction rather than a single-subject scenario. In dyadic interactions the facial expressions of the two interactants naturally interact in a sequence of cause and effect relations. This approach would capture effects such as mirroring or synchrony [14, 121].

As mentioned before, models such as HMM or linear chain CRF enforce temporal consistency of the labelling locally, only capturing correlations between consecutive frames. Finding longer term correlations is very challenging, to a large degree due to the inherent limitations of the most widely-known ML approaches. Some models have been proposed that are capable of capturing and using pairwise long-term potentials efficiently. For example, the Long-Short Term Memory NN (LSTM-NN) is a type of recurrent neural network that is capable of capturing both long and short term dependencies [147]. This model has been widely exploited in the audio community, and some works exist on audio-visual works using LSTM-NN. For example, [43] studies vocal outburst from an AV point of view, while [149] actually target multimodal emotion recognition using LSTM-NN. Other ML methods, such as the recently proposed Continuous Conditional Neural Fields [9], could be exploited within the context of expression recognition problems to explore the potential of harnessing long-term temporal correlations.

The aforementioned methods still rely on capturing pairwise correlations. That is to say, they only consider connections between two time stamps, rather than considering correlations among larger groups of variables. Capturing higher-order potentials is a different and yet again understudied aspect of the modelling of the dynamics. The inability of the most common ML tools for harnessing higher-order potentials is again to blame here. However, recent advances in ML (e.g., [67]) are beginning to open the door to using new sources of information resulting from analysing more than pairwise potentials. In fact, some very initial attempts have been proposed by e.g. Wang et al. [144], although in these works the higher-order correlations captured are still only at a frame level. Exploring the possible designs of higher-order potentials (i.e., defining exactly what should be captured and how to computationally model them) is a very interesting future challenge.

From Momentary to Higher-Level While the understanding of dynamics is paradigmatic, it is actually a specific instance of a general situation. The overwhelming majority of research to this date has focused on a momentary analysis of the facial expressions. This disregards information related to higher-level understanding of the scene, such as the context, the interaction type, the personality of the subjects, etc. If all of this information is to be harnessed into a single model, then advanced ML capable of incorporating higher-order potentials should be used.

Facial expression recognition can understood within the context of Social Signal Processing (SSP) applications. It is then part of a multi-modal problem that integrates heterogeneous cues into some higher-level understanding of the scene [140, 141]. An ideal system would integrate audio analysis (including sentiment analysis, speaker identification, etc), other different forms of video analysis (human body pose estimation, action recognition, head pose, head nods detection, etc), and even tools for natural language processing. In this context, considering facial expression recognition as an isolated problem is not fully satisfactory. Integrating learning and inference into the same system, and using feedback from other components of the system, would be a more natural way of tackling the individual problems.

Computational Efficiency It is becoming clear that algorithms capable of running on devices with low computational capabilities, most notably mobile phones, will have high-impact opportunities (see Sect. 4 below). A notable example of this trend is the IntraFace software,[4] which allows for face analysis on a mobile platform. This includes running a face alignment algorithm (that of [151]) in real time. The use of geometric features allows for inexpensive facial expression recognition, but their modelling capability is limited. Producing algorithms capable of analysing the face appearance using the computational resources provided by a generic mobile platform is both a challenge and a very interesting research direction with very important practical implications.

4 Opportunities

In this section we review different exciting opportunities, exploring the potential of applying current and imminent state-of-the-art algorithms for facial expression technology to practical problems. The state of maturity reached by the field means that long-heralded opportunities are suddenly becoming possible at sufficient reliability levels, while new opportunities are now being envisaged as creatives and industrial forces are taking interest in the facial expression recognition technology. Together, there is an exciting market for these technologies to develop. In here we group the opportunities into "umbrella" criteria: medical conditions, HCI and virtual agents, data analytics, biometrics, and implicit labelling.

Behaviomedics Medical applications of automatic facial expression analysis methods have received increasing attention due to the potential societal impact of such an endeavour. This interest is seen both in the academic and funding sides, a trend reflected in the number of current research papers, and on the number of projects targeting these problems. An interesting observation regarding the latter is the priorities set on the EU Horizon 2020 funding programme, where *Societal challenges* is one of the three core themes or "pillars", and *Health, demographic change and well-being* is defined as one of only seven specific calls within the societal challenges pillar.

A wide range of medical conditions, for instance depression or anxiety, produce distinctive alterations on the behavioural patterns observed on an individual. It is

---

[4]http://www.humansensing.cs.cmu.edu/intraface/.

then reasonable to consider the automatic analysis of the behaviour, potentially over long periods of time, as a potentially effective mechanism for early detection of such conditions. This was formalised by Valstar as Behaviomedics [131]:

Behaviomedics - The application of automatic analysis and synthesis of affective and social signals to aid objective diagnosis, monitoring, and treatment of medical conditions that alter one's affective and socially expressive behaviour.

Given the current range of state-of-the-art performances, what can be hoped to achieve now is systems that aid doctors in diagnosis and monitoring, for example current behaviomedical systems could be used to filter the cases that require the attention of a trained clinician in an efficient manner. The advantages of such systems are threefold. Firstly, many patients are either unaware of their condition or do not actively seek the help of a clinician to tackle them. Widening the reach of these services to patients that might otherwise not receive treatment is thus a fundamental target. Secondly, the use of long-term and fine-grained monitoring in a pervasive and passive manner can result in a much richer understanding of the progress of the condition and the specific behavioural idiosyncrasy of the patient. Thirdly, it would result in a more effective management of the (limited) time of the clinicians.

The systematic analysis of behavioural cues means that this problem requires techniques from both the Affective Computing and Social Signal Processing communities. It is thus a multi-modal problem in nature which requires the integration of the automatic analysis of, among others, facial expressive, body pose and hand gesture, and audio information. There is in fact a long-standing tradition in the Affective Computing field of considering medical applications, already present in very early and foundational works within the field [33, 102]. It is however only recently that many such problems started to be considered as feasible potential applications of affective computing and social signal processing techniques, which has helped to better define and systematise this family of applications [131].

Following Valstar [131], we distinguish between three groups of medical applications: mood and anxiety disorders, neuro-developmental disorders, and pain estimation. Mood and anxiety disorders encompass a wide range of different mental disorders. Notable examples include depressive disorders, bipolar disorders or substance-induced disorders among others [3]. Neuro-developmental disorders include again a wide variety of disorders such as autistic spectrum disorder, schizophrenia, foetal alcohol spectrum disorder, Down syndrome or attention deficit hyperactivity disorder. Finally, pain estimation is also considered. While pain is a symptom rather than a condition in itself, it is common to use pain as an indicator in medical settings, e.g. for clinicians controlling rehabilitation exercises or for judging the severity of an injury, making it a very interesting practical problem that is well suited for facial expression analysis [6, 75].

Obtaining a consistent and objective ground truth for such tasks is very challenging even for trained clinicians. The diagnosis and the evaluation of the severity of depression are for example typically assessed based on self-report questionnaires [7, 162]. The inherent subjectivity of this kind of measurement suggests that an approach based on objectively-measured behaviour tracked for extended periods of time might add valuable information with diagnostic potential. Similarly, while pain estimation is easy to elicit, and it is associated to some level of communicative intent [120] (thus being conveyed through a distinct and clearly visible signal), it is unclear how to produce an objective encoding of the expressed signal into a numeric scale representing the pain level objectively. An attempt at producing a systematic measuring system was made by Craig and Patrick [27] by making use of the facial AU coding system, while other works have conducted further experiments along these lines [158].

Most of these applications are however targeted at the analysis of the behaviour of one single individual considered in isolation with their environment. In particular, the analysis of the relations between individuals with the aim of detecting some diagnosable pathological behaviour, or even for the improvement of interpersonal relations, has been out of the scope. Such constraints are likely to be a result of the current state of the art, which is only now starting to focus on the modelling of (typically dyadic) interactions. It is however likely that a stage of maturity of interaction modelling techniques will bring applications to automating interventions such as counselling (e.g. marriage, or family counselling) and mediation. These are however blue-sky thinking applications right now, and their viability will directly depend on the quality of the research outcome for the next 5 years.

Data Analytics Some of the applications of automatic facial expression recognition that are currently attracting wide interest from industry are related to the analysis of a large volume of visual data. The aim is in this case to produce an easily understandable statistical summarisation of the content. The most notable example is the automatic analysis of marketing and publicity [85, 86]. Several start-ups are currently focusing on the use of automatic facial expression analysis to evaluate the effectiveness (in terms of the reaction of the viewer) of a marketing campaign. It is for example possible to measure factors such as the level of engagement or infer the emotions elicited during a screening sessions to a large audience, or for example to retrieve the reactions to ads shown through the internet to individuals while using their personal computers or mobile phones. The reactions are then summarised in a report that can be easily analysed by marketing experts without having to resort to hours of video visualisations and annotation.

While marketing studies are a prominent application in terms of the interest shown by the industrial sector so far, similar studies can be carried out for a wider range of applications. For example, [83] focused on the prediction of voting preferences based on the reaction of the screened subjects to a political debate. The work in [90] focused instead on predicting the ratings of a movie based on the behaviour of the audience, while [122] targeted instead the task of automatically measuring the level of engagement of subjects watching television. While these applications are better tackled through a multi-modal perspective (e.g. the body pose can play a key role), the analysis of facial expressions is a key modality.

Applications other than those aforementioned can be easily envisioned. Many of them are similarly mutli-modal in nature and have facial expression analysis as a component that needs to be integrated into a multi-modal framework. One example of these applications is the analysis of group interactions [46, 82]. Specific applications with high-impact industrial applications would be the analysis of dominance within a group [51], the analysis of the cohesion within a group [50], or automatically measuring the level of engagement of individuals [148]. The creation of such tools would for example allow the automatic and non-intrusive analysis of the group dynamics at the workplace, potentially transforming our understanding of group dynamics, the way working groups are configured and constructed, and the way each individual is evaluated with regards to the final outcome achieved by the group. Other potential (also multi-modal) applications could relate to the training of individuals in regards to their behaviour to optimise their performance under certain social circumstances. This could result in automatic tools for personalised training targeting for instance the improvement of public speaking abilities [11] or for the preparation of job interviews [12, 91].

Human-Computer Interaction HCI has traditionally been regarded as one of the main applications of facial expression recognition. Researchers have often cited the need for algorithms capable of endowing computers with the ability to interact in a more natural way with humans, much closer to the way that humans interact with each other [102, 157]. It is common to envision the future of human-computer interaction as moving away from being centred around peripherics such as the mouse and keyboards. Instead the interaction should move towards a more natural, often passive and pervasive approach, where computers can automatically detect and interpret your non-verbal cues (with facial expressions among them), and react to them. Little of this early promise has however been materialised to date. This might be due to the technical challenges, but also because of the lack of specific materialisation of these high-level concepts into specific interaction patterns.

One application that has actually achieved some level of success is that of Virtual Agents (VA) [111]. VA represent an (anthropomorphic) embodiment of the computer, which enables the creation of more natural Human-Computer interactions. VA require both the ability to analyse and synthesise facial expressions and more generally expressive behaviour. Thus, an underlying methodological challenge is the understanding of the "non-verbal semantic and syntactic rules". It is to this end possible to construct a generative model capable of capturing the decay of intensity of expressions with time, and the complex temporal interaction between expressions so that the virtual agent can produce realistic facial expressions [34]. It has also been argued that the use of mirroring of behaviour is an important part of human-human interaction, aimed at creating empathy [100]. It is thus interesting to endow VA with the capability to read facial expressions (among other relevant behaviour) in order to introduce similar mirroring mechanisms in the human-computer interaction.

A very related topic is that of creating expressive and socially-aware robots. The creation of robots endowed with emotional awareness and social intelligence, capable of communicating and behaving naturally with humans while respecting socials, has been a long-standing aim for researchers [31, 45]. The embodiment aspect characteristic of virtual agents is similarly present in this case, although in a more physical manner. The synthesis of facial expressions represents however a strong dissimilarity. Facial expressions are in this case more complicated to synthesise and thus the usefulness of their analysis relies on the capability of the robot to infer human emotions [32]. The physical dimension of robots, as opposed to the non-physical nature of virtual agents, means that the former are more likely to be seen as personal objects and be understood from the consumer goods perspective. Developing algorithms mimicking bonding processes could result in a new perspective on the relation between the robots and their owners/users [69].

A final important application is that of driver assistance, where facial expression analysis has in this case a direct application to improve driver's safety. The strong interest shown from the automotive industry has lead to a wide variety of systems for detecting driver drowsiness. Approaches falling within Computer Vision include the use of Near Infra-Red images [55], face analysis to detect driver drowsiness [143], or the use of facial expression analysis as a cue to infer when the driver is driving recklessly [52]. While there has been a long-standing interest in this problem, the nature of the data, with frequent large illumination variations, has driven the attention within the research community towards NIR imagery. This kind of images can be used to perform inexpensive gaze estimation, from which attention and drowsiness can be inferred. The recent state-of-the-art advancements have resulted in a significant boost on the performance of automatic facial expression recognition under varying illumination (mostly due to the in-the-wild face alignment algorithms). These advancements might result on a surge of commercial applications with this aim.

Throughout the years, and probably fuelled by the need to justify the importance of the associated research lines, several other applications of facial expression recognition have been proposed within the sphere of HCI. Examples of these are the use of facial expressions within the computer games industry, the use of facial expressions and emotional awareness to control the environment in a pervasive manner (a typical scenario is a system capable of automatically adapting the music to your mood), or the use of expressions to control computers (e.g., increase the font size when the subject is tired). Many of these aspects are however unlikely to result in practical systems, let aside commercial applications, given the very specific application scenarios and their relative lack of practical interest.

Assisting Behaviour Understanding Research While categorical approaches traditionally focus on the detection of the prototypical facial expressions, the same approach can be applied to infer any target expression directly from the input data. The only difference is the absence of universality of the expression, which affects the link between facial expression (understood as a sign) and an emotion. For example, while pain estimation can be achieved with one such approach, it is likely that different people express pain with different facial expressions. Conversely, a smile will likely be interpreted as pain since the sign used to denote pain often involves letting the eyelids droop, and stretching the mouth corners [4, 103]. Instead, facial AU approaches first extract a set of facial Action Units (signs) and then these signs can be interpreted at a later stage. The latter approach has the disadvantage of adding further complexity to the problem. AU detection is more complex than facial expression recognition. Working with AU has however a twofold advantage. Firstly, the interpretation layer, in which the sign (the facial expression) is interpreted can include contextual information or other cues in a seamless manner. Secondly, it is possible to interpret the composition of the facial expression.

The interpretability of the sign that led to the detection is of particular interest when the aim is to understand how a person can express certain cue non-verbally. This is of interest mostly for two reasons: it facilitates the realistic synthesis of facial expressions, and enables behavioural scientists and psychologists to conduct studies on the way humans express themselves and how these signs are perceived by other humans. While AU can offer powerful cues for psychologists and behavioural scientists to conduct quantitative analysis, the annotation of AU throughout a corpus from which to extract statistically significant conclusions is an extremely tedious, resource-intensive and time-consuming process. As a consequence, one of the widely extended arguments to justify research on automatic facial Action Units analysis is the creation of automatic labelling tools for supporting the research of behavioural and psychological scientists. Off-the-shelf software that can be run to produce such labelling is largely absent from the literature. Some efforts have been made publicly available, such as the Computer Expression Recognition Toolbox (CERT) [71]. While more tools with improved reliable and ease of use are necessary, the main drawback in this sense might be the absence of a semi-automatic tool. Even state-of-the-art facial AU detection algorithms are not reliable enough as to be applied as a tool without manual intervention. A semi-automatic annotation tool would instead produce some off-the-shelf output as a starting point. Then the user would have the option of correcting some of the prediction errors through an easy-to-use interface or to introduce a small number of subject-specific or scenario-specific manual annotations within the training set and produce new or refined results. This loop between manual correction and automatic re-fitting can be iterated for as long as necessary until the target data is annotated with acceptable reliability according to the criterion of the users of the tool (typically psychologists or behavioural scientists).

Implicit Tagging Given the exponential growth of the amount of multimedia digital data both at public repositories (e.g. youtube) or private ones (e.g. facebook), how to effectively and efficiently search through this content is an increasingly important problem. One such mechanism is the creation of *tags*, which is a type of metadata useful for retrieval based on content. It is for example nowadays customary to tag images within facebook with the names of the people on it. However, this manual tagging is labour intensive and in the majority of cases users are not interested in carrying it out. One can then resort to automatic tagging. Computer Vision tools can then be used to analyse the data through algorithms tasked with automatically assigning relevant tags. Facial expression recognition can play a role within this framework and be used to associate multimedia content with the associated emotions.

A third option has very recently become one of the areas of application of facial expression recognition: implicit tagging [117–119]. This problem refers to the association of tags to multimedia data based on the spontaneous reactions of users while watching the content. This reaction is measured automatically based on their facial expressions [145] or even based on their physiological reactions [65]. The wide availability of built-in sensors within devices capable of multimedia reproduction makes this an interesting and effortless way of tagging content. The set of tags involved do not necessarily correspond to prototypical emotions, and applications such as flagging inappropriate behaviour, to assessing the interest of multimedia content (e.g., in a virtual class) could be envisioned as applications of this range of techniques.

Deceit Detection Some work in the facial expression analysis literature has focused on the detection of posed expressions. These are characterised both by distinct appearance and, fundamentally, in terms of their dynamics [25, 132]. Training people to control their facial expressions and, in particular, to be able to mimic spontaneous (truthful) facial expressions is possible and even common (e.g., actor's training). Micro-expressions are instead involuntary and hard to control, and they can correspond to what Ekman and Friesen described as a leakage clue of deception [39]. The rationale here is that, during a deception episode, a subject will try to conceal his emotions. However, small clues of this concealment can *leak out* and result in small observable facial expressions. While this theory seems promising, we should be cautious about its usefulness and prominence. Micro-expressions correlate to some concealment, which is not equivalent to a downright lie. That is to say, they might indicate that there is more to the story than what is being told, but not that the part of the story told is actually true. Thus it could be seen as a "early flag" sign so that further checks could be conducted. Similarly, it is one of a number of physiological signals that could correlate with deception [35, 96, 128, 146]. One of the main potential advantages of micro-expressions respect to other physiological signals is that they are non-intrusive and non-invasive, and that (theoretically) they can be detected using cheap hardware, such as a standard webcam.

It is because of these considerations that the Computer Vision community has very recently explored the automatic detection of micro-expressions. The first datasets have been created (e.g. [153]) and some works have started performing quantitative performance measurements [101, 150]. These results are however a starting point, and further research and improved performances are necessary to turn this approach into a viable practical option. Due to the very short time span of micro-expressions, existing datasets use cameras with a high frame rate and high spatial resolution, and there is little head pose variation. Very precise face registration seems also necessary in practise. Given the very faint signal of the facial expressions compared to other sources of variation (illumination, identity or head pose), exploring which features capture the necessary information to allow for effective learning seems like a very reasonable next step. Whether it is possible to detect micro-expressions using off-the-shelf hardware (i.e., standard cameras) is another open question. It is also unknown how well we can achieve the end application, i.e., to automatically detect deceit, based on the current level of performance for micro-expression detection, or even based on manually-annotated micro-expressions.

5 Conclusions

Automatic Facial Expression Recognition has reached a state of maturity where it can now start to be reliably deployed in real-world applications. In particular the elements of the pipeline that can be considered the pre-processing steps have reached a point where they are highly accurate and robust to real-world variations in the data. Face detection has reached this point some time ago already with the advent of the Viola and Jones Face Detector [142], and has since been improved further to a point where it can now be expected to work in most practical situations. More recently face alignment has made major strides, propelled forward with the introduction of regression-based facial point localisation [134]. This was followed by the Cascade regression (e.g. SDM [151]), and finally fine-tuned to the point of perfection by works such as Project-Out Cascaded Regression by Tzimiropoulos [129]. One could safely argue that these components are now ready to be used reliably in all sorts of real-world applications.

Interestingly the actual facial expression analysis component of the pipeline has not seen the same jump towards robustness and accuracy. Certainly, the detection of the six basic emotions, and small numbers of discrete expressions in general, can be considered close to being solved. But as pointed out these expressions are not frequently displayed and are thus of limited value. Most Action Units on the other hand are still not reliably detectable, nor are the affective dimensions valence and arousal. This is not due to a lack of good ideas in this field, but instead mainly due to a lack of high-quality data recorded in realistic, natural conditions.

With the proliferation of multimedia content on social networks, ubiquitous sensors carried around and used to collect ever more natural scenes, this is bound to change. When scientists figure out how to use a sufficient amount of this data efficiently, probably through semi-supervised, transfer, multi-task, or unsupervised learning, so too will facial expression recognition become a readily applicable technology.

For decades, research works in this field have started with the same dry statements of what massive impact automatic facial expression recognition will have on wide-ranging domains such as medicine, security, marketing, and HCI. Excitingly, we believe that we are finally on the verge of making true on these promises!

Acknowledgements The work of Dr. Valstar and Dr. Martinez is funded by European Union Horizon 2020 research and innovation programme under grant agreement No. 645378. The work of Dr. Valstar is also supported by MindTech Healthcare Technology Co-operative (NIHR-HTC).

References

1. T. Almaev, M. Valstar, Local Gabor binary patterns from three orthogonal planes for automatic facial expression recognition, in *Affective Computing and Intelligent Interaction* (2013)
2. Z. Ambadar, J.F. Cohn, L.I. Reed, All smiles are not created equal: morphology and timing of smiles perceived as amused, polite, and embarrassed/nervous. J. Nonverbal Behav. **33**, 17–34 (2009)
3. American Psychiatric Association, *Diagnostic and Statistical Manual of Mental Disorders (DSM)*, 5th edn. (American Psychiatric Association, Washington, 2013)
4. A.B. Ashraf, S. Lucey, J.F. Cohn, T. Chen, Z. Ambadar, K.M. Prkachin, P.E. Solomon, The painful face - pain expression recognition using active appearance models. Image Vis. Comput. **27**(12), 1788–1796 (2009)
5. A. Asthana, S. Zafeiriou, S. Cheng, M. Pantic, Incremental face alignment in the wild, in *Computer Vision and Pattern Recognition* (2014)
6. M.S. Aung, S. Kaltwang, B. Romera-Paredes, B. Martinez, A. Singh, M. Cella, M. Valstar, H. Meng, A. Kemp, M. Shafizadeh, A.C. Elkins, N. Kanakam, A. de Rothschild, N. Tyler, P.J. Watson, A.C. de C. Williams, M. Pantic, N. Bianchi-Berthouze, The automatic detection of chronic pain-related expression: requirements, challenges and a multimodal dataset. Trans. Affect. Comput. In Press
7. M.R. Bagby, A.G. Ryder, D.R. Schuller, M.B. Marshall, The Hamilton depression rating scale: has the gold standard become a lead weight? Am. J. Psychiatry **161**, 2163–2177 (2004)
8. T. Baltrušaitis, P. Robinson, L.P. Morency, 3D constrained local model for rigid and non-rigid facial tracking, in *Computer Vision and Pattern Recognition* (2012)
9. T. Baltrusaitis, P. Robinson, L.P. Morency, Continuous conditional neural fields for structured regression, in *European Conference on Computer Vision* (2014), pp. 593–608
10. T. Baltrušaitis, M. Mahmoud, P. Robinson, Cross-dataset learning and person-specific normalisation for automatic action unit detection, in *Facial Expression Recognition and Analysis Challenge Workshop* (2015)
11. L.M. Batrinca, G. Stratou, A. Shapiro, L. Morency, S. Scherer, Cicero - towards a multimodal virtual audience platform for public speaking training, in *International Conference on Intelligent Virtual Agents* (2013), pp. 116–128
12. T. Baur, I. Damian, P. Gebhard, K. Porayska-Pomsta, E. Andre, A job interview simulation: Social cue-based interaction with a virtual character, in *International Conference on Social Computing* (2013), pp. 220–227
13. J. Bazzo, M. Lamar, Recognizing facial actions using Gabor wavelets with neutral face average difference, in *Automatic Face and Gesture Recognition* (2004)
14. S. Bilakhia, A. Nijholt, S. Petridis, M. Pantic, The MAHNOB mimicry database - a database of naturalistic human interactions. Pattern Recogn. Lett. **66**, 52–61 (2015)
15. M.B. Blaschko, C.H. Lampert, Learning to localize objects with structured output regression, in *European Conference on Computer Vision* (2008)
16. L. Bourdev, J. Malik, Poselets: body part detectors trained using 3d human pose annotations, in *International Conference on Computer Vision* (2009)
17. H. Brugman, A. Russel, Annotating multimedia/multi-modal resources with ELAN, in *International Conference on Language Resources and Evaluation* (2004)
18. X.P. Burgos-Artizzu, P. Perona, P. Dollár, Robust face landmark estimation under occlusion, in *International Conference on Computer Vision* (2013), pp. 1513–1520
19. X. Cao, Y. Wei, F. Wen, J. Sun, Face alignment by explicit shape regression, in *Computer Vision and Pattern Recognition* (2012), pp. 2887–2894
20. S. Cheng, S. Zafeiriou, A. Asthana, M. Pantic, 3D facial geometric features for constrained local models, in *International Conference on Image Processing* (2014)

21. S. Chew, P. Lucey, S. Lucey, J. Saragih, J. Cohn, S. Sridharan, Person-independent facial expression detection using constrained local models, in *Automatic Face and Gesture Recognition* (2011), pp. 915–920
22. W.S. Chu, F. Zhou, F. De la Torre, Unsupervised temporal commonality discovery, in *European Conference on Computer Vision* (2012)
23. I. Cohen, N. Sebe, A. Garg, L.S. Chen, T.S. Huang, Facial expression recognition from video sequences: temporal and static modeling. Comput. Vis. Image Underst. **91**(1–2), 160–187 (2003)
24. J.F. Cohn, P. Ekman, Measuring facial actions, in *The New Handbook of Methods in Nonverbal Behavior Research*, ed. by J.A. Harrigan, R. Rosenthal, K. Scherer (Oxford University Press, New York, 2005), pp. 9–64
25. J. Cohn, K. Schmidt, The timing of facial motion in posed and spontaneous smiles. Int. J. Wavelets Multiresolution Inf. Process. **2**(2), 121–132 (2004)
26. R. Cowie, E. Douglas-Cowie, S. Savvidou, E. McMahon, M. Sawey, M. Schröder, FEEL-TRACE: an instrument for recording perceived emotion in real time, in *ISCA Tutorial and Research Workshop on Speech and Emotion* (2000)
27. K.D. Craig, C.J. Patrick, Facial expression during induced pain. J. Pers. Soc. Psychol. **48**(4), 1080–1091 (1985)
28. N. Dalal, B. Triggs, Histograms of oriented gradients for human detection, in, *Computer Vision and Pattern Recognition* (2005), pp. 886–893
29. M. Dantone, J. Gall, G. Fanelli, L.J.V. Gool, Real-time facial feature detection using conditional regression forests, in *Computer Vision and Pattern Recognition* (2012), pp. 2578–2585
30. C. Darwin, *The Expression of the Emotions in Man and Animals* (John Murray, London, 1872)
31. K. Dautenhahn, Getting to know each other – artificial social intelligence for autonomous robots. Robot. Auton. Syst. **16**(2), 333–356 (1995)
32. K. Dautenhahn, Socially intelligent robots: dimensions of human–robot interaction. Philos. Trans. R. Soc. B **362**(1480), 679–704 (2007)
33. K. Dautenhahn, I. Werry, Towards interactive robots in autism therapy: background, motivation and challenges. Pragmat. Cogn. **12**(1), 1–35 (2004)
34. F. de Rosis, C. Pelachaud, I. Poggi, V. Carofiglio, B.D. Carolis, From Greta's mind to her face: modelling the dynamics of affective states in a conversational embodied agent. Int. J. Hum. Comput. Stud. **59**(1–2), 81–118 (2003)
35. B.M. DePaulo, J.J. Lindsay, B.E. Malone, L. Muhlenbruck, K. Charlton, H. Cooper, Cues to deception. Psychol. Bull. **129**(1), 74 (2003)
36. A. Dhall, R. Goecke, S. Lucey, T. Gedeon, Collecting large richly annotated facial-expression databases from movies. IEEE MultiMedia **19**(3), 34–41 (2012)
37. X. Ding, W.S. Chu, F.D. la Torre, J.F. Cohn, Q. Wang, Facial action unit event detection by cascade of tasks, in *International Conference on Computer Vision* (2013)
38. S. Du, Y. Tao, A. Martinez, Compound facial expressions of emotion. Proc. Natl. Acad. Sci. **111**(15), 1454–1462 (2014)
39. P. Ekman, W.V. Friesen, Nonverbal leakage and clues to deception. Psychiatry **32**(1), 88–106 (1969)
40. P. Ekman, W. Friesen, Constants across cultures in the face and emotion. J. Pers. Soc. Psychol. **17**, 124–129 (1971)
41. P. Ekman, W.V. Friesen, *Facial Action Coding System: A Technique for the Measurement of Facial Movement* (Consulting Psychologists, Palo Alto, 1978)
42. P. Ekman, W. Friesen, J.C. Hager, in *Facial Action Coding System* (A Human Face, Salt Lake City, 2002)
43. F. Eyben, S. Petridis, B. Schuller, G. Tzimiropoulos, S. Zafeiriou, M. Pantic, Audiovisual classification of vocal outbursts in human conversation using long-short-term memory networks, in *International Conference on Acoustics, Speech and Signal Processing* (2011), pp. 5844–5847

44. P. Felzenszwalb, R. Girshick, D. McAllester, D. Ramanan, Object detection with discriminatively trained part-based models. Trans. Pattern Anal. Mach. Intell. **32**(9), 1627–1645 (2010)
45. T. Fong, I. Nourbakhsh, K. Dautenhahn, A survey of socially interactive robots. Robot. Auton. Syst. **42**(3), 143–166 (2003)
46. D. Gatica-Perez, Automatic nonverbal analysis of social interaction in small groups: a review. Image Vis. Comput. **27**(12), 1775–1787 (2009)
47. A. Gudi, H.E. Tasli, T.M. den Uyl, A. Maroulis, Deep learning based FACS action unit occurrence and intensity estimation, in *Facial Expression Recognition and Analysis Challenge* (2015)
48. H. Gunes, B. Schuller, Categorical and dimensional affect analysis in continuous input: current trends and future directions. Image Vis. Comput. **31**(2), 120–136 (2013)
49. T. Hassner, S. Harel, E. Paz, R. Enbar, Effective face frontalization in unconstrained images, in *Computer Vision and Pattern Recognition* (2015)
50. H. Hung, D. Gatica-Perez, Estimating cohesion in small groups using audio-visual nonverbal behavior. Trans. Multimedia **12**(6), 563–575 (2010)
51. H. Hung, Y. Huang, G. Friedland, D. Gatica-Perez, Estimating dominance in multi-party meetings using speaker diarization. IEEE Trans. Audio Speech Lang. Process. **19**(4), 847–860 (2011)
52. M.E. Jabon, J.N. Bailenson, E. Pontikakis, L. Takayama, C. Nass, Facial expression analysis for predicting unsafe driving behavior. IEEE Pervasive Comput. **10**(4), 84–95 (2011)
53. V. Jain, E. Learned-Miller, FDDB: a benchmark for face detection in unconstrained settings. Technical Report UM-CS-2010-009, University of Massachusetts, Amherst (2010)
54. S. Jaiwand, B. Martinez, M. Valstar, Learning to combine local models for facial action unit detection, in *Facial Expression Recognition and Analysis Challenge, in conj. with Face and Gesture Recognition* (2015)
55. Q. Ji, X. Yang, Real-time eye, gaze, and face pose tracking for monitoring driver vigilance. Real-Time Imaging **8**(5), 357–377 (2002)
56. H. Jia, A.M. Martinez, Support vector machines in face recognition with occlusions, in *Computer Vision and Pattern Recognition* (2009), pp. 136–141
57. B. Jiang, M.F. Valstar, M. Pantic, Action unit detection using sparse appearance descriptors in space-time video volumes, in *Automatic Face and Gesture Recognition* (2011), pp. 314–321
58. B. Jiang, B. Martinez, M. Pantic, Parametric temporal alignment for the detection of facial action temporal segments, in *British Machine Vision Conference* (2014)
59. B. Jiang, B. Martinez, M.F. Valstar, M. Pantic, Decision level fusion of domain specific regions for facial action recognition, in *International Conference on Pattern Recognition* (2014)
60. B. Jiang, M.F. Valstar, B. Martinez, M. Pantic, Dynamic appearance descriptor approach to facial actions temporal modelling. Trans. Cybern. **44**(2), 161–174 (2014)
61. B. Jiang, B. Martinez, M. Pantic, Automatic analysis of facial actions, a survey. Trans. Affect. Comput. (under review)
62. S. Kaltwang, O. Rudovic, M. Pantic, Continuous pain intensity estimation from facial expressions, in *Advances in Visual Computing* (Springer, Heidelberg, 2012), pp. 368–377
63. S. Kaltwang, S. Todorovic, M. Pantic, Latent trees for estimating intensity of facial action units, in *Computer Vision and Pattern Recognition* (2015)
64. M. Kipp, ANVIL - a generic annotation tool for multimodal dialogue, in *European Conference on Speech Communication and Technology* (2001), pp. 1367–1370
65. S. Koelstra, I. Patras, Fusion of facial expressions and EEG for implicit affective tagging. Image Vis. Comput. **31**(2), 164–174 (2013)
66. S. Koelstra, M. Pantic, I. Patras, A dynamic texture based approach to recognition of facial actions and their temporal models. Trans. Pattern Anal. Mach. Intell. **32**(11), 1940–1954 (2010)
67. N. Komodakis, Efficient training for pairwise or higher order CRFs via dual decomposition, in *Computer Vision and Pattern Recognition* (2011), pp. 1841–1848

68. A. Krizhevsky, I. Sutskever, G.E. Hinton, Imagenet classification with deep convolutional neural networks, in *Advances in Neural Information Processing Systems* (2012)
69. I. Leite, G. Castellano, A. Pereira, C. Martinho, A. Paiva, Empathic robots for long-term interaction. Int. J. Soc. Robot. **6**(3), 329–341 (2014)
70. G. Littlewort, M.S. Bartlett, I. Fasel, J. Susskind, J. Movellan, Dynamics of facial expression extracted automatically from video, in *Image and Vision Computing* (2004), pp. 615–625
71. G. Littlewort, J. Whitehill, T. Wu, I.R. Fasel, M.G. Frank, J.R. Movellan, M.S. Bartlett, The computer expression recognition toolbox (CERT), in *Automatic Face and Gesture Recognition* (2011), pp. 298–305
72. M. Liu, S. Shan, R. Wang, X. Chen, Learning expressionlets on spatio-temporal manifold for dynamic facial expression recognition, in *Computer Vision and Pattern Recognition* (2014), pp. 1749–1756
73. P. Liu, S. Han, Z. Meng, Y. Tong, Facial expression recognition via a boosted deep belief network, in *Computer Vision and Pattern Recognition* (2014)
74. P. Lucey, J.F. Cohn, I. Matthews, S. Lucey, S. Sridharan, J. Howlett, K.M. Prkachin, Automatically detecting pain in video through facial action units. Trans. Syst. Man Cybern. B **41**(3), 664–674 (2011)
75. P. Lucey, J.F. Cohn, K.M. Prkachin, P.E. Solomon, I. Matthews, Painful data: the UNBC-McMaster shoulder pain expression archive database, in *Automatic Face and Gesture Recognition* (2011)
76. M. Lyons, S. Akamatsu, M. Kamachi, J. Gyoba, Coding facial expressions with Gabor wavelets, in *Automatic Face and Gesture Recognition* (1998)
77. A. Maalej, B.B. Amor, M. Daoudi, A. Srivastava, S. Berretti, Shape analysis of local facial patches for 3D facial expression recognition. Pattern Recogn. **44**(8), 1581–1589 (2011)
78. B. Martinez, M.F. Valstar, L21-based regression and prediction accumulation across views for robust facial landmark detection. Image Vis. Comput. In press
79. B. Martinez, M.F. Valstar, X. Binefa, M. Pantic, Local evidence aggregation for regression based facial point detection. Trans. Pattern Anal. Mach. Intell. **35**(5), 1149–1163 (2013)
80. M. Mathias, R. Benenson, M. Pedersoli, L. van Gool, Face detection without bells and whistles, in *European Conference on Computer Vision* (2014)
81. D. Matsumoto, More evidence for the universality of a contempt expression. Motiv. Emot. **16**, 363–368 (1992)
82. I. McCowan, D. Gatica-Perez, S. Bengio, G. Lathoud, M. Barnard, D. Zhang, Automatic analysis of multimodal group actions in meetings. Trans. Pattern Anal. Mach. Intell. **27**(3), 305–317 (2005)
83. D. McDuff, R. El Kaliouby, E. Kodra, R. Picard, Measuring voter's candidate preference based on affective responses to election debates, in *Affective Computing and Intelligent Interaction* (2013), pp. 369–374
84. D. McDuff, R. Kaliouby, T. Senechal, A, Amr, J.F. Cohn, R. Picard, Affectiva-MIT facial expression dataset (AM-FED): naturalistic and spontaneous facial expressions collected in-the-wild, in *Computer Vision and Pattern Recognition Workshop* (2013), pp. 881–888
85. D. McDuff, R. El Kaliouby, T. Senechal, D. Demirdjian, R. Picard, Automatic measurement of ad preferences from facial responses gathered over the internet. Image Vis. Comput. **32**(10), 630–640 (2014)
86. D. McDuff, R. Kaliouby, J. Cohn, R. Picard, Predicting ad liking and purchase intent: large-scale analysis of facial responses to ads. Trans. Affect. Comput. **6**, 223–235 (2015)
87. G. McKeown, I. Sneddon, Modeling continuous self-report measures of perceived emotion using generalized additive mixed models. Psychol. Methods **19**(1), 155–74 (2014)
88. G. McKeown, M. Valstar, R. Cowie, M. Pantic, M. Schroder, The semaine database: annotated multimodal records of emotionally colored conversations between a person and a limited agent. IEEE Trans. Affect. Comput. **3**, 5–17 (2012). doi:http://doi.ieeecomputersociety.org/10.1109/T-AFFC.2011.20
89. L. Morency, I. de Kok, J. Gratch, Context-based recognition during human interactions: automatic feature selection and encoding dictionary, in *International Conference on Multimodal Interaction* (2008), pp. 181–188

90. R. Navarathna, P. Lucey, P. Carr, E. Carter, S. Sridharan, I. Matthews, Predicting movie ratings from audience behaviors, in *IEEE Winter Conference on Applications of Computer Vision* (2014), pp. 1058–1065
91. L.S. Nguyen, A. Marcos-Ramiro, M.M. Romera, D. Gatica-Perez, Multimodal analysis of body communication cues in employment interviews, in *International Conference on Multimodal Interaction* (2013), pp. 437–444
92. M.A. Nicolaou, H. Gunes, M. Pantic, Output-associative RVM regression for dimensional and continuous emotion prediction. Image Vis. Comput. **30**(3), 186–196 (2012)
93. M.A. Nicolaou, V. Pavlovic, M. Pantic, Dynamic probabilistic CCA for analysis of affective behaviour and fusion of continuous annotations. Trans. Pattern Anal. Mach. Intell. **36**(7), 1299–1311 (2014)
94. T. Ojala, M. Pietikainen, D. Harwood, A comparative study of texture measures with classification based on featured distribution. Pattern Recogn. **29**(1), 51–59 (1996)
95. J. Orozco, B. Martinez, M. Pantic, Empirical analysis of cascade deformable models for multi-view face detection. Image Vis. Comput. **42**, 47–61 (2015)
96. I. Pavlidis, N.L. Eberhardt, J.A. Levine, Human behaviour: seeing through the face of deception. Nature **415**(6867), 35–35 (2002)
97. P. Perakis, G. Passalis, T. Theoharis, I. Kakadiaris, 3D facial landmark detection under large yaw and expression variations. Trans. Pattern Anal. Mach. Intell. **35**(7), 1552–1564 (2013)
98. S. Petridis, M. Pantic, Audiovisual discrimination between laughter and speech, in *International Conference on Acoustics, Speech and Signal Processing* (2008), pp. 5117–5120
99. S. Petridis, B. Martinez, M. Pantic, The MAHNOB laughter database. Image Vis. Comput. **31**(2), 186–202 (2013)
100. J.H. Pfeifer, M. Iacoboni, J.C. Mazziotta, M. Dapretto, Mirroring others' emotions relates to empathy and interpersonal competence in children. NeuroImage **39**(4), 2076–2085 (2008)
101. T. Pfister, X. Li, G. Zhao, M. Pietikäinen, Recognising spontaneous facial micro-expressions, in *International Conference on Computer Vision* (2011), pp. 1449–1456
102. R.W. Picard, *Affective Computing* (MIT, Cambridge, 1997)
103. K.M. Prkachin, P.E. Solomon, The structure, reliability and validity of pain expression: evidence from patients with shoulder pain. Pain **139**, 267–274 (2008)
104. O. Rudovic, M. Pantic, Shape-constrained Gaussian process regression for facial-point-based head-pose normalization, in *International Conference on Computer Vision* (2011), pp. 1495–1502
105. J.A. Russell, A circumplex model of affect. J. Pers. Soc. Psychol. **39**, 1161–1178 (1980)
106. G. Sandbach, S. Zafeiriou, M. Pantic, Binary pattern analysis for 3D facial action unit detection, in *The British Machine Vision Conference* (2012)
107. G. Sandbach, S. Zafeiriou, M. Pantic, Markov random field structures for facial action unit intensity estimation, in *International Conference on Computer Vision Workshop* (2013)
108. J.M. Saragih, S. Lucey, J.F. Cohn, Deformable model fitting by regularized landmark mean-shift. Int. J. Comput. Vis. **91**(2), 200–215 (2011)
109. E. Sariyanidi, H. Gunes, A. Cavallaro, Automatic analysis of facial affect: a survey of registration, representation and recognition. Trans. Pattern Anal. Mach. Intell. **37**(6), 1113–1133 (2015)
110. K. Scherer, P. Ekman, *Handbook of Methods in Nonverbal Behavior Research* (Cambridge University Press, Cambridge, 1982)
111. M. Schröder, E. Bevacqua, R. Cowie, F. Eyben, H. Gunes, D. Heylen, M. ter Maat, G. pain, S. Pammi, M. Pantic, C. Pelachaud, B. Schuller, E. de Sevin, M.F. Valstar, M. Wöllmer, Building autonomous sensitive artificial listeners. Trans. Affect. Comput. **3**(2), 165–183 (2012)
112. T. Senechal, V. Rapp, H. Salam, R. Seguier, K. Bailly, L. Prevost, Facial action recognition combining heterogeneous features via multi-kernel learning. IEEE Trans. Syst. Man Cybern. B **42**(4), 993–1005 (2012)
113. T. Sha, M. Song, J. Bu, C. Chen, D. Tao, Feature level analysis for 3D facial expression recognition. Neurocomputing **74**(12–13), 2135–2141 (2011)

114. C. Shan, S. Gong, P. McOwan, Facial expression recognition based on local binary patterns: a comprehensive study. Image Vis. Comput. **27**(6), 803–816 (2009)
115. P.E. Shrout, J.L. Fleiss, Intraclass correlations: uses in assessing rater reliability. Psychol. Bull. **86**(2), 420–428 (1979)
116. T. Simon, M.H. Nguyen, F.D.L. Torre, J. Cohn, Action unit detection with segment-based SVMs, in *Computer Vision and Pattern Recognition* (2010), pp. 2737–2744
117. M. Soleymani, M. Pantic, Human-centered implicit tagging: overview and perspectives, in *International Conference on Systems, Man, and Cybernetics* (2012), pp. 3304–3309
118. M. Soleymani, J. Lichtenauer, T. Pun, M. Pantic, A multimodal database for affect recognition and implicit tagging. Trans. Affect. Comput. **3**(1), 42–55 (2012)
119. M. Soleymani, M. Larson, T. Pun, A. Hanjalic, Corpus development for affective video indexing. Trans. Multimedia **16**(4), 1075–1089 (2014)
120. M.J.L. Sullivan, P. Thibault, A. Savard, R. Catchlove, J. Kozey, W.D. Stanish, The influence of communication goals and physical demands on different dimensions of pain behavior. Pain **125**(3), 270–277 (2006)
121. X. Sun, J. Lichtenauer, M. Valstar, A. Nijholt, M. Pantic, A multimodal database for mimicry analysis, in *Affective Computing and Intelligent Interaction* (2011), pp. 367–376
122. M. Takahashi, M. Naemura, M. Fujii, S. Satoh, Estimation of attentiveness of people watching TV based on their emotional behaviors, in *Affective Computing and Intelligent Interaction* (2013), pp. 809–814
123. H. Tang, T. Huang, 3D facial expression recognition based on properties of line segments connecting facial feature points, in *Automatic Face and Gesture Recognition* (2008)
124. E. Taralova, F. De la Torre, M. Hebert, Motion words for video, in *European Conference on Computer Vision* (2014)
125. D. Tax, M.F. Valstar, M. Pantic, E. Hendrix, The detection of concept frames using clustering multi-instance learning, in *International Conference on Pattern Recognition* (2010), pp. 2917–2920
126. Y. Tong, J. Chen, Q. Ji, A unified probabilistic framework for spontaneous facial action modeling and understanding. Trans. Pattern Anal. Mach. Intell. **32**(2), 258–273 (2010)
127. F. Tsalakanidou, S. Malassiotis, Real-time 2D+3D facial action and expression recognition. Pattern Recogn. **43**(5), 1763–1775 (2010)
128. P. Tsiamyrtzis, J. Dowdall, D. Shastri, I. Pavlidis, M. Frank, P. Ekman, Imaging facial physiology for the detection of deceit. Int. J. Comput. Vis. **71**(2), 197–214 (2007)
129. G. Tzimiropoulos, Project-out cascaded regression with an application to face alignment, in *Computer Vision and Pattern Recognition* (2015), pp. 3659–3667
130. G. Tzimiropoulos, M. Pantic, Gauss-Newton deformable part models for face alignment in-the-wild, in *Computer Vision and Pattern Recognition* (2014), pp. 1851–1858
131. M. Valstar, Automatic behaviour understanding in medicine, in *Workshop on Roadmapping the Future of Multimodal Interaction Research, including Business Opportunities and Challenges, RFMIR@ICMI* (2014), pp. 57–60
132. M. Valstar, M. Pantic, Fully automatic recognition of the temporal phases of facial actions. IEEE Trans. Syst. Man Cybern. B **42**(1), 28–43 (2012)
133. M. Valstar, I. Patras, M. Pantic, Facial action unit detection using probabilistic actively learned support vector machines on tracked facial point data, in *Computer Vision and Pattern Recognition Workshops* (2005)
134. M.F. Valstar, B. Martinez, X. Binefa, M. Pantic, Facial point detection using boosted regression and graph models, in *Computer Vision and Pattern Recognition* (2010), pp. 2729–2736
135. M.F. Valstar, M. Mehu, B. Jiang, M. Pantic, K. Scherer, Meta – analysis of the first facial expression recognition challenge. IEEE Trans. Syst. Man Cybern. B **42**(4), 966–979 (2012)

136. M. Valstar, B. Schuller, K. Smith, T. Almaev, F. Eyben, J. Krajewski, R. Cowie, M. Pantic, AVEC 2014: 3D dimensional affect and depression recognition challenge, in *International Workshop on Audio/Visual Emotion Challenge* (2014), pp. 3–10
137. M.F. Valstar, T. Almaev, J.M. Girard, G. McKeown, M. Mehu, L. Yin, M. Pantic, J.F. Cohn, FERA 2015 - second facial expression recognition and analysis challenge, in *Automatic Face and Gesture Recognition Workshop* (2015)
138. L. van der Maaten, E. Hendriks, Action unit classification using active appearance models and conditional random fields. Cogn. Process. **13**(2), 507–518 (2012)
139. L. van der Maaten, M. Chen, S. Tyree, K.Q. Weinberger, Learning with marginalized corrupted features, in *International Conference on Machine Learning* (2013), pp. 410–418
140. A. Vinciarelli, M. Pantic, H. Bourlard, Social signal processing: survey of an emerging domain. Image Vis. Comput. **27**(12), 1743–1759 (2009)
141. A. Vinciarelli, M. Pantic, D. Heylen, C. Pelachaud, I. Poggi, F. D'Errico, M. Schröder, M.: Bridging the gap between social animal and unsocial machine: a survey of social signal processing. Trans. Affect. Comput. **3**(1), 69–87 (2012)
142. P. Viola, M.J. Jones, Robust real-time face detection. Int. J. Comput. Vis. **57**(2), 137–154 (2004)
143. E. Vural, M. Cetin, A. Ercil, G. Littlewort, M. Bartlett, J. Movellan, Drowsy driver detection through facial movement analysis, in *IEEE International Conference on Human-Computer Interaction* (2007), pp. 6–18
144. Z. Wang, Y. Li, S. Wang, Q. Ji, Capturing global semantic relationships for facial action unit recognition, in *International Conference on Computer Vision* (2013), pp. 3304–3311
145. S. Wang, Z. Liu, Y. Zhu, M. He, X. Chen, Q. Ji, Implicit video emotion tagging from audiences' facial expression. Multimedia Tools Appl. **74**(13), 4679–4706 (2015)
146. G. Warren, E. Schertler, P. Bull, Detecting deception from emotional and unemotional cues. J. Nonverbal Behav. **33**(1), 59–69 (2009)
147. F. Weninger, Introducing CURRENNT: the munich open-source CUDA recurrent neural network toolkit. J. Mach. Learn. Res. **16**, 547–551 (2015)
148. J. Whitehill, Z. Serpell, Y. Lin, A. Foster, J.R. Movellan, The faces of engagement: automatic recognition of student engagement from facial expressions. Trans. Affect. Comput. **5**(1), 86–98 (2014)
149. M. Wöllmer, A. Metallinou, F. Eyben, B. Schuller, S.S. Narayanan, Context-sensitive multimodal emotion recognition from speech and facial expression using bidirectional LSTM modeling, in *Interspeech* (2010), pp. 2362–2365
150. Q. Wu, X. Shen, X. Fu, The machine knows what you are hiding: an automatic micro-expression recognition system, in *Affective Computing and Intelligent Interaction* (2011), pp. 152–162
151. X. Xiong, F. De la Torre, Supervised descent method and its applications to face alignment, in *Computer Vision and Pattern Recognition* (2013)
152. J. Yan, Z. Lei, D. Yi, S.Z. Li, Learn to combine multiple hypotheses for accurate face alignment, in *International Conference on Computer Vision Workshop* (2013), pp. 392–396
153. W. Yan, Q. Wu, Y. Liu, S. Wang, X. Fu, CASME database: a dataset of spontaneous micro-expressions collected from neutralized faces, in *Automatic Face and Gesture Recognition* (2013)
154. J. Yan, X. Zhang, Z. Lei, S.Z. Li, Face detection by structural models. Image Vis. Comput. **32**(10), 790–799 (2014)
155. P. Yang, Q. Liu, D.N. Metaxas, Boosting encoded dynamic features for facial expression recognition. Pattern Recogn. Lett. **30**(2), 132–139 (2009)
156. X. Yu, Z. Lin, J. Brandt, D. Metaxas, Consensus of regression for occlusion-robust facial feature localization, in *European Conference on Computer Vision* (2014), pp. 105–118
157. Z. Zeng, M. Pantic, G. Roisman, T.S. Huang et al., A survey of affect recognition methods: audio, visual, and spontaneous expressions. Trans. Pattern Anal. Mach. Intell. **31**(1), 39–58 (2009)

158. X. Zhang, L. Yin, J.F. Cohn, Three dimensional binary edge feature representation for pain expression analysis, in *Automatic Face and Gesture Recognition* (2015)
159. G. Zhao, M. Pietikainen, Dynamic texture recognition using local binary patterns with an application to facial expressions. Trans. Pattern Anal. Mach. Intell. **29**(6), 915–928 (2007)
160. L. Zhong, Q. Liu, P. Yang, B. Liu, J. Huang, D.N. Metaxas, Learning active facial patches for expression analysis, in *Computer Vision and Pattern Recognition* (2012), pp. 2562–2569
161. X. Zhu, D. Ramanan, Face detection, pose estimation, and landmark localization in the wild, in *Computer Vision and Pattern Recognition* (2012), pp. 2879–2886
162. M. Zimmerman, I. Chelminski, M. Posternak, A review of studies of the Hamilton depression rating scale in healthy controls: implications for the definition of remission in treatment studies of depression. J. Nerv. Ment. Dis. **192**(9), 595–601 (2004)

Exaggeration Quantified: An Intensity-Based Analysis of Posed Facial Expressions

Harish Bhaskar, Davide La Torre, and Mohammed Al-Mualla

Abstract Posed facial expressions are characterized by deliberate and often exaggerated behaviours that usually fail to generalize to the complexity of expressive exhibition of human emotion in real-life. The detailed understanding of exaggeration and its quantification that defines each facial expression as a feature vector in the face space will provide deep insight to the growing interest in analysing posed facial expressions. In this paper, an attempt to quantify the intensity of facial expressions via estimating exaggeration as the deviation in the combined relationship between correlated landmark points that characterize the deformable face model, is made. Such relationships that underpin the discrimination of different facial expressions using exaggeration measurements is based on novel geometric morphometric inspired feature descriptors together with enhanced appearance models that seed a cascaded Support Vector Machine (SVM) classifier. In addition to demonstrating the superiority of the proposed method by applying it to images containing a wide range of expressions using standard datasets; results also distinguish different important facial landmarks for classifying that expression from every other expression.

H. Bhaskar (✉)
Visual Signal Analysis and Processing (VSAP) Research Center, Khalifa University
of Science Technology and Research, Abu Dhabi, United Arab Emirates
e-mail: harish.bhaskar@kustar.ac.ae

D. La Torre
Department of Applied Mathematics and Sciences, Khalifa University of Science,
Technology and Research, P.O.Box 127788, Abu Dhabi, United Arab Emirates

Department of Economics, Management and Quantitative Methods,
University of Milan, 20122 Milan, Italy
e-mail: davide.latorre@kustar.ac.ae

M. Al-Mualla
Khalifa University of Science Technology and Research, Abu Dhabi, United Arab Emirates
e-mail: almualla@kustar.ac.ae

© Springer International Publishing Switzerland 2016
M. Kawulok et al. (eds.), *Advances in Face Detection and Facial Image Analysis*,
DOI 10.1007/978-3-319-25958-1_5

1 Introduction and Related Work

Facial expressions are considered a visual manifestation of the affective psychological state and the cognitive ability of a human mind, when engaged in a non-verbal
communication [1]. Facial expressions are crucial to understand these more subtle
signals as a larger part of the communication process. While detecting facial
expressions is already complex, their interpretation can vary largely. Therefore,
attaching semantics such as emotions to such expressions require being accustomed
to cultural, familial, and business backgrounds. The automatic recognition of facial
expressions has received significant research attention in the recent years due to
its wide range of applications in emotion analysis [1], biometrics [2] and Human
Computer Interaction (HCI) [3], among others. Facial expression recognition
and analysis techniques can been classified in a number of different ways. This
includes: (a) message and sign-based approaches that interpret facial expressions
from an observer's inference point-of-view; (b) micro-level and macro-level facial
expression analysis based on suppression of effects such as the emotional artefacts
and (c) spontaneous versus posed (SVP) facial expressions characterised by the
exaggeration of each expression category. Facial expressions can also be distinguished based on the type of expression being recognized. Most research efforts
in the literature have been based on the analysis of basic expressions and some non-
basic expressions that requires recognizing finer changes in expressions [4, 5].

Message-based and sign-judgement approaches are holistic techniques that combine information from various regions of the face to decipher the cognitive state that
may be expressed. Such methods heavily rely on a priori information from human
perception and context for accurate facial expression recognition [6]. The various
changes in facial behaviour have traditionally been parametrized and modelled using
action units (AU) that are characterised by muscle actions either individually or
in groups. For example, facial expression recognition using AU was developed by
Tian et al. in [7] where the automatic recognition of six upper face AU's and ten
lower face AU's was targeted. However, the method was found limited to frontal
views alone and therefore required an extension into AU recognition for profile
views for real-time applications. Based on the study of AU's, a comprehensive
system for measuring all facial movements was developed in the form of the Facial
Action Coding System (FACS) [8]. FACS taxonomy consisted of 44 unique AU's
defined by gray-level variations between expressions and using electrical activity
of the underlying facial muscles. In addition to AU's, FACS also defines action
descriptors that illustrate facial movements. In [8], FACS represented facial actions
that vary in intensity measured at three distinct levels (X, Y and Z) which was more
recently updated into five levels ranging from A to E. Several combinations of AU's
and actions descriptors have also been described in the literature. The reliability of
FACS as descriptors for face analysis was based on several parameters including
precision of measurement of onsets, peaks, offsets, and changes in AU intensity. In
an attempt to study the impact of human perception on AU, inter-observer agreement
between AU's was studied in [9] and was found to have variations. Despite, the

detailed literature within the message and sign based facial expression analysis, most methods were found to be limited in several ways including: (a) difficulties in handling a blend of multiple emotions or expressions, (b) generalization to different image sources and applications and (c) attaching labels to expressions that may result in unwarranted inference [10].

Macro-level facial expressions are the most common among expressions that can be found during daily interactions. On the other hand, micro-expressions occur when there exists conscious or unconscious concealing or repression of human emotion. While macro-expressions are easy to detect and perform inference with, micro-expressions are difficult to detect and requires more training as their recognition could have misconceptions in interpretations. Macro-expression recognition techniques in the literature have been classified into three categories. First, methods that utilize point-model representations of the face to study the inter-dynamics of these points temporally over the video sequence can be considered. One of the main challenges in this category of macro-expression recognition emanates from poor illumination conditions that can cause inconsistent and inaccurate detection of key features that are required for robust classification. Second, a dense representation of the face is considered for macro-expression recognition. Finally, the third category is a combination of both the aforementioned classes. On the other hand, the state-of-the-art in micro-expression detection is focused on more localised study of certain regions of the face. For example, in [11] the upper and lower halves of the face have been considered separately for micro-expression detection. The conclusion of [11] was further supported by other recent studies such as in [12], where localized analysis of the face has permitted reliable recognition of micro-expressions. In a similar study by [21], the discriminative power of the dynamics of eyelid, cheek, and lip corner movements were capitalized for distinguishing between spontaneous and posed enjoyment smiles. Further, not much research has been possible within automatic recognition of micro-expressions, mainly because these expressions are displayed very quickly, mostly correctively and are difficult to elicit and capture [13, 14].

Finally, SVP expression categorization is considered. Posed expressions are artificial expressions that are different from spontaneous expressions in terms of their appearance, intensity, symmetry, timing and other temporal characteristics. Furthermore, posed-expression studies in the literature have revealed that high exaggeration is a common characteristics among such category of expressions [15–17]. The literature in SVP expression classification is divided into global and local approaches. Global methods aim to extract global facial features such as shape, geometry, appearance and motion and are used in conjunction with a trained classifier for SVP expression classification [18, 19]. On the other hand, local approaches advocate a two-staged process, one where the class of expression is first recognized/or assumed known and then posed expressions is identified within the selected class. Several research studies have been conducted analysing individual classes of facial expressions including, smiles [20, 21], eyebrow actions [23] and pain [24, 25]. The work within SVP expressions analysis is dominated by the use of distinctive and unique feature sets including the typical combination of geometric and appearance features [19]. For example, in distinguishing between SVP smiles on the BBC and Cohn-Kanade databases, the method of [26] has reported nearly 85 and 91 % accuracy respectively. Motion features on tracked landmark points have been fused into a multi-model system for distinguishing between SVP expressions in [20]. The use of appearance feature such as the Gabor wavelets [25, 27] has been used to differentiate spontaneous from posed pain. In the study by Bhaskar and Al-Mualla [28], the combination of geometric features using Euclidean Distance Matrix Analysis (EDMA) and appearance features using Gabor wavelet features has been shown to be valuable for SVP expression classification. In addition, the fusion of geometric deviations estimated using reflection symmetry with appearance quality through structural similarity has also been demonstrated for SVP expression recognition in [29].

All facial expression analysis systems are composed of three main modules: (1) face detection, facial feature tracking and registration, (2) feature extraction and (3) supervised or unsupervised classification. The initial step of face detection is usually performed using the most common Viola and Jones face detection algorithm [30]. Following this, facial features are localised to either facilitate the registration of face instances in a video or for feature extraction. Face feature localization can be accomplished using Parametrized Appearance Models (PAMs) such as the Eigen tracking [31], Active Shape [32, 33] or Appearance Models [34] and Morphable Models [35]. In the next step, geometric and appearance feature descriptors that encapsulate the underlying facial expression are often extracted for the classification of facial expressions. While geometric features contains facial shape information, appearance descriptors capture appearance variations such as dynamic textures among feature points [36]. Despite advances, there exists no concrete evidence whether geometric or appearance descriptors are better for expression classification. For example, in [36] appearance feature based on texture variations coded for specific AU's has been shown to outperform geometric shape features. Other examples of appearance features include, Gabor wavelets [27, 37, 38], shape features [39], and Independent Component Analysis (ICA) [40] based representations. However, more recently, the combination of geometric and appearance features have resulted in accurate classification of facial expressions [28, 29]. In addition, other texture features such as the LBP [41] and invariant feature descriptors such as the SIFT [42] have also been explored for face expression recognition. Global features such as optical flow and dynamic textures [41] also have a place in the face analysis literature. Finally, the classification of extracted features can operate in both supervised and unsupervised manner. Algorithms such as the nearest neighbour classifier [43], Neural network [44], Adaboost Classifier [45] and the Support Vector Machine (SVM) classifier [22, 28, 29, 39] are all among popular classification strategies in the literature. Recent work has also focused on the use of cascaded classifiers for spontaneous and posed expression classification [28].

Despite recent efforts, the autonomous recognition of facial expressions and their classification either as spontaneous or posed is complicated due to changes in pose and scale, illumination variations, and occlusion. In addition, there is an increasing need for research on non-frontal and head-motion invariant facial expression detection. Detection of micro-expressions, where building models that are complicated by inherently subtle variations in expressions are also highly demanded. Finally, much research attention is also required for modelling temporal dynamics of facial movements thereby facilitating continuous facial expression recognition.

In this paper, an aggregated 2-stage classifier model for the detection of SVP expressions in still images using combined geometric and appearance feature sets is proposed. During stage 1 of classification, geometric descriptors inspired by Lele and Richtsmeier [46] including an enhanced definition of Form Difference Matrix (*FDM*), Growth Matrix (*GM*) and Shape Difference Matrix (*SDM*) is proposed. These descriptors are extracted using autonomously selected influential facial landmark points and are combined with a weighted structural similarity based appearance quality indicators for facial expression recognition. Further, in an extended stage 2, a combination of geometric descriptors based on reflective symmetry and appearance features using complex wavelets structural similarity are used to discriminate spontaneous from posed expressions within that class of facial expression identified in stage 1.

One key novelty of the proposed method is the combination of novel descriptors using complementary geometric and appearance features for quantifying exaggeration and hence classification of SVP expressions. Results suggest that using the proposed distinctive and unique feature sets for the individual stages (1 & 2) of classification can improve the accuracy of facial expression recognition and hence robustly identify posed expressions in comparison to a joint classification model. Although previous work in [28] has already demonstrated the use of form matrices as geometric descriptors, the selection of influential landmark points and hence the computation of SM can be considered novel in this study. Further, the use of a weighted component structural similarity for appearance quality modelling for facial expression classification is new under the present context. Finally, the choice of reflection symmetry based geometric features in stage 2 classification not alone encapsulates intensity variations but also identifies anti-symmetry in facial expressions that allows the accurate recognition of posed expressions.

2   Proposed Method

A generic block diagram of the proposed SVP expressions classification framework is illustrated in Fig. 1. As mentioned above, the framework incorporates an aggregated classification process. During stage 1 of classification, a multi-class (one-against-all) SVM classifier is trained for facial expression recognition. Further, in stage 2, multiple two-class SVM classifiers, one for each expression, are trained for separating spontaneous from posed expressions. Note that the facial expression recognized in stage 1 of the framework is cascaded onto stage 2 and only that SVM classifier for the expression identified in stage 1 is used for detecting posed expressions in stage 2. The main distinguishing aspect of stages

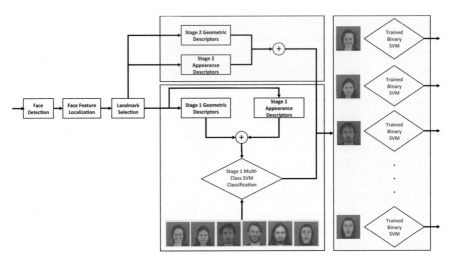

Fig. 1 The proposed posed expression recognition framework using the 2-stage aggregated classification model using combined geometric descriptors and appearance indicators

1 and 2 are the geometric and appearances descriptors chosen for classification. In stage 1, feature descriptors are chosen such that the intrinsic global geometric and appearance differences between different facial expressions are captured. However, during stage 2, combined features that can quantify exaggeration in terms of the intensity of expression and anti-symmetry within that expression, are chosen.

Face detection and facial feature localization are the fundamental steps in the proposed expression recognition framework. In the proposed method, following successful face detection using the Viola-Jones face detector [30], a face model is represented using a set of N landmark points, $\mathbf{X} = \{x_1, x_2, \ldots, x_N\}$. Therefore, in the next step, the optimal set of landmark points $\mathbf{X}^*$, that best fits the face image is obtained by minimizing the posterior probability:

$$p(\mathbf{X}|\mathbf{I}) \propto p(\mathbf{I}|\mathbf{X})p(\mathbf{X}) \qquad (1)$$

A number of different techniques are well known for such optimization including the popular Active Shape Model (ASM) [32] and Active Appearance Model (AAM) [34]. In this paper, the combined shape models of [33] has been used. The first step in [33] begins by initializing the point locations at $\mathbf{X}_0^*$ and further, iteratively selecting candidates $\mathrm{argmax}_Y\, p(\mathbf{I}|\mathbf{Y})p(\mathbf{Y}|\mathbf{X})$ until regularization that updates the landmark points to the optimal set $\mathbf{X}^*$ is obtained. This technique is known for deriving an approximate solution $\mathbf{Y}$ from a restrictive set of optimal candidates for each x_i efficiently using Markov Random Fields (MRF) inference scheme.

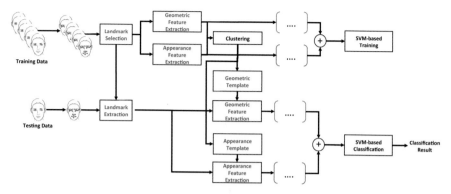

Fig. 2 Block diagram of the stage-1 classification model for facial expression classification illustrating the training and testing processes of the framework

2.1 Stage 1: Face Expression Recognition

Facial expression recognition is attempted during stage 1 classification of the proposed framework. A detailed block diagram of this stage 1 classification is presented in Fig. 2. The process in stage 1 begins with the extraction of geometric descriptors based on Form Matrix (FM). Further, the geometrical differences between each expression and its neutral counterpart is exploited to determine a subset of the most influential landmark points. In addition, novel geometric descriptors and appearance features using a weighted structural similarity method are deduced across those selected landmark points and further, jointly used for classification.

2.1.1 Training

During training, the training data in the form of (Γ_1, ℓ_1), (Γ_2, ℓ_2), (Γ_3, ℓ_3), ..., (Γ_M, ℓ_M), are prepared by extracting joint descriptors $\Gamma_{1,2,...,M}$ from all M training images. The corresponding expression class label ℓ, where $\ell \in 1, 2, .., 6$ are concatenated to the joint descriptors for each image and together are presented to a multi-class SVM model. A brief description of the various steps during training is described in the algorithm below. For every image on the training set,

- Extract the optimal set of landmark points $\mathbf{X}^*$ using the method described earlier in Sect. 2.
- Using the preliminary geometric features mentioned in Sect. 2.2, transform the optimal set of landmark points $\mathbf{X}^*$ into a selected subset of influential landmark points represented as $\hat{\mathbf{X}}^*$. Using a simple majority voting approach across the all considered facial expressions, a subset of influential landmark points is generated and stored.

- Extract geometric descriptors Γ_g consisting of *FDM*, *GM* and *SDM* as described in Sect. 2.2.1 using those selected landmark points.
- Extract appearance descriptors Γ_a using the weighted structural similarity method as described in Sect. 2.2.2 at those chosen landmark points.

After the geometric and appearance descriptors for all the training images have been successfully extracted, the training data consisting of the joint descriptor $\Gamma = \{\Gamma_g, \Gamma_a\}$ and the class label ℓ for all training is presented as $\{(\Gamma_1, \ell_1), (\Gamma_2, \ell_2), (\Gamma_3, \ell_3), \ldots, (\Gamma_M, \ell_M)\}$ to a multi-class SVM classifier for training. Further, geometric and appearance templates along with the mean shape of all facial expressions are generated using a simple clustering algorithm and stored for the detection (or testing) step.

2.1.2 Testing

Given a trained face expression classification model and the necessary geometric and appearance templates, recognition of face expressions on test data is initiated. During testing, the following steps are followed for the accurate recognition of the facial expression on each test image. On each test image,

- Extract the optimal set of landmark points $\mathbf{X}^*$ using the method described earlier in Sect. 2.
- Extract geometric descriptors Γ_g consisting of *FDM*, *GM* and *SDM* according to the method described in Sect. 2.2.1 using those selected landmark points and the geometric template generated from the training step, as a reference.
- Extract appearance descriptors Γ_a using the weighted structural similarity method as described in Sect. 2.2.2 at those chosen landmark points, using the appearance template from training as a reference.
- Present the joint descriptor $\Gamma = \{\Gamma_g, \Gamma_a\}$ along with the trained SVM model to the SVM classifier described in Sect. 2.2.3 for face expression recognition.

2.2 Stage 1 Landmark Selection

In order to describe the technique for reducing the optimal set of landmark points into a selected subset of influential landmarks, some background into Form Matrix (FM) representation is introduced. Consider that S is equivalent to the space of all faces represented by N landmarks which belong to the 2D Euclidean space, then according to the study by Lele and Richtsmeier [46] and Bhaskar and Al-Mualla [28], an Euclidean Distance Matrix (EDM) representation of S consists of all the inter-landmark distances and can easily be formulated as a FM of the form:

$$FM(s) = \begin{bmatrix} 0 & d(1,2) & d(1,3) & . . & d(1,K) \\ d(2,1) & 0 & d(2,3) & . . & d(2,K) \\ . & . & . & . . & . \\ . & . & . & . . & . \\ d(K,1) & d(K,2) & d(K,3) & . . & 0 \end{bmatrix} \qquad (2)$$

Such FM has been known to encapsulate all the relevant information about the geometric form of the face. However, it was reported in [28] that not all landmark points retain this geometric form exclusively. Empirically, it was shown in [28] that certain landmark point are more influential to the classification of facial expressions than others. Therefore, it has become necessary to automatically extract a selected subset of those most influential (K, where $K \leq N$) landmark points dubbed as $\hat{\mathbf{X}}^*$ from the original optimal set $\mathbf{X}^*$.

In this context, the mean form representation defined in [28] as FDM is used to compare the geometry of any facial expression against its neutral counterpart using:

$$FDM(t,n) = \frac{FM_{ij}(t)}{FM_{ij}(n)} \qquad (3)$$

where $i,j = 1,2,3 \ldots N$, $FM_{ij}(t)$ is the FM corresponding to the target expression t and $FM_{ij}(n)$ is the FM of its neutral expression counterpart. That is, FDM is a simple ratio of the distances between landmarks i and j in the target expression t to the same distance in the neutral expression n.

In order to determine the most influential landmark points as in [46], the $N(N-1)/2$ off-diagonal elements are placed into a column vector in a rank-ordered manner. The influential landmark points can be identified as those that presented the largest spread of element values, i.e. which corresponded to the extreme elements of the column vectors of the sorted off-diagonals. This proposition could be validated by considering the distribution of the FDM values of the influential landmarks that showed higher skew than others. Thus, the optimal set of K landmark points $\mathbf{X}^*$ are transformed into $\hat{\mathbf{X}}^*$.

During empirical study of this selection procedure, it was noticed that the K influential landmark points thus chosen, using the process mentioned above, were different for different facial expression. Therefore, a simple majority voting approach was used across all facial expressions to reduce the landmark points from the original size of $N = 22$ to $K = 12$ consisting of the corners of the eyes, eyebrow and mouth.

2.2.1 Stage 1 Geometric Descriptors

Geometric features play an important role in encapsulating shape variation information between facial expressions. In order to adequately model all such variations, descriptors such as the FDM, GM and SDM are extracted using the $\hat{\mathbf{X}}^*$ landmark points. *FDM* is re-estimated as a feature vector according to the definition in Eq. (3). Note, while computing *FDM* and *SDM*, the neutral expression that was used during training phase is replaced by the geometric template (obtained from training) during the testing phase.

As further extensions, the absolute growth matrix (GM) is redefined as the deviation from the (global) mean of all expressions using:

$$GM(t, \hat{\mu}) = \frac{FM_{ij}(t)}{FM_{ij}(\hat{\mu})} \tag{4}$$

where $FM_{ij}(t)$ refers to the FM of the target expression and $FM_{ij}(\hat{\mu})$ is the FM of the mean of all forms of expressions. Further, a shape difference matrix (SDM) that estimates and compares the mean shape of the landmark points with respect to its neutral and mean forms is formulated as,

$$SDM(t, n, \hat{\mu}) = \frac{SM_{ij}(t) - SM_{ij}(n)}{SM_{ij}(t) - SM_{ij}(\hat{\mu})} \tag{5}$$

where *SM* represents the shape matrix or the average size of the expression and is computed as the ratio of FM of the expression against a constant scaling factor δ which in this case is the inter-ocular distance.

$$SM_{ij}(.) = \frac{FM_{ij}(.)}{\delta} \tag{6}$$

Finally, the geometric descriptors are composed into a single vector of the form,

$$\Gamma_g = [FDM(t, n), GM(t, \mu), SDM(t, n, \hat{\mu})] \tag{7}$$

2.2.2 Stage 1 Appearance Descriptors

While the overall shape of a facial expression is important for classification, appearance cues have been proven to be equally important, if not influential. Although the literature reports common interest in the use of Gabor-type [47] or Local Binary Patterns (LBP)-based [48] texture features for appearance modelling, in a recent work by Bhaskar et al. [29] it has been shown that appearance quality indicators using Structural Similarity Index (SSIM) can also be useful in the present context. According to [49, 50], the SSIM measure between two image patches *A* and *B* can be defined as,

$$
\begin{aligned}
S(A, B) &= S_1(A, B)S_2(A, B)S_3(A, B) \\
&= \left[\frac{2\mu_A\mu_B + \gamma_1}{\mu_A{}^2 + \mu_B{}^2 + \gamma_1} \right] \left[\frac{2\sigma_A\sigma_B + \gamma_2}{\sigma_A^2 + \sigma_B^2 + \gamma_2} \right] \left[\frac{\sigma_{AB} + \gamma_3}{\sigma_A\sigma_B + \gamma_3} \right],
\end{aligned} \tag{8}
$$

where,

$$\mu_A = \frac{1}{N_l}\sum_{i=1}^{N_l} A_i, \quad \sigma_{AB} = \frac{1}{N_l - 1}\sum_{i=1}^{N_l}(A_i - \mu_A)(B_i - \mu_B), \quad \sigma_A = \sqrt{\sigma_{AA}}, \quad \text{etc. (9)}$$

and γ_k is a small positive constant that contributes to numerical stability. In this paper, image patches A are generated around a pre-defined neighbourhood of each selected landmark points in $\hat{\mathbf{X}}^*$ and image patched B are corresponding patches from the Neutral expression counterpart (during training phase) or from the appearance template learnt from training (during the test phase).

One novelty of this paper evolves around the proposed alternative formulation for the appearance indicators using SSIM, where the above indexes S_1, S_2, and S_3 are combined in a unique manner as follows,

$$\Gamma_a = \lambda_1 S_1(A, B) + \lambda_2 S_2(A, B) + \lambda_3 S_3(A, B) \tag{10}$$

where λ_1, λ_2, and λ_3 are three trade-off parameters, $\lambda_1 + \lambda_2 + \lambda_3 = 1$. Here, the effect of the individual components of the SSIM can be analysed and an appropriately weighted combination (from training) can be chosen to represent a robust appearance quality indicator. As in the previous studies of [50, 51], S_1 in Eq. (10) measures the similarity between the means of A and B by comparing the local patch luminance or brightness value. on the other hand, S_2 measures the similarity in contrast while, S_3, if $\gamma_3 = 0$, coincides with the correlation $\Xi(A, B)$ between A and B and measures similarity in patch structures. Note that $-1 \leq S(A, B) \leq 1$, and $S(A, B) = 1$ if and only if $A = B$.

2.2.3 Stage 1 Classification

Stage 1 classification model is set to recognize the class of expression given a target face expression. The joint descriptor $\Gamma = \{\Gamma_g, \Gamma_a\}$ is used for training and classification. The SVM classifier attempts to solve the optimisation problem by constructing multiple two-class rules where the kth function $w_k^T \phi(\Gamma_i) + b_k$ separates the training vectors of the class k from other vectors by minimising the objective function:

$$\min \frac{1}{2}\sum_{k=1}^{6} w_k^T w_k + C \sum_{i=1}^{M}\sum_{k \neq \ell_i} \epsilon_i^k \tag{11}$$

subject to the constraints,

$$w_{\ell_i}^T \phi(\Gamma_i) + b_{\ell_i} \geq w_k^T \phi(\Gamma_i) + b_k + 2 - \epsilon_i^k \tag{12}$$

where $\epsilon_i^k \geq 0$, $i = \{1, 2, \ldots, M\}$, $k \in \{1, 2, \ldots, 6\}\backslash \ell_i$ and ϕ is that function that maps the joint features into a higher dimensional space, where it is assumed that the data is linearly or non-linearly separable, C is the penalisation factor for training errors, b the bias vector and ξ the slack variable. The final decision function is thus defined as:

$$argmax_{k=\{1,2,\ldots,6\}}(w_k^T \phi(\Gamma) + b_k) \tag{13}$$

Using the above defined decision function, target expressions are classified into one of the six classes of expression.

2.3   Stage 2 Classification: Posed Expression Recognition

During stage 2 of the aggregated classification process, SVP expression recognition is performed within the class of facial expression detected in stage 1. That is, for each facial expression category, a two-class SVM classifier is trained with relevant features for discriminating SVP expressions within that class. The block diagram in Fig. 3 illustrates the stage 2 of classification.

The fundamental elements of stage 2 classification are similar to those used during stage 1 classification, such as the joint feature space. However, the individual

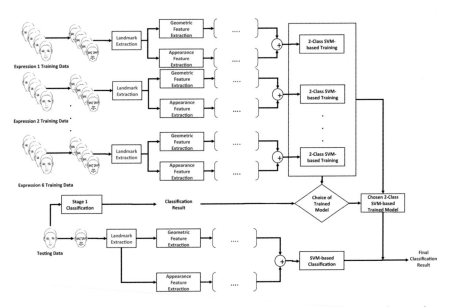

Fig. 3 Block diagram of the stage-2 classification model for spontaneous versus posed expression classification (in turn posed expression recognition) illustrating the training and testing processes of the framework using joint geometric and appearance cues

geometric and appearance descriptors chosen for quantifying exaggeration of expressions are different to those used in stage 1. At some level, the features in stage 1 capture coarser characteristics that demonstrate specific deviations of different classes of expressions from the neutral counterpart. However, in stage 2, features are chosen such that the intensity of the expression and anti-symmetry are highlighted to represent exaggeration and thus distinguish SVP expressions.

2.3.1 Training

In the training phase of stage 2 classification, six different SVP expression classifiers are trained for recognizing posed expressions corresponding to each facial expression category. Training data for the individual classes of facial expression in the form of $\{(\zeta_1^{\ell}, \omega_1), (\zeta_2^{\ell}, \omega_2), (\zeta_3^{\ell}, \omega_3), \ldots, (\zeta_W^{\ell}, \omega_W)\}$, are prepared by extracting joint descriptors $\zeta_{1,2,\ldots,W}^{\ell}$ from all W training images for the different classes of facial expressions $\ell \in \{1, 2, \ldots, 6\}$. The corresponding expression class label ω representing either spontaneous or posed classes, where $\omega \in \{-1, +1\}$ are concatenated to the joint descriptors for each image and together are presented to the respective SVM classification model for that facial expression. The steps in training for stage 2 classification includes, for every image in the training set for a chosen class of facial expression ℓ,

- Extract the optimal set of landmark points $\mathbf{X}^*$ using the method described earlier in Sect. 2.
- Extract geometric descriptors of *FDM*, *GM* and *SDM* according to the method described in Sect. 2.3.3 for those selected landmark points generated from the training step in stage 1 classification. In addition, using reflective symmetry constraints, estimate the form matrix representation of *FM_H* based on the technique described in Sect. 2.3.3. Further, the mean of all facial expressions $\hat{\mu}_{\ell}$ for each category ℓ is generated and stored.
- Extract appearance descriptors using the complex wavelets based structural similarity method as described in Sect. 2.3.4 to generate a form matrix representation of the appearance deviations *FM_S*.
- Fuse the geometric descriptors ζ_{id} consisting of *FDM*, *GM*, *SDM* together with a joint representation of *FM_H* and *FM_S* dubbed as ζ_{as} to form the joint geometric and appearance descriptors ζ^{ℓ}.

After extracting such geometric and appearance descriptors from all the training images for the individual category of facial expressions, the different SVP expression classifiers are trained using $\{(\zeta_1^{\ell}, \omega_1), (\zeta_2^{\ell}, \omega_2), (\zeta_3^{\ell}, \omega_3), \ldots, (\zeta_W^{\ell}, \omega_W)\}$.

2.3.2 Testing

When a test image arrives for recognition, stage 1 classification is performed to label the test image with the class of the facial expression ℓ. Then, the corresponding

SVP expression classifier pre-trained for that class ℓ is chosen for posed expression recognition. For recognition on the test image,

- Extract the optimal set of landmark points $\mathbf{X}^*$ using the method described earlier in Sect. 2.
- Extract geometric descriptors of *FDM*, *GM* and *SDM* according to the method described in Sect. 2.3.3 for those selected landmark points generated from the training step in stage 1 classification. In addition, using reflective symmetry constraints, estimate the form matrix representation of *FM$_H$* based on the technique described in Sect. 2.3.3.
- Extract appearance descriptors using the complex wavelets based structural similarity method as described in Sect. 2.3.4 to generate a form matrix representation of the appearance deviations *FM$_S$*.
- Present the joint descriptor ζ^ℓ along with the trained SVP expression classifier to the SVM classifier described in Sect. 2.2.3 for posed expression recognition.

2.3.3 Stage 2 Geometric Descriptors

The detection of the intensity of a facial expression is crucial towards analysing exaggeration. Therefore, in stage 2, geometric descriptors that encapsulate variation in the intensity of a given facial expression using FM features of the following types are generated similar to Sect. 2.2.1.

$$\bar{FDM}(t, \hat{\mu}_\ell) = \frac{FM_{ij}(t)}{FM_{ij}(\hat{\mu}_\ell)} \tag{14}$$

$$\bar{GM}(t, \hat{\mu}_\ell) = \frac{FM_{ij}(t)}{FM_{ij}(\hat{\mu}_\ell)} \tag{15}$$

$$\bar{SDM}(t, \hat{\mu}_\ell, \hat{\mu}) = \frac{SM_{ij}(t) - SM_{ij}(\hat{\mu}_\ell)}{SM_{ij}(t) - SM_{ij}(\hat{\mu})} \tag{16}$$

where $\hat{\mu}_\ell$ represents the mean of that of class of expression and $\hat{\mu}$ represents the mean template of all classes of expression as generated during stage 1 classification. Therefore, these descriptors indicate the extent of deviation of the target expression (for example smile) from the (local) mean of its class (happy).

Further to the intensity of expressions, anti-symmetry can also be indicative of exaggeration and hence posed expressions. Therefore, as discussed in the work of [29], a reflection symmetry inspired computation of geometric descriptors is detailed in this section. According to [29], the optimal set of landmark points are split into three groups based on a reflective symmetry criteria as in $\mathbf{X}^* = \{\hat{\mathbf{X}}_s^*, \hat{\mathbf{X}}_l^*, \hat{\mathbf{X}}_r^*\}$. Further, Delaunay triangulation is applied and the vertices of L triangle pairs satisfying reflective symmetry constraints are chosen as $\{p_1, p_2, \ldots, p_L\}$. and $\{q_1, q_2, \ldots, q_L\}$ respectively. The geometric deviations between the left and right parts of the face is measured using Hausdorff distance between sets $\mathbf{P}$ and

$\mathbf{Q}$ of the landmark points using the corresponding triangle pairs. A form matrix representation of the facial expression using K triangle pairs is of the form,

$$
FM_H = \begin{bmatrix} H(p_1, q_1) & H(p_1, q_2) & H(p_1, q_3) & .. & H(p_1, q_K) \\ H(p_2, q_1) & H(p_2, q_2) & H(p_2, q_3) & .. & H(p_2, q_K) \\ . & . & . & .. & . \\ . & . & . & .. & . \\ H(p_K, q_1) & H(p_K, q_1) & H(p_K, q_3) & .. & H(p_K, q_K) \end{bmatrix} \tag{17}
$$

where $H(\mathbf{P}, \mathbf{Q})$ represents the Hausdorff distance between $\mathbf{P}$ and $\mathbf{Q}$ and is measured as the maximum distance of a set $\mathbf{P}$ to the nearest point in the other set $\mathbf{Q}$. Mathematically,

$$
H(\mathbf{P}, \mathbf{Q}) = \max\{h(\mathbf{P}, \mathbf{Q}), h(\mathbf{Q}, \mathbf{P})\} \tag{18}
$$

where

$$
h(\mathbf{P}, \mathbf{Q}) = \max_{p \in \mathbf{P}}\{\min_{q \in \mathbf{Q}}\{d(p, q)\}\} \tag{19}
$$

It is important to note that the term h is oriented (or in other words asymmetrical), which means that

$$
h(\mathbf{P}, \mathbf{Q}) \neq h(\mathbf{Q}, \mathbf{P}) \tag{20}
$$

As aforementioned, reflection symmetry is critical to the computation of both geometric and appearance features during stage 2 of classification. Hence, a brief description of this procedure is detailed as below. The fundamental aim of reflection symmetry in face images is to partition the optimal set of landmark points into $\hat{\mathbf{X}}^* = \{\mathbf{X}_s^*, \mathbf{X}_l^*, \mathbf{X}_r^*\}$. In order to do this, the following steps are adopted,

- Determine the optimal transformation T^* that maps every point pair in the optimal set $\{x_i^* = (u_i, v_i)\}$ into the transform domain characterized by points $\{\Upsilon_j = (m_j, \psi_j)\}$.
- For example, if p and q are the point-pair under consideration from the 2D space of optimal landmark points, then T is simply the line that passes through the midpoint $m = \frac{p+q}{2}$ with normal direction $p - q$.
- Thus, the transform domain in m and ϕ will host each unique pair votes between landmark points in the optimal set $\mathbf{X}^*$.
- The largest cluster $\Upsilon_j^{\hat{c}}$ can be identified as the peak of this distribution of votes, as in $\Upsilon_j^{\hat{c}} = \max_j \Upsilon_j^c$. These clusters Υ_j^c are generated using the Density Based Clustering algorithm DBSCAN [52].
- An inverse mapping of each $\Upsilon_j \in \Upsilon_j^{\hat{c}}$ presents a unique set of N_s points that are landmark points from $\mathbf{X}_s^*$ that lie on the line of symmetry.

- Thus, the optimal set of landmark points $\mathbf{X}^*$ are partitioned into set, $\{\mathbf{X}_s^*, \mathbf{X}_l^*, \mathbf{X}_r^*\}$, where $N = N_s + N_l + N_r$, $N_l = N_r$, $\mathbf{X}_l^*$ represents landmark points on the left of the line of symmetry and $\mathbf{X}_l^*$ are landmarks points on the right of line of symmetry.

2.3.4 Stage 2 Appearance Descriptors

In contrast to the weighted structural similarity appearance modelling method as detailed in Sect. 2.2.2, during stage 2 of classification, the complex wavelets SSIM (CW-SSIM) metric proposed in [53] is used as appearance indicators. This choice of metric is primarily motivated due to the sensitivity of the original structural similarity index to non-structural distortions [53]. The CW-SSIM index thus used is anticipated to be able to capture both variations in the intensity of expressions and anti-symmetry. As in [29, 53], the CW-SSIM is formulated as follows. Given two sets of complex wavelet coefficients $\mathbf{c}_u = \{c_{u,i} | i = 1, 2, \ldots, M\}$ and $\mathbf{c}_v = \{c_{v,i} | i = 1, 2, \ldots, M\}$ extracted at the same spatial location (u, v) in the same wavelet sub-bands of the two images being compared, the local CW-SSIM index is defined as,

$$\hat{S}(\mathbf{c}_u, \mathbf{c}_v) = \frac{2|\sum_{i=1}^{M} c_{u,i} c_{v,i}^*| + \kappa}{\sum_{i=1}^{M} |c_{u,i}|^2 + \sum_{i=1}^{M} |c_{v,i}|^2 + \kappa} \tag{21}$$

where c^* denotes the complex conjugate of c and κ is a small positive stabilizing constant. The CW-SSIM index ranges between 0 to 1, where 0 indicates maximum structural distortion. The global CW-SSIM index $\hat{S}(A, B)$ is computed as the mean of all local CW-SSIM values $\bigcup \hat{S}(\mathbf{c}_u, \mathbf{c}_v)$ over the entire wavelet sub-bands.

Since, each triangle pair as illustrated in Sect. 2.3.3, consists of ρ image patches representing the facial feature point, the overall CW-SSIM can be factorized as $\prod_{\wp=1}^{\rho} \hat{S}(a_{\wp,.}, \Re(b)_{\wp,.})$, where $\hat{S}$ represents the CW-SSIM. Therefore, a similar form matrix for CW-SSIM based appearance quality indicator using K triangle pairs can be estimated using,

$$FM_S = \begin{bmatrix} \prod_{\wp=1}^{\rho} \hat{S}(a_{\wp,1}, \Re(b)_{\wp,1}) & \prod_{\wp=1}^{\rho} \hat{S}(a_{\wp,1}, \Re(b)_{\wp,2}) & \cdots & \prod_{\wp=1}^{\rho} \hat{S}(a_{\wp,1}, \Re(b)_{\wp,K}) \\ \prod_{\wp=1}^{\rho} \hat{S}(a_{\wp,2}, \Re(b)_{\wp,1}) & \prod_{\wp=1}^{\rho} \hat{S}(a_{\wp,2}, \Re(b)_{\wp,2}) & \cdots & \prod_{\wp=1}^{\rho} \hat{S}(a_{\wp,2}, \Re(b)_{\wp,K}) \\ \cdot & \cdot & \cdots & \cdot \\ \cdot & \cdot & \cdots & \cdot \\ \prod_{\wp=1}^{\rho} \hat{S}(a_{\wp,K}, \Re(b)_{\wp,1}) & \prod_{\wp=1}^{\rho} \hat{S}(a_{\wp,K}, \Re(b)_{\wp,2}) & \cdots & \prod_{\wp=1}^{\rho} \hat{S}(a_{\wp,K}, \Re(b)_{\wp,K}) \end{bmatrix} \tag{22}$$

2.3.5   Integration of Descriptors

As aforementioned, the features describing exaggeration combines both intensity deviation geometric measurements, dubbed as ζ_{id} and the integrated geometric and appearance descriptor for anti-symmetry detection referred as ζ_{as}. While $\zeta_{id} = \{F\bar{D}M, G\bar{M}, S\bar{D}M\}$ as computed in Sect. 2.3.3, ζ_{as} is a seamless integration of both geometric and appearance features as below.

In stage 2 classification, geometric descriptors in FM_H and appearance indicators in FM_S are integrated into ζ_{as} that represents the fused feature for anti-symmetry detection in the form,

$$
\zeta_{as} = \begin{bmatrix} H(p_1,q_1), \prod_{\wp=1}^{\rho} \hat{S}_{(a_{\wp,1}, \Re(b)_{\wp,1})} & H(p_1,q_2) & \cdots & H(p_1,q_K) \\ \prod_{\wp=1}^{\rho} \hat{S}_{(a_{\wp,2}, \Re(b)_{\wp,1})} & H(p_2,q_2), \prod_{\wp=1}^{\rho} \hat{S}_{(a_{\wp,2}, \Re(b)_{\wp,2})} & \cdots & H(p_2,q_K) \\ & \cdot & \cdots & \\ & \cdot & \cdots & \cdot \\ \prod_{\wp=1}^{\rho} \hat{S}_{(a_{\wp,K}, \Re(b)_{\wp,1})} & \prod_{\wp=1}^{\rho} \hat{S}_{(a_{\wp,K}, \Re(b)_{\wp,2})} & \cdots & H(p_K,q_K), \prod_{\wp=1}^{\rho} \hat{S}_{(a_{\wp,K}, \Re(b)_{\wp,K})} \end{bmatrix} \tag{23}
$$

Further, the overall feature ζ if formed by combining ζ_{as} with ζ_{id}. Thus $\zeta = \{\zeta_{id}, \zeta_{as}\}$ is presented as an input to the two-class SVM classification model for posed expression recognition.

2.3.6   Stage 2 Classification

The stage 2 SVM classification model can be formulated as an optimization problem of the following form,

$$
\min \frac{1}{2} w_k^T w_k + C_k \sum_{i=1}^{W} \sum_{k \neq \omega_i} \epsilon_i^k \tag{24}
$$

subject to the constraints:

$$
\rho_i^k (w_k^T \phi(\zeta_i) + b_k) \geq 1 - \epsilon_i^k \tag{25}
$$

where $\epsilon_i^k \geq 0$ and $i = \{1, 2, \ldots, W\}$. C_k is the penalisation factor for training errors, b_k the bias vector and ϵ the slack variable.

$$
f(\zeta) = sign(w_k^T \phi(\zeta_i) + b_k) \tag{26}
$$

ϕ is a function that non-linearly maps the combined geometric and appearance features to a higher dimensional space defined by a positive kernel $h(\zeta_i, \zeta_j)$. The function used as the SVM kernel where α is the degree of the polynomial function is defined as:

$$
h(\zeta_i, \zeta_j) = (\zeta_i^T \zeta_j + 1)^{\alpha} \tag{27}
$$

and the Radial Basis Function (RBF) kernel:

$$h(\zeta_i, \zeta_j) = exp(-\gamma||\zeta_i - \zeta_j||^2) \tag{28}$$

where is γ is the spread of the Gaussian function.

3   Results and Discussion

In this section, experiments and evaluation of the proposed framework in comparison with competing baseline methods are discussed. Performance evaluation using both qualitative and quantitative analysis is conducted for the two stages of classification. In addition, the effect of selecting influential landmark points as against using all landmark points towards classification is also demonstrated. Further, the complementary advantages of the individual geometric and appearance feature sets towards both the recognition of facial expressions and posed expressions is also reported. Finally, tests were conducted to validate the 2-stage aggregated classifier and compare it against the direct classification of SVP expressions using manually labelled face expression classes.

3.1   Dataset Description

In this work, samples from the publicly available MUG facial expression dataset [54], Cohn-Kanade (CK) dataset [55], extended Cohn-Kanade (CK+) dataset [56] and JAFFE dataset [37, 57] were collectively chosen for training and testing purposes. The MUG database consists of image sequences elaborating six different facial expressions from 52 subjects. The images in the database are all 896 × 896 pixels of the subjects frontal face against a blue background. The Cohn-Kanade datasets consists a total of 593 sequences from 123 subjects recorded at a resolution of 640 × 480 pixels. In addition to conventional facial expressions, the extended Cohn-Kanade dataset also contains spontaneous expressions of 122 smiles recorded from 66 subjects. Finally, the Japanese Female Facial Expression (JAFFE) Database contains 213 images of seven facial expressions (six basic facial expressions and one neutral).

Since the proposed work relies on quantifying exaggeration in facial expressions, images where at least two out of three FACS labellers identified exaggeration in facial expressions were chosen as positive samples and those where none identified exaggeration, were considered negative samples. After appropriate labelling of positive and negative samples a total of 1200 examples from across all the databases were chosen for analysis. The composition of this set included 200 samples, consisting of 100 spontaneous and 100 posed expressions, from each of the six basic expressions: Anger (An), Disgust (Di), Fear (Fe), Happy (Ha), Sadness (Sa), Surprise (Su) and corresponding samples of the Neutral (Nu) expression for training. Note that the Nu expression category has been excluded for testing purposes as it is used within the framework for building the geometric and appearance templates and measuring expression deviations using FM. In addition to this, 200 samples of smiles were chosen from the original and extended Cohn-Kanade datasets for a case study on smile expressions, consisting of 75 positive and 125 negative samples. It is important to note that the samples considered under the "happy" category of expressions and the smile expressions used in the case study are mutually exclusive (or non-overlapping).

3.2 Performance Evaluation

Further to the choice of an appropriate dataset, different evaluation criteria were considered for measuring performance and benchmarking the proposed technique against baselines. Qualitative and quantitative evaluations are usually common in facial expression analysis. Although qualitative evaluation is simply performed through visual inspection of results, various metrics are also available for quantitative analysis. In this paper, classification accuracy, precision, recall and F-measure metrics have been chosen for the evaluation of the framework. Given the results of classification as a confusion matrix, the aforementioned statistics can be computed for each class label. For example, during SVP expression classification, metrics can be provided for both spontaneous and posed classes individually. However, since this study focuses on posed expression recognition, only the results on the posed class are provided.

Recall represents the proportion of actual positives, which were predicted positive. Recall is synonymous to the True Positive Rate (TPR) or sensitivity.

$$Recall = TPR = \frac{TP}{TP + FN} \tag{29}$$

where, TP denotes the number of true positives and FN represents the number of False Negatives. On the other hand, precision (or positive predictive value) is the proportion of predicted positives which were actual positives and is measured as,

$$precision = \frac{TP}{TP + FP} \tag{30}$$

where, FP represents the number of False Positives. The combined criteria, F-measure is estimated as the harmonic mean between precision and recall using,

$$F - measure = \frac{2}{1/precision + 1/recall} \tag{31}$$

Finally, the overall classification accuracy is calculated as

$$Accuracy = \frac{TP + TN}{TP + TN + FP + FN} \tag{32}$$

where, TN denotes the number of True Negatives.

3.3 Baseline Methods

In order to benchmark the performance of the proposed framework, two competing baseline strategies were chosen to assess the two stages of classification respectively. For comparison of stage 1 classification, the methods of [47] and [48] have been chosen. According to the technique described in [47], a combination of geometric descriptors together with Gabor filter banks based appearance features have been utilized for face expression recognition. On the other hand, the method in [48], selects salient patches for extracting Local Binary Patterns (LBP) features and using Linear Discriminant Analysis (LDA) and SVM attains facial expression classification. Both chosen methods are similar to the proposed method in a manner that they extract and combine geometric and appearance features from detected facial landmark points for face expression classification. Therefore, comparisons with these methods shall highlight the impact of the selection of influential landmark points and the usefulness of the FM-based geometric features together with the weighted structural similarity appearance indicators.

Further, to benchmark stage 2 classification, the methods described in [23] and [29] have been chosen. The comparison with technique in [29] will fundamentally assess the impact of the aggregation process of classification for posed expression recognition as against SVP expression classification using manually labelled face expression classes. Further, the technique in [23] will help evaluate the case study on smile expressions.

3.4 Experimental Analysis

3.4.1 Stage 1 Evaluation

In stage 1 classification, all 1200 samples are considered towards analysis. The samples are divided into a training and testing sets consisting of 850 and 350 randomly chosen samples respectively. To the 850 chosen training samples, corresponding Nu expression counterparts are chosen for generating the templates and estimating FDM during training. The evaluation procedure is repeated in order to obtain fivefold validation results and negate any bias that may have occurred due to random sampling. In Table 1, the results of face expression classification of the proposed

Table 1 Classification accuracy (%) results of stage 1 classification for identifying the class of expression

Method	An	Di	Fe	Ha	Sa	Su
PCFIL	90.1	87.3	84.8	98.6	85.7	83.4
PGIL	83.5	82.9	77.2	91.6	84.3	80.2
PAIL	76.6	74.2	74.8	77.3	76.5	73.1
PCFAL	88.6	85.2	80.8	96.4	82.2	79.6
Baseline 1 [47]	85.3	81.1	77.5	93.2	81.8	78.6
Baseline 2 [48]	87.3	78.1	76.5	88.9	82.6	81.2
Baseline 3 [28]	84.6	80.5	75.8	90.3	80.1	79.2

Comparative results of the proposed method using combined features against baseline methods of [47] and [48]. Here, PCFIL indicates the proposed method with combined feature sets using influential landmarks, PGIL and PAIL represents proposed method with geometric descriptors and appearance indicators, and PCFAL proposed method using all landmark points

model compared against some of its own variations and other competing baseline methods [47] and [48] are presented. In general, it can be observed from Table 1 that the proposed method with combined feature sets using influential landmarks (PCFIL) demonstrates superior classification accuracy across all expression classes. Therefore, it is possible to conclude that the joint descriptors chosen in this study are more accurate in distinguishing the fundamental facial expressions and are robust towards reducing false classifications.

In addition to the generic classification results of stage 1 classification reported as PCFIL in Table 1, the results of the proposed method with geometric descriptors (PGIL) and appearance indicators (PAIL) independently are also presented. The results of the multi-class SVM classifier indicates the advantage of using combined geometric and appearance feature vectors as against using them individually. By building the combined descriptors the mutual complementarity of the two descriptors are merged thereby resulting in superior discrimination of the basic facial expressions. In addition, the geometric descriptors emerge as a stronger cue for the classification of the fundamental facial expressions in comparison to its appearance counter-part. One novelty of the proposed work is the selection of influential landmark points for the extraction of the joint geometric and appearance descriptors. Therefore, in order to validate this simplification of the feature space, experiments comparing the classification rates of different expressions using the selected subset of influential landmark points against all the landmark points (PCFAL) are also presented in Table 1. It is clear that the selection of influential landmark points has a positive impact towards the classification accuracy at reduced computational costs.

One of the key parameters that contributes to the novel appearance descriptors proposed in this paper, are the weights of the individual components of structural similarity. Here in Table 2, different combinations of weights have been tested and reported. In addition, Table 2 also presents the classification accuracy of both the equally weighted (row 7) and non-weighted (row 8) schemes in comparison

Table 2 Classification accuracy (%) results of stage 1 classification using only the weighted appearance indicators at different choices of λ_1, λ_2 and λ_3

λ_1	λ_2	λ_3	An	Di	Fe	Ha	Sa	Su
0.5	0.0	0.5	66.2	63.1	64.8	63.6	64.4	60.9
0.5	0.1	0.4	68.3	66.8	60.4	64.4	67.3	64.5
0.5	0.2	0.3	71.4	72.5	71.8	70.6	73.2	69.7
0.5	0.3	0.2	77.8	74.1	76.2	75.3	76.8	71.4
0.5	0.4	0.1	76.7	75.0	73.8	74.2	74.6	70.8
0.5	0.5	0.0	69.5	68.8	66.9	67.6	68.1	67.2
0.334	0.333	0.333	73.2	71.7	72.6	73.9	74.1	69.8
$S_1.S_2.S_3$			70.2	69.5	70.6	71.4	69.6	65.9

Table 3 Classification accuracy (%) results of stage 1 classification for distinguishing spontaneous from posed expressions

	PCCCF		SVPCF		Bhaskar et al. [29]		Bhaskar and Al-Mualla [28]	
Expression	Sp	Po	Sp	Po	Sp	Po	Sp	Po
An	81.2	80.6	73.4	78.6	80.7	79.6	78.2	77.9
Di	86.5	88.9	79.5	85.6	81.8	83.2	79.8	80.4
Fe	80.1	82.4	75.2	77.9	76.5	75.8	74.6	72.1
Ha	90.5	92.7	88	86.8	86.6	89	85.5	82.9
Sa	83.6	82.8	73.7	76.5	80.3	81.4	78.6	77.5
Su	85.5	87	87.2	86.3	84.1	85.4	83.9	85.1

Comparative results of spontaneous vs. posed classification of expressions using conventional multi-class classifier. Here, PCCCF represents the proposed cascaded classification method using combined features and SVPCF is the multi-class classification of SVP expression classification using combined features

to the others. The results clear highlights the benefit of the weighted structural similarity mechanism, while it is also apparent that of the individual components of structural similarity, S_1 plays the most dominant role followed by S_2 and then by S_3.

3.4.2   Stage 2 Evaluation

In stage 2, SVP expression classification is performed by training each individual expression classifier using 150 samples consisting of 75 spontaneous and 75 posed expressions. These 150 samples are drawn from the same set of 1050 training samples as used during stage 1 classification. The mean classification accuracy results of the stage 2 SVP expression classification using the aggregated binary classifiers are presented in Table 3 for each individual expression.

The superior results of the proposed cascaded classification method using combined features (PCCCF) are indicative of the power of the aggregated classifier as the detection of posed expression within the already detected class of facial expression is more accurate than against all expressions in general. In other words, it is clear that the results of the proposed method using the 2-stage aggregated classier is superior to its multi-class SVP expression classification using combined features (SVPCF) counterpart. The results indicate that by obtaining an initial label for the class of the facial expression and then classifying SVP expressions within the identified category of facial expression produces better detection rates than otherwise.

In addition to assessing the performance of the stage 2 classification using classification accuracy, the metrics of precision, recall, and f-scores for the individual facial expressions comparing the proposed method to the baseline method [29] are illustrated in Table 4. Experimental results in Table 4 indicate that in most expressions the general accuracy of the proposed method is better than the baseline strategy. However one common failure mode could be observed due to the measurement of deviations in landmark points across the perpendicular axes to the plane of symmetry. Under this condition, the constraints set to detect anti-symmetry through the measurement of reflection symmetry fails to generalize. Hence, the reduction in the overall performance. In addition to the quantitative results already presented, some qualitative results of the SVP expression classification are illustrated in Fig. 4.

The impact of the chosen geometric and appearance descriptors for stage 2 classification are evaluated using the smile expression as a case study. Experimental results comparing the proposed method to the baseline strategy of [29] are presented in Table 5. The trend observed regarding the superiority of the geometric descriptors against the appearance indicators in stage 1 remains consistent in stage 2 classification as well.

Finally, the results of smile detection using the Cohn-Kanade datasets comparing the proposed framework and relevant baseline techniques are presented in Table 6. Although not directly comparable, our results are 1.2 % higher than the nearest counterpart in [23]. We acknowledge that this comparison cannot be direct due the differences in the composition of images chosen between the baseline and ours. However, these results present a general idea of the superiority of the proposed method.

4   Conclusions

This chapter proposed a novel method for the recognition of posed facial expressions using joint geometric and appearance descriptors and a 2-stage aggregated classifier. The usefulness of the proposed descriptors and the cascading classifiers has both been validated through systematic experimentation using public datasets. The initial facial expression decision during the first stage classification not alone reduces the decision space during SVP expression classification but also restricts the number of false positives that can be generated while all expressions are analysed all at once. The future work of this research will focus on extending the joint descriptors to analyse video sequences for continuous tracking of facial expressions.

Table 4 Results of the proposed method compared with the baseline method of [29] for posed expression recognition using the metrics of precision, recall and F-measure

Expression	Precision			Recall			F-measure		
	Proposed	Bhaskar et al. [29]	Bhaskar and Al-Mualla [28]	Proposed	Bhaskar et al. [29]	Bhaskar and Al-Mualla [28]	Proposed	Bhaskar et al. [29]	Bhaskar and Al-Mualla [28]
An	0.8788	0.7941	0.8478	0.9063	0.8438	0.8612	0.8923	0.8182	0.8544
Di	0.8400	0.7667	0.7261	0.9130	0.8846	0.8190	0.8750	0.8214	0.7698
Fe	0.8571	0.9333	0.8872	0.7826	0.8235	0.8073	0.8182	0.8750	0.8454
Ha	0.8889	0.8750	0.8550	0.9600	0.8750	0.8812	0.9231	0.8750	0.8679
Sa	0.8621	0.9167	0.8586	0.8929	0.8462	0.9022	0.8772	0.8800	0.8799
Su	0.8485	0.8000	0.9126	0.8750	0.8571	0.8492	0.8615	0.8276	0.8798
All	0.7977	0.7016	0.7665	0.8571	0.8375	0.8265	0.8263	0.7635	0.7954

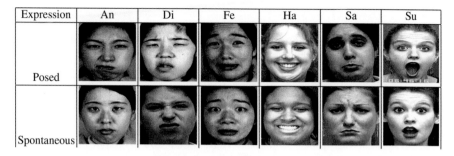

Expression	An	Di	Fe	Ha	Sa	Su
Posed						
Spontaneous						

Fig. 4 Illustration of the classified examples from the posed and spontaneous categories using the proposed algorithm. Note that the examples in row 1, columns 4, 5 and 6 and row 2, columns 2, 4, 5 and 6 are protected under Jeffrey Cohn [55, 56]

Table 5 Results of posed smile detection using the proposed method compared to the baseline method of [29] using the performance metrics of precision, recall and F-measure

Method	Precision	Recall	F-measure
Proposed	0.9323	0.9440	0.9371
Bhaskar et al. [29]	0.8843	0.8912	0.8875

Table 6 Results of spontaneous vs. posed smile classification using the proposed method compared against the baseline techniques of [28] and [23]

Method	Spontaneous	Posed	Rate (%)
Proposed	66/75	118/125	92
Bhaskar et al. [29]	60/75	111/125	85.5
Valstar et al. [23]	58/70	109/114	90.8

References

1. D. Matsumoto, H.S. Hwang, *Reading Facial Expressions of Emotion*. Psychological Science Agenda Science Brief (American Psychological Association, 2011), http://www.apa.org/science/about/psa/2011/05/facial-expressions.aspx
2. S. Tulyakov, T. Slowe, Z. Zhang, V. Govindaraju, Facial expression biometrics using tracker displacement features, in *IEEE Conference on Computer Vision and Pattern Recognition (CVPR)* (2007), pp. 17–22
3. F. Dornaika, B. Raducanu, Facial expression recognition for HCI applications, in *Encyclopedia of Artificial Intelligence*, vol. II (IGI Global, London, 2008), pp. 625–631
4. V. Beek, D. Yolanda, J. Semon, Decoding basic and non-basic facial expressions and depressive symptoms in late childhood and adolescence. J. Nonverbal Behav. **32**(1), 53–64 (2008)
5. V. Bettadapura, Face expression recognition and analysis: the state of the art. Technical Report (2012), http://arxiv.org/abs/1203.6722
6. A. Dhall, Context based facial expression analysis in the wild, in *2013 Humaine Association Conference on Affective Computing and Intelligent Interaction (ACII)* (2013), pp. 636–641

7. Y. Tian, T. Kanade, J. Cohn, Recognizing action units for facial expression analysis. IEEE Trans. Pattern Anal. Mach. Intell. **23**(2), 97–115 (2001)
8. P. Ekman, W.V. Friesen, *The Facial Action Coding System: A Technique for the Measurement of Facial Movement* (Consulting Psychologists Press, San Francisco, 1978)
9. M.A. Sayette, J.F. Cohn, J.M. Wertz, M.A. Perrott, D.J. Parrott, A psychometric evaluation of the facial action coding system for assessing spontaneous expression. J. Nonverbal Behav. **25**, 167–186 (2001)
10. F. De la Torre, J.F. Cohn, Facial expression analysis, in *Guide to Visual Analysis of Humans: Looking at People* (Springer, New York, 2011)
11. W-J. Yan, S-J. Wang, G. Zhao, X. Fu, Quantifying micro-expressions with constraint local model and local binary pattern, in *Computer Vision - ECCV 2014 Workshops*. Lecture Notes in Computer Science, vol. 8925 (Springer, Heidelberg, 2014), pp. 296–305
12. M. Shreve, S. Godavarthy, D. Goldgof, S. Sarkar, Macro- and micro-expression spotting in long videos using spatio-temporal strain, in *Proceedings of the International Conference on Automatic Face and Gesture Recognition (FG)* (2011)
13. M.A. Shreve, Automatic macro- and micro-facial expression spotting and applications, Ph.D. Thesis, University of Southern Florida, 2013
14. M. Shreve, S. Godavarthy, V. Manohar, D. Goldgof, S. Sarkar, Towards macro- and micro-expression spotting in video using strain patterns, in *Workshop on Applications of Computer Vision (WACV)* (2009), pp. 1–6
15. B. Fasel, J. Luettin, Recognition of asymmetric facial action unit activities and intensities, in *Proceedings of the International Conference on Pattern Recognition (ICPR)* (2000)
16. S. Mitra, Y. Liu, Local facial asymmetry for expression classification, in *Proceedings of the IEEE International Conference on Computer Vision and Pattern Recognition (CVPR)*, vol. 2 (2004), pp. 889–894
17. Y. Liu, K.L. Schmidt, J.F. Cohn, S. Mitra, Facial asymmetry quantification for expression invariant human identification. J. Comput. Vis. Image Underst. **91**(1–2), 138–159 (2003)
18. M. Kyperountas, A. Tefas, I. Pitas, Salient feature and reliable classifier selection for facial expression classification. Pattern Recogn. **43**(3), 972–986 (2010)
19. L. Zhang, D.W. Tjondronegoro, V. Chandran, Geometry vs appearance for discriminating between posed and spontaneous emotions, in *Neural Information Processing*. Lecture Notes in Computer Science, vol. 7064 (Springer, Heidelberg, 2011), pp. 431–440
20. M.F. Valstar, H. Gunes, M. Pantic, How to distinguish posed from spontaneous smiles using geometric features, in *Proceedings of the 9th International Conference on Multimodal Interfaces (ICMI)* (2007)
21. H. Dibeklioglu, A.A. Salah, T. Gevers, Are you really smiling at me? Spontaneous versus posed enjoyment smiles, in *Proceedings of the European Conference on Computer Vision (ECCV)* (2012), pp. 525–538
22. M.S. Bartlett, G. Littlewort, C. Lainscsek, I. Fasel, J. Movellan, Machine learning methods for fully automatic recognition of facial expressions and facial actions, in *IEEE International Conference on Systems, Men and Cybernetics* (2004), pp. 592–597
23. M.F. Valstar, M. Pantic, Z. Ambadar, J.F. Cohn, Spontaneous vs. posed facial behavior: automatic analysis of brow actions, in *Proceedings of the ACM International Conference on Multimodal Interfaces* (2006), pp. 162–170
24. G.C. Littlewort, M.S. Bartlett, K. Lee, Automatic coding of facial expressions displayed during posed and genuine pain. Image Vis. Comput. **27**(12), 1797–1803 (2009)
25. M. Bartlett, G. Littlewort, E. Vural, K. Lee, M. Cetin, A. Ercil, J. Movellan, Data mining spontaneous facial behavior with automatic expression coding, in *Verbal and Nonverbal Features of Human-Human and Human-Machine Interaction* (Springer, Heidelberg, 2008), pp. 1–20
26. H. Dibeklioglu, R. Valenti, A. Salah, T. Gevers, Eyes do not lie: spontaneous versus posed smiles, in *Proceedings of the ACM International Conference on Multimedia* (2010), pp. 703–706

27. S.M. Lajevardi, M. Lech, Facial expression recognition using neural networks and log-gabor filters, in *Proceedings of the Digital Image Computing: Techniques and Applications (DICTA)* (2008), pp. 77–83

28. H. Bhaskar, M. Al-Mualla, Spontaneous Vs. posed facial expression analysis using deformable feature models and aggregated classifiers, in *Proceedings of the International Conference on Information FUSION* (2013)

29. H. Bhaskar, D. La Torre, M. Al-Mualla, *Posed Facial Expression Detection using Reflection Symmetry and Structural Similarity (2015) - 218-228 - ICIAR 2015*, ed. by A. Campilho, M. Kamel. Lecture Notes in Computer Science, vol. 9164 (Springer, 2015), pp 218–228

30. P. Viola, M.J. Jones, Robust real-time face detection. Int. J. Comput. Vis. **57**(2), 137–154 (2004)

31. M.J. Black, A.D. Jepson, Eigentracking: Robust matching and tracking of objects using view-based representation. Int. J. Comput. Vis. **26**(1), 63–84 (1998)

32. T.F. Cootes, D. Cooper, C.J. Taylor, J. Graham, Active shape models - their training and application. J. Comput. Vis. Image Underst. **61**(1), 38–59 (1995)

33. P.A. Tresadern, H. Bhaskar, S.A. Adeshina, J.C. Taylor, T.F. Cootes, Combining local and global shape models for deformable object matching, in *Proceedings of the British Machine Vision Conference (BMVC)* (2009), pp. 1–12

34. T.F. Cootes, G.J. Edwards, C.J. Taylor, Active appearance models, in *Proceedings of the European Conference on Computer Vision (ECCV)*. LNCS, vol. 2 (Springer, Heidelberg, 1998), pp. 484–498

35. V. Blanz, T. Vetter, A morphable model for the synthesis of 3D faces, in *Proceedings of SIGGRAPH* (1999)

36. A.R. Rivera, J.A.R. Castillo, O. Chae, Recognition of face expressions using local principal texture pattern, in *IEEE International Conference on Image Processing (ICIP)* (2012), pp. 2609–2612

37. M.J. Lyons, S. Akemastu, M. Kamachi, J. Gyoba, Coding facial expressions with gabor wavelets, in *3rd IEEE International Conference on Automatic Face and Gesture Recognition* (1998), pp. 200–205

38. S. Dongcheng, C. Fang, D. Guangyi, Facial expression recognition based on gabor wavelet phase features, in *Proceedings of the International Conference on Image and Graphics (ICIG)* (2013), pp. 520–523

39. S. Jain, H. Changbo, J.K. Aggarwal, Facial expression recognition with temporal modeling of shapes, in *Proceedings of the IEEE International Conference on Computer Vision Workshop* (2011), pp. 1642–1649

40. I. Buciu, C. Kotropoulos, I. Pitas, ICA and Gabor representation for facial expression recognition, in *Proceedings of the International Conference on Image Processing*, vol. 2 (2003), pp. 855–858

41. Z. Guoying, M. Pietikainen, Dynamic texture recognition using local binary patterns with an application to facial expressions. IEEE Trans. Pattern Anal. Mach. Intell. **29**(6), 915–928 (2007)

42. H. Soyel, H. Demirel, Improved SIFT matching for pose robust facial expression recognition, in *Proceedings of the International Conference on Automatic Face and Gesture Recognition and Workshops (FG 2011)* (2011), pp. 585–590

43. A.S. Md. Sohail, P. Bhattacharya, Classification of facial expressions using K-nearest neighbor classifier, in *Computer Vision/Computer Graphics Collaboration Techniques*. Lecture Notes in Computer Science, vol. 4418 (Springer, Heidelberg, 2007), pp. 555–566

44. A. Graves, C. Mayer, M. Wimmer, J. Schmidhuber, B. Radig, Facial expression recognition with recurrent neural networks, in *Proceedings of the International Workshop on Cognition for Technical Systems* (2008)

45. E. Owusu, Y. Zhan, Q.R. Mao, A neural-AdaBoost based facial expression recognition system. Expert Syst. Appl. **41**(7), 3383–3390 (2014)

46. S. Lele, J.T. Richtsmeier, Euclidean distance matrix analysis: confidence intervals for form and growth differences. Am. J. Phys. Anthropol. **98**(1), 73–86 (1995)

47. A.K.K. Bermani, A.Z. Ghalwash, A.A.A. Youssif, Automatic facial expression recognition based on hybrid approach. Int. J. Adv. Comput. Sci. Appl. **3**(11), 102–107 (2012)
48. S.L. Happy, A. Routray, Automatic facial expression recognition using features of salient facial patches. IEEE Trans. Affect. Comput. **6**(1), 1–12 (2015)
49. Z. Wang, A.C. Bovik, Mean squared error: love it or leave it? A new look at signal fidelity measures. IEEE Signal Process. Mag. **26**(1), 98–117 (2009)
50. Z. Wang, A.C. Bovik, H.R. Sheikh, E.P. Simoncelli, Image quality assessment: from error visibility to structural similarity. IEEE Trans. Image Process. **13**(4), 600–612 (2004)
51. M.J. Wainwright, O. Schwartz, E.P. Simoncelli, Natural image statistics and divisive normalization: modeling non-linearity and adaptation in cortical neurons. in *Probabilistic Models of the Brain: Perception and Neural Function* (MIT Press, Cambridge, 2002), pp. 203–222
52. M. Ester, H-P. Kriegel, J. Sander, X. Xu, A density-based algorithm for discovering clusters in large spatial databases with noise, in *Proceedings of the Second International Conference on Knowledge Discovery and Data Mining (KDD-96)* (AAAI Press, Portland, 1996), pp. 226–231
53. Y. Gao, A. Rehma, Z. Wang, CW-SSIM based image classification, in *Proceedings of the IEEE International Conference on Image Processing (ICIP)* (2011), pp. 1249–1252
54. N. Aifanti, C. Papachristou, A. Delopoulos, The MUG facial expression database, in *Proceedings of the 11th International Workshop on Image Analysis for Multimedia Interactive Services (WIAMIS)* (2010)
55. T. Kanade, J.F. Cohn, Y. Tian, Comprehensive database for facial expression analysis, in *Proceedings of the Fourth IEEE International Conference on Automatic Face and Gesture Recognition (FG)* (2000), pp. 46–53
56. P. Lucey, J.F. Cohn, T. Kanade, J. Saragih, Z. Ambadar, I. Matthews, The extended Cohn-Kande dataset (CK+): a complete facial expression dataset for action unit and emotion-specified expression, in *Proceedings of the Third IEEE Workshop on Computer Vision and Pattern Recognition for Human Communicative Behavior Analysis* (2010)
57. M.N. Dailey, C. Joyce, M.J. Lyons, M. Kamachi, H. Ishi, J. Gyoba, G.W. Cottrell, Evidence and a computational explanation of cultural differences in facial expression recognition. Emotion **10**(6), 874–893 (2010)

Method of Modelling Facial Action Units Using Partial Differential Equations

Hassan Ugail and Nur Baini Ismail

Abstract In this paper we discuss a novel method of mathematically modelling facial action units for accurate representation of human facial expressions in 3-dimensions. Our method utilizes the approach of Facial Action Coding System (FACS). It is based on a boundary-value approach, which utilizes a solution to a fourth order elliptic Partial Differential Equation (PDE) subject to a suitable set of boundary conditions. Here the PDE surface generation method for human facial expressions is utilized in order to generate a wide variety of facial expressions in an efficient and realistic way. For this purpose, we identify a set of boundary curves corresponding to the key features of the face which in turn define a given facial expression in 3-dimensions. The action units (AUs) relating to the FACS are then efficiently represented in terms of Fourier coefficients relating to the boundary curves which enables us to store both the face and the facial expressions in an efficient way.

1 Introduction

Human faces can be represented using both two-dimensional (2D) and three-dimensional (3D) geometry. It is true to say that in computer graphics, facial modelling and animation is one of the most challenging areas. This is due to the fact that the human face is an extremely complex geometric form [1–3].

The most common approaches in facial modelling and animation are geometric manipulation and those based on image manipulations [4]. Approaches based on geometric manipulations include key-frame expression interpolation [5], pseudo-muscle modelling including physics based modelling [6] and parametrisation techniques [7]. Image manipulations include image morphing between photographic images [8], texture manipulations [9] and image blending [10].

H. Ugail (✉) • N.B. Ismail
Centre for Visual Computing, School of Engineering and Informatics,
University of Bradford, Bradford BD7 1DP, UK
e-mail: h.ugail@bradford.ac.uk; N.B.Ismail@student.bradford.ac.uk

© Springer International Publishing Switzerland 2016
M. Kawulok et al. (eds.), *Advances in Face Detection and Facial Image Analysis*,
DOI 10.1007/978-3-319-25958-1_6

129

As stated previously, much work have been done in modelling human faces in 2D and 3D. One of the most popular models among 3D parametric face models is the 3D morphable model of Blanz and Vetter [11]. This parametric model has a low dimensional 3D face shape and texture. Further, the parametric description of the face is based on the principles that underlie in solving a mathematical optimization problem. Since Blanz and Vetter's pioneering work, a number of approaches have been adopted to develop sophisticated fitting algorithms of morphable models. For example, the work presented in [12] used a Lambert model for simplification of the optimization. Further work presented in [13] and [14] show improved optimization efficiency by using an input image. However, even with the existing methods, the iteration process for optimization in matching algorithms for morphable models is very time consuming and computational inefficient.

In computer animation, mainly six facial expressions are used for facial animation and facial simulations, i.e. anger, joy, fear, sad, disgust and surprise [15]. Further, there are some studies in this area whereby the Facial Action Coding Systems (FACS) based on the facial muscle movements are utilised e.g. [7, 16, 17].

Action units within the FACS are classified according to the location of the given muscle on the face and the type of action involved. Within the definitions of FACS the human facial muscles are divided into two parts namely those that are in the upper face and those that are in the lower face. The upper face comprises the muscles around eyebrows, forehead and eyelids. In contrast, muscles around the mouth and lips affect lower part of the face [16].

Although, FACS was initially designed for psychologists and behavioural scientists to understand facial expressions and behaviour, it has been recently adapted to visual communication, teleconferencing, computer graphics, etc. [18]. Thus, today FACS is widely acceptable for use in computer graphics and computer based simulations.

Conventional methods in Computer-Aided Design to generate parametric surface tend to be based on splines such as B-Splines and Non-Uniform Rational B-Splines (NURBS). Much work in modelling and animation of facial expressions has been based on B-Spline and NURBS. Hoch et al. [19], for example, used a canonical model for animating facial expressions by fitting splines to the scan data. They manually positioned the control points of the chosen splines on the geometry of the face accordingly so as to create the right action units. This process is usually quite tedious. Thus, it is not practical for complex facial animations. Meanwhile, Huang and Yan [20, 21] used NURBS for modelling and animating facial expressions by positioning a control polygon based on the facial anatomy. They used a fuzzy set formulation for simulating the features of facial muscles. Various expressions were simulated by changing the weights or positioning the control points. However, this approach requires a higher resolution face model for realistic animation which would in turn increase the computational cost.

The use of Partial Differential Equations (PDEs) for efficiently modelling human face and facial features have been discussed previously. PDEs can be used to generate complex geometry by posing the geometry generation as a mathematical boundary boundary-value problem. Due to the efficiency in parameterising complex geometry the use of PDEs for modelling faces is a plausible idea [2].

For example, the work in [22] used a 3D scan of a human face to efficiently represent and generate facial geometry. Later, the work in [23] applied simple mathematical transformations for generating PDE faces for various facial expressions. Given that the boundary conditions of the chosen PDE are used to describe the key feature points of the geometric object, it is thought that integrating FACS on these boundaries conditions would be a way forward for modelling facial expressions efficiently.

Thus, the aim of this work was to develop a PDE based mathematical formulation for representing action units for a human face. The approach we have adopted here for generating FACS is by mapping the boundary curves (in other words the boundary conditions of the chosen PDE) with key features of the face. There are many benefits in using a formulation of this nature. These include the generation of smooth PDE surface for the faces with associated FACS muscle structure, generation of facial expressions in real time as well as efficient storage of facial data.

The rest of the paper is organized as follows. In Sect. 2 we introduce the PDE method for surface generation. Then in Sect. 3, the process for modelling neutral face configuration using the PDE method is discussed. In Sect. 4, the methodology for generating PDE faces with given action units are discussed and this is illustrated with relevant examples. Results as well as the accuracy of the methodology we have presented in this paper are discussed in Sect. 5. Finally we conclude this work in Sect. 6.

2 The PDE Method for Parametric Surfaces

The PDE method for parametric surfaces is based upon by solving a suitably posed boundary-value problem where an elliptic PDE is solved subject to a given set of boundary conditions [2].

A PDE surface is a parametric surface patch $\underline{X}(u, v)$, defined as a function of two parameters u and v on a finite domain $\Omega \subset R^2$, by specifying boundary data around the edge region of $\partial\Omega$. Typically the boundary data are specified in the form of $\underline{X}(u, v)$ and a number of its derivatives on $\partial\Omega$. Moreover, this approach regards the coordinates of the (u, v) point as a mapping from that point in Ω to a point in the physical space. To satisfy these requirements the surface $\underline{X}(u, v)$ is regarded as a solution of a PDE based on the bi-harmonic equation $\nabla^4 = 0$ namely,

$$\left(\frac{\partial^2}{\partial u^2} + a^2 \frac{\partial^2}{\partial v^2}\right)^2 \underline{X}(u, v) = 0. \tag{1}$$

Here the boundary conditions on the function $\underline{X}(u, v)$ and its normal derivatives $\frac{\partial X}{\partial n}$ are imposed at the edges of the surface patch. The parameter a (where $a \neq 0$) is a special design parameter which controls the relative smoothing of the surface in the u and v directions.

Note that the Eq. (1) is known as a form of the well known Biharmonic equation. The Biharmonic equation can model phenomena related to solid and fluid mechanics as well as stress/strain analysis problems in engineering analysis. Further, there are great many solution techniques to solve PDEs of the nature described in Eq. (1). These range from analytical to numerical methods [24–26].

2.1 Solution of the PDE

As stated above, there exist many methods to determine the solution of Eq. (1) ranging from analytic solution techniques to sophisticated numerical methods. For the work described here restricting to periodic boundary conditions a closed form analytic solution of Eq. (1) is utilised.

Choosing the parametric region to be $0 \leq u \leq 1$ and $0 \leq v \leq 2\pi$, the periodic boundary conditions can be expressed as, $\underline{X}(0, v) = \underline{P}_1(v)$, $\underline{X}(1, v) = \underline{P}_2(v)$, $\underline{X}_u(0, v) = \underline{d}_1(v)$ and $\underline{X}_u(1, v) = \underline{d}_2(v)$.

Note that the boundary conditions $\underline{P}_0(v)$ and $\underline{P}_1(v)$ define the edges of the surface patch at $u = 0$ and $u = 1$ respectively. Using the method of separation of variables, the analytic solution of Eq. (1) can be written as,

$$\underline{X}(u, v) = \underline{A}_0(u) + \sum_{n=1}^{\infty} [\underline{A}_n(u) \cos(nv) + \underline{B}_n(u) \sin(nv)], \qquad (2)$$

where

$$\underline{A}_0 = \underline{a}_{00} + \underline{a}_{01}u + \underline{a}_{02}u^2 + \underline{a}_{03}u^3, \qquad (3)$$

$$\underline{A}_n = \underline{a}_{n1}e^{anu} + \underline{a}_{n2}ue^{anu} + \underline{a}_{n3}e^{-anu} + \underline{a}_{n4}ue^{-anu}, \qquad (4)$$

$$\underline{B}_n = \underline{b}_{n1}e^{anu} + \underline{b}_{n2}ue^{anu} + \underline{b}_{n3}e^{-anu} + \underline{b}_{n4}ue^{-anu}, \qquad (5)$$

where $\underline{a}_{00}, \underline{a}_{01}, \underline{a}_{02}, \underline{a}_{03}$ $\underline{a}_{n1}, \underline{a}_{n2}, \underline{a}_{n3}, \underline{a}_{n4}, \underline{b}_{n1}$ $\underline{b}_{n2}, \underline{b}_{n3}$ and $\underline{b}_{n4}$ are vector constants, whose values are determined by the imposed boundary conditions at $u = 0$ and $u = 1$.

For a general set of boundary conditions, in order to define the various constants in the solution, it is necessary to Fourier analyse the boundary conditions and identify the various Fourier coefficients. If the boundary conditions can be expressed exactly in terms of a finite Fourier series, the solution given in Eq. (2) will also be finite. However, this is often not possible, in which case the solution will be the infinite series given Eq. (2).

The technique for finding an approximation to $\underline{X}(u, v)$ is based on the sum of the first few Fourier modes and a 'remainder term', i.e.,

$$\underline{X}(u, v) \simeq \underline{A}_0(u) + \sum_{n=1}^{N} [\underline{A}_n(u) \cos(nv) + \underline{B}_n(u) \sin(nv)] + \underline{R}(u, v), \qquad (6)$$

where N is usually small (e.g. $N < 10$) and $\underline{R}(u, v)$ is a remainder function defined as,

$$\underline{R}(u, v) = \underline{r}_1(v)e^{wu} + \underline{r}_2(v)e^{wu} + \underline{r}_3(v)e^{-wu} + \underline{r}_4(v)e^{-wu}, \qquad (7)$$

where $\underline{r}_1$, $\underline{r}_2$, $\underline{r}_3$, $\underline{r}_4$ and w are obtained by considering the difference between the original boundary conditions and the boundary conditions satisfied by the function,

$$\underline{F}(u, v) = \underline{A}_0(u) + \sum_{n=1}^{N} [\underline{A}_n(u) \cos(nv) + \underline{B}_n(u) \sin(nv)]. \qquad (8)$$

An important point to note here is that although the solution is approximate this new solution technique guarantees that the chosen boundary conditions are exactly satisfied since the remainder function $\underline{R}(u, v)$ is calculated by means of the difference between the original boundary conditions and the boundary conditions satisfied by the function $\underline{F}(u, v)$ [27].

Note, with the above solution form one could define the position and derivative boundary conditions at the edges of $u = 0$ and $u = 1$ to generate a typical surface patch. Such a surface patch can be controlled by the position vectors (defined by the position boundary conditions) and derivative vectors (defined by the derivative boundary conditions).

However, for practical reasons, in order for a given surface patch to pass through all the four boundary conditions one could make a slight adjustment to the way the boundary conditions are defined as described below.

Choosing the parametric region to be $0 \le u \le 1$ and $0 \le v \le 2\pi$, the periodic boundary conditions can be expressed as in terms of just positional conditions as, $\underline{X}(0, v) = \underline{P}_1(v)$, $\underline{X}(1, v) = \underline{P}_2(v)$, $\underline{X}_a(a, v) = \underline{P}_a(v)$, and $\underline{X}_b(b, v) = \underline{P}_b(v)$.

The above does not affect the core of the PDE formulation and without loss of generality, PDE surface representation with these curves can be obtained [28]. For example, in order to create the vase with a closed base, we can use seven boundary curves. These seven boundary curves will generate two different patches; one patch for the vase body and the other patch for the base of the vase. Using four consecutive boundary curves accordingly (e.g. 1, 2, 3, 4 and 4, 5, 6, 7) means the adjacent PDE patch shares a common boundary curve (i.e. in this case curve 4). This ensures positional continuity across PDE patches. We show this by way of an example in Fig. 1. Figure 1a shows the boundary curves (1, 2, 3, 4, 5, 6 and 7) that are used to generate the vase shape in Fig. 1b.

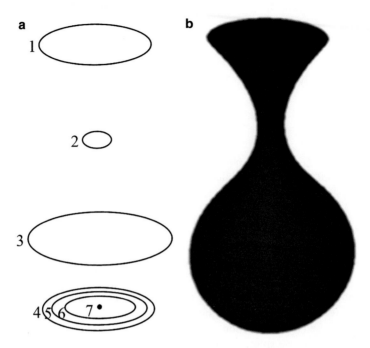

Fig. 1 Example of a PDE surface. (**a**) Boundary curves, (**b**) PDE surface of a vase shape

3 Method of Facial Modelling

The basic idea we have used for representing facial geometry is to identify and utilise a number of curves across the face which define facial geometry accurately. For this purpose we one can obtain such curve data from real face measurements such as the use of a 3D scanning device.

For the purpose of this work, we have utilized a database of dynamic 3D FACS developed at the Department of Computer Science at the University of Bath [29]. This database contains ten subjects, consisting of four males and six females. All subjects performed between 19 and 97 different AUs both individually and in combination with a total of 519 AU sequences. The peak expression frame of each sequence has been manually FACS coded by certified FACS experts.

The 3D data in the database are available in .OBJ file format for the 3D spatial data along with .BMP file format for UV color texture map data, for each frame of the dynamic data. The OBJ data consists of the two pod half-face meshes joined together. A database of this nature has been highly useful for us to develop and verify our work described in this paper. In particular, we have utilised this database to extract boundary curves in order to realistically describe the face as well as develop the PDE based AUs.

Within the 3D FACS database, there are ten subjects performing single action units and a combination of action units. From the sequences of 3D mesh data available, the neutral face for each person was chosen. The chosen neutral configuration acts as a template of PDE face for each person. Then, the template is used for extracting boundary curves for a given action unit.

The obtained 3D mesh data from the database would require pre-processing prior to that boundary curves extraction procedure. A raw surface mesh may have a different alignment and orientation. This pre-processing step is important to have a standardized and normalized mesh. The procedure is performed using Autodesk Maya and its scripting language, MEL script.

To start with the 3D mesh data is scaled to a normalized coordinate system and then rotated around to put the face in the correct orientation. Here correct orientation means that the facial mesh should face the xy Cartesian coordinate plane. When the tip of the nose is located the mesh is translated so the nose tip is located at origin. Next, the nose bridge and the center of the nose bottom are located manually. Based on the nose tip, nose bridge and nose bottom, the facial mesh can be rotated by aligning the mesh parallel to yz Cartesian coordinate plane. This task is performed automatically within Maya using a MEL script specifically written for this purpose.

3.1 Boundary Curve Extraction

The most important task when shape modelling using the PDE formulation is to make available a set of suitable boundary curves that represent the object in question. For the purpose of human face representation we can ensure curves are extracted so as important features of the face are accurately represented. These include facial feature such as nose, mouth and the eye areas. Previous PDE face models (e.g. [22, 28, 30]) have used 28 boundary curves across the face to accurately represent the face. This formulation of boundary curves produced nine different PDE surface patches where each of surface patches is generated by four consecutive boundary curves. Note that all nine PDE surface patches share common boundary curves to guarantee positional continuity between all the generated patches.

Referring to the previous work on PDE face representation we have re-formulated the way curves are extracted. This is necessary due to the fact that our prime concern here is not just to represent the face but also the underlying muscle structure of the face. Thus, in this work, the boundary curves are based on the facial anatomy representing the geometric aspects of facial muscle distribution and the movement of muscles of the face. In order to position points to form boundary curves, we determine a set of characteristic points that correspond to feature points corresponding to the muscles as defined by FACS.

For this purpose, vertices on the original mesh are located corresponding to the related muscle and muscle distribution. Once the points were selected, the MEL script was used to export them to an OBJ file. The boundary curves extracted for a typical face are shown in Fig. 2b. We found that a total of 31 boundary curves is sufficient to accurately represent a face, as shown in Fig. 2a.

It is important to mention that all 31 extracted boundary curves must be in closed form to ensure that the correct PDE face is produced. The generating boundary

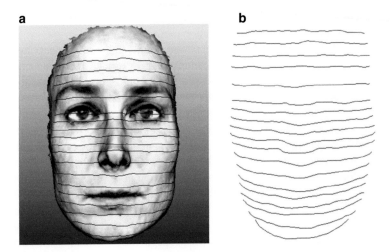

Fig. 2 A neutral face and position of extracted curves, (**a**) a face from the FACS database, (**b**) generating boundary curves

curves shown in Fig. 2b are not in closed form as the mesh image only contains the image of front face. Thus the MEL script is used to close the curves by reading the OBJ file and adding fictitious points by reflecting the coordinate to opposite side.

Following this, we then represent the points of the curves in terms of their corresponding Fourier series. If we approximate the boundary curves using a finite Fourier series such as,

$$f_i(v) = \underline{a}_{0u} + \sum_{n=1}^{5} [\underline{A}_{ni} \cos(nv) + \underline{B}_{ni} \sin(nv)], \tag{9}$$

where $i = 1, 2, 3, 4$.

This procedure enables us to store the facial information in the form of Fourier coefficients corresponding to the curves representing the face. i.e. taking Eq. (9), we can represent facial information in form,

$$M_f = \begin{bmatrix} a_{01} \ a_{11} \dots a_{51} \ b_{11} \dots b_{51} \\ a_{02} \ a_{12} \dots a_{52} \ b_{12} \dots b_{52} \\ a_{03} \ a_{13} \dots a_{53} \ b_{13} \dots b_{53} \\ a_{04} \ a_{14} \dots a_{54} \ b_{14} \dots b_{54} \end{bmatrix}. \tag{10}$$

With this formulation a full PDE face is generated by four sets of consecutive boundary curves resulting in ten continuous facial surface patches as shown in Fig. 3. Figure 3a shows the neutral PDE face with ten different patches. The corresponding PDE face with a texture applied on it is shown in Fig. 3b.

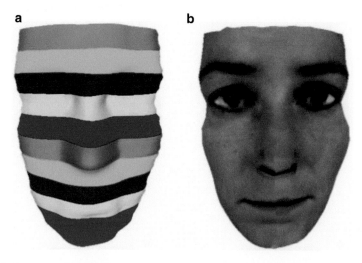

Fig. 3 PDE generated neutral fac, (**a**) original PDE face with patches shown, (**b**) Textured PDE neutral face

4 Modeling Action Units Using PDE Formulation

Previously in [23] it has been shown how simple mathematical transformations can be utilised to create generic facial expression, by identifying the boundary curves and points in these curves that are affected by each expression.

In this work we utilise the facial expression data from the 3D FACS database in order to produce more accurate facial expressions. For this purpose we use the FACS formulation. Here FACS is adopted to control the modification of the neutral PDE face to a generic PDE face with a given action unit. By adjusting the position of the boundary curves related to a given action unit, the generated PDE face is created with that given action unit.

The process to create a given action unit from neutral PDE face representation with a given action unit is carried out by aligning the neutral mesh and the action unit mesh in the same position. Key features of the two meshes must be positioned so that they nearly overlap for facilitating the correct correspondence between two surfaces. Once the mesh for a given action unit is properly aligned, the Mel script reads the points of neutral configuration of boundary curves from external file and file the closest point for all set of points from action unit mesh image. Then, the script will automatically close the boundary curves and write the OBJ file based on the closest point found by the script. The process was repeated for all 31 boundary curves on the mesh with the given action unit.

If we refer M_f as the matrix of Fourier coefficients representing the natural pose of the face and M_g corresponding to the matrix of Fourier coefficients for a given action unit then for the given action we store the action unit data as,

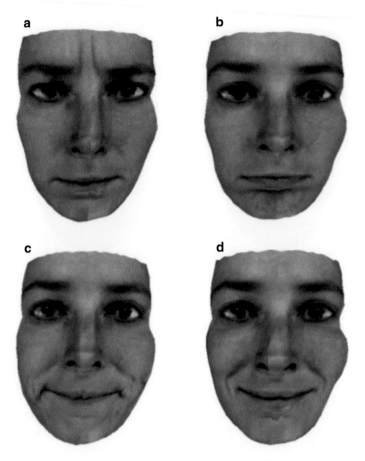

Fig. 4 PDE generated facial action units, (**a**) AU4, (**b**) AU17, (**c**) AU14 and (**d**) AU12

$$M_{AU} = M_f - M_g \tag{11}$$

Once we have computed all the necessary M_{AU}, (which we have done by utilising the data from the 3D FACS database) then a given FACS can be created using by adding M_{AU} to M_f, the neutral pose of the face.

Thus, in order to generate a PDE face with a given action unit, the coefficients M_{AU} are added to neutral Fourier coefficients, M_f, and Eq. (1) is solved to generated the PDE face with the given action unit. Figure 4 shows some example of generic PDE faces with action units. Figure 4a–d respectively represent AU4 (brow lowerer), AU17 (chin raiser), AU14 (dimpler) and AU12 (lip corner puller).

5   Results and Analysis

The PDE based FACS approach proposed here has been implemented using C# and Mel scripting. Once the relevant FACS are created for the PDE geometry we have then tested our FACS generation capacity for the other facial data in the database.

Apart from the efficiency through which we can generate facial feature via the PDE formulation it has an added advantage in that we can store facial expressions very efficiently. For instance, facial images found in the 3D FACS database range in mesh density between 15K and 16K vertices. The file size of the mesh in the database varies from 3 to 5 MB. In contrast to this, by only storing small amounts of parametric data in terms of Fourier coefficients, the PDE formulation enables us to store the facial data very efficiently. In fact, a typical facial expression can be stored using only 50 kB data, which include storage of the both the matrices M_{AU} and M_f. It is important to highlight that given M_{AU} and M_f the given facial expression corresponding to the given set of action units can be created to any given resolution using the pseudo-analytic solution given in Eq. (6).

A natural question one would ask is the accuracy of the facial data we are able to represent using the PDE formulation when compared to raw scanned data, for example. To demonstrate the accuracy of our data representation with respect to the raw data we overlaid the generic PDE mesh and the mesh from the database as illustrated in Fig. 5. Here, the green surface represents the PDE generated face which is placed on the corresponding facial data from the database.

Further, we created three cut-planes through the facial data as shown in Fig. 5. Note that, cut-plane 1 is a vertical plane that lies from head to the chin. This vertical cut-plane is not located at the centre of the face as the centre of the face does

Fig. 5 Original face mesh and the corresponding PDE face overlaid along with three different cut-planes across the faces

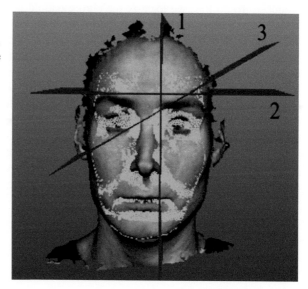

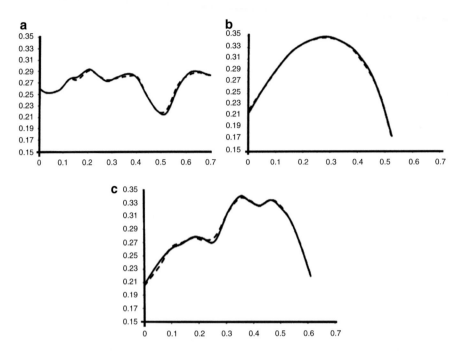

Fig. 6 Accuracy comparison between original and PDE generated facial mesh. Graphs showing three different cut-planes for a neutral face. (**a**) Vertical, (**b**) horizontal and (**c**) diagonal cut-planes

not effect the action unit movement. Cut-plane 2 and cut-plane 3 are somewhat randomly placed where the aim is to measure the accuracy of data the PDE method produces. To do this, we identify the corresponding points on both the faces (the PDE face and the corresponding face from the database) on each of the cut-planes and compare their differences. The results for neutral face are shown in Fig. 6.

The solid line and the dotted lines, respectively, in the graphs shown in Fig. 6 represent the original mesh taken from database and generic PDE generated face. From Fig. 6, it is clear that PDE face is very close to the original mesh. The mean error for vertical, horizontal and diagonal planes are within the 10^{-4} range.

Finally we look at the errors in the PDE generated face for action units in comparison with the corresponding meshes from the database. In particular, we look at the errors for AU4 and AU14. This is illustrated in Figs. 7 and 8. Again from the graphs shown in Figs. 7 and 8 we note that the PDE generated facial expressions are in close agreement with the real expressions found in the database.

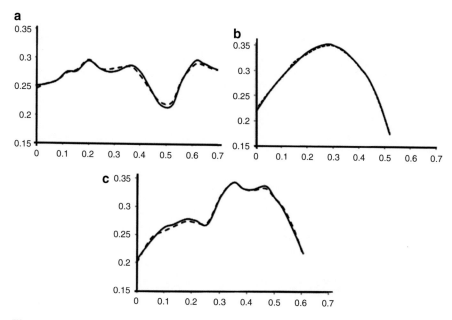

Fig. 7 Accuracy comparison between original and PDE generated facial mesh. Graphs showing three different cut-planes for a face with AU4. (**a**) Vertical, (**b**) horizontal and (**c**) diagonal cut-planes

6 Conclusions

In this paper we have described an approach to model action units on human faces based on Partial Differential Equations. In particular, we show how a boundary-value approach to solving a Biharmonic type PDE can be utilised to efficiently model facial action units and store the facial expression data in a mathematically compressed form. This work has also demonstrated that the PDE based approach is not only flexible but also accurately represent human facial expression data.

Work is currently under way to utilise this formulation so as it can be used to define a wide variety of realistic facial expressions. Such expression can then be used for realistic facial animations whereby subtle changes in expressions among individual faces can also be taken into account.

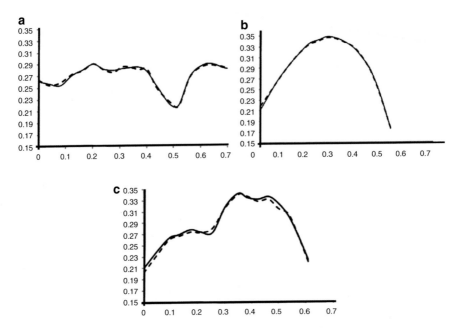

Fig. 8 Accuracy comparison between original and PDE generated facial mesh. Graphs showing three different cut-planes for a face with AU14. (**a**) Vertical, (**b**) horizontal and (**c**) diagonal cut-planes

References

1. I.A. Essa, A.P. Pentland, Coding, analysis, interpretation, and recognition of facial expressions. IEEE Trans. Pattern Anal. Mach. Intell. **19**(7), 757–763 (1997)
2. H. Ugail, *Partial Differential Equation for Geometric Design* (Springer, London, 2011)
3. Y. Sheng, P. Willis, G. Castro, H. Ugail, Facial geometry parameterisation based on partial differential equations. Math. Comput. Model. **54**(5–6), 1536–1548 (2011)
4. J. Noh, U. Neumann, A survey of facial modeling and animation techniques. USC Technical Report (1998), pp. 99–705
5. F.I. Parke, Computer generated animation of faces, in *ACM 72, Proceedings of the ACM Annual Conference*, vol. 1 (1972), pp. 451–457
6. D. Terzopoulus, K. Waters, Physically-based facial modeling. J. Vis. Comput. Animat. **1**(4), 73–80 (1990)
7. F.I. Parke, Parameterized models for facial animation. IEEE Comput. Graph. Appl. **2**(9), 61–68 (1982)
8. T. Beier, S. Neely, Feature-based image metamorphosis. ACM SIGGRAPH Comput. Graph. **26**, 35–42 (1992)
9. M. Oka, K. Tsutsui, A. Ohba, Y. Karauchi, T. Tago, Real-time manipulation of texture-mapped surfaces. ACM Comput. Graph. **21**(4), 181–188 (1987)
10. Y. He, Y. Zhao, D. Jiang, H. Sahli, Speech driven photo-realistic face animation with mouth and jaw dynamics, in *Signal and Information Processing Association Annual Summit and Conference (APSIPA), 2013 Asia-Pacific*. IEEE, Kaohsiung (ACM, 2013), pp. 1–4

11. V. Blanz, T. Vetter, A morphable model for the synthesis of 3D faces, in *Proc. 26th Annu. Conf. Comput. Graph. Interact. Tech., SIGGRAPH 99* (1999), pp. 187–194
12. L. Zhang, S. Member, D. Samaras, Face recognition from a single training image under arbitrary unknown lighting using spherical harmonics. IEEE Trans. Pattern Anal. Mach. Intell. **28**(3), 351–363 (2006)
13. B. Moghaddam, J. Lee, H. Pfister, R. Machiraju, Model-based 3D face capture with shape-from-silhouettes, in *IEEE Work Analysis and Modelling of Faces and Gesture Recognition* (2003), pp. 20–27
14. S. Romdhani, T. Vetter, Estimating 3D shape and texture using pixel intensity, edges, specular highlights, texture constraints and a prior, in *IEEE Comput. Soc. Conf. Comput. Vis. Pattern Recognit.*, vol. 2 (2005), pp. 20–27
15. H. Yu, O.G.B. Garrod, P.G. Schyns, Perception-driven facial expression synthesis. Comput. Graph. **36**(3), 152–162 (2012)
16. S. Villagrasa, A.S. Sánchez, Face! 3D facial animation system based on facs, in *IV Iberoamerican Symposium in Computer Graphics - SIACG 2009* (2009), pp. 202–207
17. A. Wojdel, L.J.M. Rothkrantz, Parametric generation of facial expressions based on FACS. Comput. Graph. Forum **24**(4), 743–757 (2005)
18. P. Havaldar, Performance driven facial animation, in *ACM SIGGRAPH 2006* (2006), pp. 1–20
19. M. Hoch, G. Fleischmann, B. Girod, Modeling and animation of facial expressions based on B-splines. Vis. Comput. **11**(2), 87–95 (1994)
20. D. Huang, H. Yan, Modeling and animation of human expressions using NURBS curves based on facial anatomy. Signal Process. Image Commun. **17**(6), 457–465 (2002)
21. D. Huang, H. Yan, NURBS curve controlled modelling for facial animation. Comput. Graph. **27**(3), 373–385 (2003)
22. E. Elyan, H. Ugail, Reconstruction of 3D human facial images using partial differential equations. J. Comput. **2**(8), 1–8 (2007)
23. G.G. Castro, H. Ugail, P. Willis, Y. Sheng, Parametric representation of facial expressions on PDE-based surfaces, in *Proc. of the 8th IASTED International Conference Visualization, Imaging, and Image Processing (VIIP 2008)* (2008), pp. 402–407
24. G.G. Castro, H. Ugail, P. Willis, I. Palmer, A survey of partial differential equations in geometric design. Vis. Comput. **24**(3), 213–225 (2008)
25. E. Elyan, H. Ugail, Interactive surface design and manipulation using PDE-method through autodesk maya plug-in, in *International Conference on Cyberworlds 2009* (2009), pp. 119–125
26. H. Ugail, Method of trimming PDE surfaces. Comput. Graph. **30**(2), 225–232 (2006)
27. M.I.G. Bloor, M.J. Wilson, Spectral approximations to PDE surfaces. Comput. Aided Des. **28**(2), 145–152 (1996)
28. Y. Sheng, P. Willis, G. Castro, H. Ugail, PDE face: a novel 3D face model, in *Proc. of the 8th IASTED International Conference Visualization, Imaging, and Image Processing (VIIP 2008)* (2008), pp. 408–415
29. D. Cosker, E. Krumhuber, A. Hilton, A FACS valid 3D dynamic action unit database with applications to 3D dynamic morphable facial modeling, in *IEEE International Conference on Computer Vision (ICCV) 2011*. IEEE, Barcelona (Elsevier, 2011), pp. 2296–2303
30. Y. Sheng, P. Willis, G. Castro, H. Ugail, PDE-based facial animation: making the complex simple, in *Advances in Visual Computing*. Lecture Notes in Computer Science, vol. 5359 (Springer, Heidelberg, Germany, 2008), pp. 723–732

Trends in Machine and Human Face Recognition

Bappaditya Mandal, Rosary Yuting Lim, Peilun Dai, Mona Ragab Sayed, Liyuan Li, and Joo Hwee Lim

Abstract Face recognition (FR) is a natural and intuitive way for human beings to identify or verify or at least get familiar and interact with other members of the community. Hence, human beings expect and endeavor to develop similar competency in machine recognition of human faces. Due to the rapid increase in computing power in recent decades and the need to automate the FR tasks for many applications, researchers from diverse areas like cognitive and computer sciences are making efforts in understanding how humans and machines recognize human faces respectively. Its application is innumerable (like access control, surveillance, social interactions, e-commerce, just to name a few). In this chapter we will review two aspects of FR: machine recognition of faces and how human beings recognize human faces. We will also discuss the recent benchmark studies, their protocols and databases for FR and psychophysical studies of FR abilities of human beings.

1 Introduction

Among many biometrics, such as finger print, palm print, ear, iris, gait, etc., face is considered to be most user-friendly and intuitive as the authentication can be performed at a distance, even without the knowledge or cooperation of the subject. The main difficulties that face recognition (FR) algorithms have to deal with are two types of variations: intrinsic factors (independent of viewing conditions) such as age and facial expressions and extrinsic factors (dependent on viewing conditions) such as pose and illumination. Large amount of work has been done over the last three decades to address these issues. Starting from the pioneering work of Eigenfaces by Pentland et al. [1] to the latest results of DeepFace by Wolf et al. [2] and DeepID by Wang et al. [3], researchers are able to reduce the recognition error rate (%) from two digits to near perfection [4].

B. Mandal (✉) • R.Y. Lim • P. Dai • M.R. Sayed • L. Li • J.H. Lim
Institute for Infocomm Research, A*STAR, 1 Fusionopolis Way, #21-01, Connexis (South Tower), Singapore 138632, Singapore
e-mail: bmandal@i2r.a-star.edu.sg

© Springer International Publishing Switzerland 2016
M. Kawulok et al. (eds.), *Advances in Face Detection and Facial Image Analysis*,
DOI 10.1007/978-3-319-25958-1_7

Although high recognition rates are reported in the academic papers, there are large gaps between the reported performance in constrained framework and their performances in the large scale unconstrained environment. In this study, we will discuss the recent face recognition vendor test (FRVT 2013) [5] conducted by the National Institute of Standards and Technology (NIST) as a third party independent evaluator of FR algorithms. We will discuss these gaps, the difference in old and new benchmark protocols and results reported in the recent benchmark study of large-scale unconstrained FR by Stan Z. Li et al. [6] on the well-known 'labeled faces in the wild' (LFW) database in 2014. Unlike numerous previous studies on LFW, this large scale unconstrained FR algorithm evaluation with new protocol using entire LFW database reveals that only 41.66 % correct verification rates (CVR) can be obtained at 0.1 % false acceptance rates (FAR) and 18.07 % open-set face identification (FI) rates at rank 1 and 1 % FAR. As these numbers show that FR problem is still largely unsolved, we will devote more attention and efforts in reviewing new invariant feature representations and learning algorithms that can advance the algorithm development for FR.

In addition to machine recognition of faces, we will review how human beings perceive human faces for recognition. We respond to faces differently from other classes of objects. Interaction involves a certain level of social cognition that needs to be adapted for each situation. Research on human face processing is now moving away from the use of static face stimuli and delving into dynamic faces to simulate a more realistic context for FR and processing. This endeavor gave rise to the formulation of two popular hypotheses by O'Toole et al. [7], in an attempt to explain the benefits that dynamic faces impart on human recognition of faces: the supplemental information hypothesis and representation enhancement hypothesis. However, both hypotheses are unable to explain how humans are able to learn and recognize a face with much fewer templates than machines. What are the possible strategies that could optimize learning of novel faces, even under challenging conditions? In this chapter, we analyze the findings for human psychophysics experiments that investigated human performance in FR across varying conditions of illumination, expression, viewing perspectives, and time lapses in age. We suggest possible FR strategies utilized by humans that could be incorporated into machines to pave the way for next-generation recognition systems.

In Sect. 2, we will discuss briefly the challenges involved in FR and its general pre-processing and normalization steps. In this section, we will also study the dynamics of FR in unconstrained environment involving emerging techniques. We will do a review on FRVT 2013 and some of the emerging databases, their findings, protocols and summary in Sect. 3. Motivated by the limitations of machine recognition of faces as discussed before, we also do a comprehensive review for dynamics in human recognition of faces in Sect. 4. It would help us to understand how we human beings solve these problems and challenges posed by machines. We also share the experimental results of human performances in dynamic FR. Finally in Sect. 5, we conclude and discuss the future trends.

2 Machine Face Recognition: Its Existing Challenges and Emerging Methods

Human face recognition plays an important role in our daily life. We utilize our visual memory to recognize an individual [8] or at least able to recall seen / unseen (familiar or unfamiliar) individual faces [9]. For humans, FR is the most natural and common way to identify and/or verify individuals. It is so intuitive and non-intrusive (without user intervention) that we aim to replicate this capability into machines. Even after four decades of intensive research in machine FR, the problem is still far from solved for large scale unconstraint FR. So what are the existing challenges that extirpate us from achieving human like high recognition rates?

2.1 Challenges for Face Recognition

For unconstrained FR, the challenging factors are: *Illumination variation, Pose and viewpoint variation, Expression variation, Aging, Scale variation, Occlusions and Motion blur.* One or in combination with others, have caused tremendous challenging problems for large scale unconstrained FR. Below we discuss each of these problems briefly.

2.1.1 Illumination Variation

A person's face appears quite different at different times throughout the course of a video capturing when it passes through underneath lights or some strong lights in certain directions. Illumination also results in self shadowing making the problem even harder. Some samples images of a person with different illumination conditions from YaleB database [10] are shown in Fig. 1. A large amount of research work has been devoted to study and alleviate this problem [11–13], however, all of them studied the problem of illumination under constrained (studio settings) environment. Very few studies are performed on real-life unconstrained illuminating conditions [14].

2.1.2 Pose and Viewpoint Variation

In natural settings, either the subject or/and viewer are moving. Capturing facial images from a stationary or moving (wearable devices like Google Glass) camera, the moving faces can lead to shots from a variety of angles causing the correspondences between pixel locations and points on the face to differ from image to image. Since human face is a 3D structure, using only 2D images to reconstruct unknown poses can become an ill-posed problem. Camera capturing human face

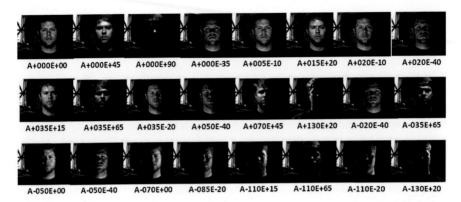

| A+000E+00 | A+000E+45 | A+000E+90 | A+000E-35 | A+005E-10 | A+015E+20 | A+020E-10 | A+020E-40 |

| A+035E+15 | A+035E+65 | A+035E-20 | A+050E-40 | A+070E+45 | A+130E+20 | A-020E-40 | A-035E+65 |

| A-050E+00 | A-050E-40 | A-070E+00 | A-085E-20 | A-110E+15 | A-110E+65 | A-110E-20 | A-130E+20 |

Fig. 1 Appearance of a person under different illuminating conditions from YaleB database. 'A+035E+15' implies that the light source direction with respect to the camera axis is at 35° azimuth ('A+035') and 15° elevation ('E+15'). (Note that a positive azimuth implies that the light source was to the *right* of the person while negative means it was to the *left*. Positive elevation implies above the horizon, while negative implies below the horizon)

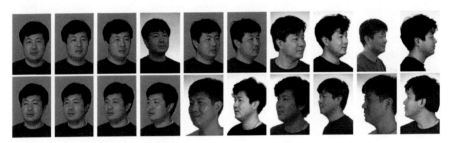

Fig. 2 Sample images from the FERET database [15] for one person with varying degree of poses

images results in in-plane or out-of-plane rotations as shown in Fig. 2. The former is a pure 2D problem and can be solved much more easily, like placing the eyes on the same horizontal axis [16]. However, the latter is very challenging and is also known as in-depth rotations. When part of a face is invisible in an image due to rotation in-depth, the facial texture is recovered from the visible side of the face using the bilateral symmetry of faces. Human face is limited to three degrees of freedom in pose, which can be characterized by pitch, roll and yaw angles. Extracting accurate face pose information in terms of these angles has always been a very challenging problem in FR literature [17, 18].

2.1.3 Expression Variation

Although all faces share the same configuration of two eyes, a nose and a mouth, forehead and cheek regions, signification in-depth deformations occur because of our expressions. Some sample images from AR database [19] are shown in Fig. 3. They pose serious problems to FR performance [20].

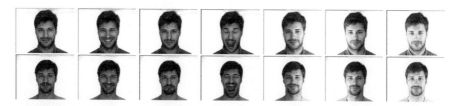

Fig. 3 Sample images from the AR database for one person with 14 different expressions

Fig. 4 Sample images of one person in different ages

2.1.4 Aging

Human face changes considerably along with aging, it gets effected in different forms at different ages. During one's younger years the cranium's shape of the face gets more effected whereas they are more effected in terms of wrinkles and other skin artifacts during one's older age. Human face also undergo growth related changes and changes arising from environmental effects that are manifested in the form of textural, color and shape variations. Some sample images of a person in different ages (from [21] database) are shown in Fig. 4. Extracting features that are invariant to large variations in ages for FR is a very challenging problem. Moreover, collecting and archiving face images across different ages in different years (decades) itself is non-trivial effort.

2.1.5 Scale Variation

Because of moving cameras and/or moving persons, face images are captured at different scales resulting in different resolutions. Some samples images captured at 1, 2, 3 and 4 m with no zooming condition using Google Glass are shown in Fig. 5. The original image resolution in Google Glass is set to 360×640 and the cropped images shown in Fig. 5 are of 150×140 dimensions each. Existing research shows that a high resolution 2D face image is better for FR than one 3D face image [22].

Fig. 5 Sample images of one person captured at 1, 2, 3 and 4 m (*left to right*) with no zooming condition using Google Glass. The face sizes are 90×90, 50×50, 36×36 and 21×21 respectively

Fig. 6 Sample images of three persons, one in normal and two partially occluded conditions (opaque glass and scarp)

Fig. 7 Sample images of three persons with motion blur captured using Google Glass. In all the cases face and eyes detections were successful [23]

2.1.6 Occlusions

Objects in the scene can block a face resulting in reducing the visible area of the face. Common cases like wearing opaque glasses can cause severe occlusions to the eyes areas. Due to occlusions the amount of face information captured is reduced, which makes the FR problem more difficult (Fig. 6).

2.1.7 Motion Blur

Either a moving face or/and a moving camera can cause motion blur. Also when the camera exposure time is set too long or the head moves rapidly, motion blur can occur. Distinctive characteristics of a face are lost when they are blur resulting in poor FR performance, such as in wearable devices [23]. Some samples images are shown in Fig. 7.

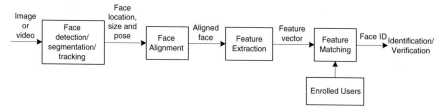

Fig. 8 A general face recognition system

2.2 Pre-processing and Normalization

In the general FR framework as shown in Fig. 8, numerous researchers perform detection of face and its features (like eyes) [16, 24]. Detected face and facial features are used for face alignment. Generally, eyes are placed on the same horizontal axis and at fixed distance (pixels) apart. A face mask is then applied to mask out the non-face portions (like the background) arising above the shoulder and below the chin. It also helps to remove the hair region which has high variations. This whole process is called pre-processing or normalization step. If facial features (like eyes) are detected wrongly then the subsequent processes may fail or the system will achieve very low recognition accuracy. The dependency between the detection precision and recognition accuracy has been studied in [25] by Kawulok et al. In recent years, face and its features detection has been improved to a very large extent. However, for unconstrained scenarios they are still challenging [26, 27]. Low-dimensional features are extracted from high-dimensional objects like face images and stored into the database. When new images (of enrolled users or imposters) are captured, they also undergo similar processes and matching is done with the features stored in the database. Finally, a match ID or non-match (unknown) outcome is given as output.

2.3 Trends in Unconstrained Face Recognition: Promising Directions

Over the past three decades researchers from diverse fields are making efforts in improving the FR algorithms. We have tried to summarize the popular or distinct algorithms that are developed over these years in Fig. 9. It is beyond the scope of this chapter to discuss each of these approaches. For details of the methodologies belonging to holistic, component and hybrid based approaches, the readers are advised to refer the FR survey paper by Zhao et al. [28]. For methodologies grouping based on three levels of taxonomy of facial features, the readers can refer to the paper by Klare et al. [29]. A recent 2014 survey on single and multimodal FR can be found in [30]. There are also a few papers that review FR across pose variations [17, 18], illumination variations [31, 32], aging [33] and forensic applications [34].

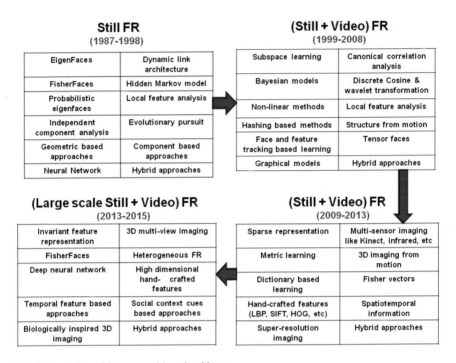

Fig. 9 Evolution of face recognition algorithms

The recent interest in FR is motivated from a few promising directions, which are (1) approaches that use the biologically motivated theory of invariance identity-preserving transformations, (2) video-based FR and (3) deep-learning based convolutional neural-network framework. Below we discuss these promising directions and a few recent successful examples.

2.3.1 Methods Using Invariance Identity-Preserving Transformations

It is evident from the recent literature reviews on unconstrained FR that in order to develop the next generation FR algorithm that can perform better FR as compared to humans and even surpass human performance, we would need more challenging databases as compared to the past. Leibo et al. [21] has tried to come up with a unconstrained FR database which is much more challenging as compared to the previously well studied labeled faces in the wild (LFW) [35] and YouTube face image (YTF) databases [14]. They named this database as subtasks of unconstrained face recognition (SUFR) [21]. Their idea is to isolate faces with specific transformation or a set of transformations for different subtasks to suppress the common computational problem of FR which is transformation invariance to various translations, illuminations, rotations and scalings. Leibo et al. [21] produced six artificial face image datasets using 3D graphics based on this concept, where each of them contains face images created using a set of transformations with various cluttered/homogenous backgrounds. Although they proposed a good idea to handle the unconstrained conditions resulted from various transformations but they are still using affine transformation. They have not included the difficult variations and deformations like face expression, along with pose and aging. So, this approach is still incomplete and cannot be used in most of the real-world conditions.

Motivated by the recent theory of transformation invariance [36], Liao et al. [37] used the SUFR database for face verification (FV) following the same idea of finding the invariance features using various transformations. A signature or invariant representation for each image is computed with respect to a group of transformations. As the inner product of the image and transformed template is the same as the inner product of the template with the same transformed image, their empirical distribution function of the inner products can be used as signature for each image. Although the authors reported a good performance of this model but it may assume some restrictions where the transformation is non-affine. For example, the authors stated that this model may work in case of non-affine transformation when it is restricted to certain nice class such as the 2D transformation mapping of the face image to its frontal view is similar to transformation of another face within the same scene. Also, as this model depends on the distribution of the inner product of an image and transformed template, it means that the transformation has to be known or can be measured in advance which is not applicable in many practical cases. Finally, it does not consider the background variations in all circumstances.

A vast number of FR research follow the algorithmic flow of *face detection* → *normalization* → *face recognition* [16, 28, 38, 39]. However, the recent theory of invariant recognition by feedforward hierarchical networks [40], like HMAX [41, 42], and other convolutional networks [43], or possibly the ventral stream, implies an alternative approach to unconstrained FR. The main idea is to remove traditional FR pipeline techniques such as face cropping, alignment and normalization and use the whole image (possibly with a face in it) for recognition. This is a biologically motivated way for performing FR as we human beings do not use normalization explicitly while recognizing a face [44].

Liao et al. [44] used a three-levels HW-Module architecture (in honor of Hubel and Wiesel's original proposal for the connectivity of V1 simple and complex cells [45]) to obtain the face signature (identity) of an individual. At level one, face is detected, nearly cropped and low level features are extracted at different positions and scales for each image in the training set. These features are stored in vectors as training templates. Then they compute dense overlapping set of windows for each test image, convolved them with training templates and applied max pooling to get new templates. Finally, matching process is done at the third layer by obtaining the dot product between these templates and the training templates and scores are computed for each test image. In that work, Liao et al. [44] tried to reduce the complexity of these processes by hashing and rank approximation using principal component analysis (PCA). They applied this model on different unconstrained databases like LFW, SUFR-W (SUFR-in the wild), LFW-J (LFW-jittered) to get state-of-the-art FR accuracy rate of 87.55 %. This accuracy is near to other popular

FR approaches that use cropping, alignment and normalization of the testing set [46]. So, they proved experimentally that their biologically plausible hierarchical model can effectively replace face detection, alignment and normalization pipelines [44], however, these techniques are of limited use with non-affine transformations.

2.3.2 Video Based Face Recognition

In the machine face recognition literature, majority of the research has focused on improving the ability of FR using static (still) face images. As pointed out in [47], this is primary because of factors such as (1) the need to constrain FR problem, so that the researchers can focus on specific type(s) of FR problem (one in combination with other, such as illumination and/or expression in AR database [19]) and assume all other factors as more or less constant, (2) computational or hardware constraints for both acquiring, processing and storing large amount of face images, (3) the large amount of legacy still face images (e.g. ID cards, mug shots) and (4) its limited availability (or sharing such as in social networks [48, 49]). Today, many of these constraints are no longer valid. A large number of researchers are working on computational, biological and cognitive aspects of FR [50–52], tackling the problem well and coming up with new model, theory and challenging unconstrained databases [53]. FR using still images has witnessed an exponential decrease in error rates [5]. Hardware devices (like digital cameras) for acquiring or capturing images are becoming less and less expensive. Availability of distributed and parallel computing has helped in processing a very large number of images. Lastly, people are very active in sharing images/videos across multiple domains, internet and channels (like social networks), hence they are more readily available as compared to the past [48, 49].

As described earlier, compared to legacy static (still) images, videos help in enhancing FR process as additional information can come from motion and other aspects such as multiple faces of different poses, expressions and illuminations. Firstly, there are techniques based on feature extraction from video input, such as [54]. These features may represent the relation between facial features or the invariant structural features that do not change under different conditions such as skin-model based and color-based approaches. Also global features are useful, such as shape of the face, skin and size or detail features of the internal face components (like eyes and nose). Secondly, there are methods based on probability density function. They deal with face images as random variables of certain probability where the similarity between images can be measured by similarity of their corresponding probability density function. Thirdly, some techniques use the dynamic variance of faces in images to enhance the face detection and identification by integrating features extracted from sequence of images like motion information.

Rowden et al. [47] proposed two techniques to fuse information from image sequences in unconstrained conditions using YouTube faces database [14]. Their multi-frame fusion deals with video as a group of single still frames. Each frame in the query video is matched to the corresponding video in the database and similarity score is computed and measured as a part of the verification process. Scores from all images are then combined by averaging, max, min and median rules. Fusion can also be done to combine matchers score before or after the multi-frame fusion using the similar rules. The former is called the multi-matcher multi-frame (MMMF), while the latter is called multi-frame multi-matcher (MFMM) technique. These techniques are tested using three commercial off-the-shelf algorithms. According to their results, the accuracy of identification using frames fusion is better than still images which means that videos are better than still images to recognize faces in unconstrained conditions. Also, fusion of more than one matchers achieves better performance. Although good performances are achieved by these techniques but matching each frame in query video with all frames in the database videos may take large computation time. Also, as the final decision is a result of combining more than one matcher scores, it may lead to failure if some of the matchers scores are very low. This could happen especially in unconstrained conditions such as low resolution and occluded face images.

Li et al. [55] proposed a technique to decrease the complexity of identification on large-scale databases by representing each subject in all relevant videos by one Eigen-PEP (probabilistic elastic part) representation with invariant length over different YouTube face videos. This representation can be used later in the matching process to make identification using joint Bayesian classifier. This approach achieves high performance identification of 85.04 % on YTF dataset and verification rate of 88.97 % on LFW dataset.

Other researchers like Chen et al. [56] tried to exploit the temporal information between video frames using joint sparse representation. They divided the database into various partitions, each partition has images of the same pose and illumination for the same face from the same video. Each video is represented by many partitions which is learned under strict sparsity to find the best representation of each face in each of the partitions. Same methodology is applied for the test images, used in the later matching step. Using this technique, the best identification rate 98.04 % is achieved on UMD dataset [57], where each subject has at least six sequences of images. This technique takes into account the illumination and pose conditions but have not exploited all the unconstrained conditions.

2.3.3 Deep-Learning Framework for Face Recognition

Recently, an emerging class of FR algorithms using large number of diverse yet labeled face images and deep neural nets (DNN) have shown promising recognition performance in unconstrained environment. The generalization capability of many machine learning tools like support vector machines (SVM), PCA, linear discriminant analysis (LDA), Bayesian interpersonal classifier tend to get saturated quickly as the volume of the training increases [58–61]. DNNs have shown to perform significantly better as compared to traditional machine learning algorithms [2] when trained with large number (millions) of diverse images, for example, images appearing in Facebook [49] at different times (and not similar appearing faces in videos). However, DNNs requires large amount of training data without which the network fails to learn and deliver impressive recognition performance. Moreover, training such massive data requires huge computational resources, like thousands of CPU cores and/or GPUs. Zhu et al. [62] trained DNNs to transform faces from different poses and illumination to frontal faces and normal illumination. They used features from the last hidden layer and transformed the faces for FR. Sun et al. [63] used multiple DNNs to learn high level face similarity features and used restricted Boltzmann machine for FV. They extracted features from a pair of face images instead from a single face.

DeepFace developed by Taigman et al. [2] has become very popular among the FR society. Primarily they have two good contributions in their work. Firstly, a 3D alignment process, where they used 3D modeling of the face based on fiducial points, that is used to warp a detected aligned 2D facial crop to a 3D frontal mode. They extracted fiducial points by using a support vector regressor trained to predict point configurations from local binary pattern (LBP) histograms based image descriptors [64]. For the alignment of faces with out-of-plane rotations, Taigman et al. used a generic 3D shape model and registered a 3D affine camera. Using these they transformed the frontal face plane of the 2D aligned crop to the image plane of the 3D shape [2].

Secondly, Taigman et al. [2] developed an efficient DNN architecture using 4 million images from 4000 persons. Face detection and localization are performed by extracting 67 fiducial points on each of the face images. Then, triangulation and frontalization are done to 3 RGB layers which are feed into 32 filters (convolutional layer 1: C1) as shown in Fig. 10. The output of this step includes 32 feature maps. M2 layer is a max pooling to get the maximum of these maps over 3×3 spatial neighborhood. Convolutional layer 3 (C3) contains filters which extract the low level features. So, C1, M2, C3 are responsible for features extraction. Three layers (L4, L5 and L6) are used to apply filter bank where every location in feature map learn a set of filters. Finally, the last two layers are connected to get the correlation between the features extracted. This DNN involves more than 120 million parameters using several locally connected layers without sharing weight, unlike the standard convolutional layers. DeepFace when applied to LFW and YTF databases achieves and an impressive accuracy rate of 97.35 % and 92.5 % respectively.

Fig. 10 DeepFace learning framework, from Taigman et al. [2] (Best viewed in *color*)

Another notable deep convolutional network (ConvNets) architecture called DeepID is developed by Sun et al. [3]. It contains four convolutional layers (with max-pooling) to extract features hierarchically. DeepID features are taken from the last hidden layer neuron activations of the ConvNets, followed by the softmax output layer indicating identity classes. Weakly aligned face image patches are used as inputs to each of the ConvNet, which extracts local low-level features. Number of extracted features gets reduced along the feature extraction hierarchy until the last hidden layer (DeepID layer) is reached. In this DeepID layer low dimensional predictive features are formed, which can predict an impressive 10,000 identity classes [3]. They have pioneered CelebFaces and CelebFaces+ face databases. The latter being a superset of the former contains 202,599 face images of 10,177 celebrities from the Internet. People in LFW and CelebFaces+ are mutually exclusive. Using their proprietary (not publicly available) databases and highly compact 160-dimensional DeepID features, they could achieve 97.45 % face verification accuracy on LFW, using weakly aligned face images [3].

Although, there are great and promising performance enhancement in these works, they still need to deal with very large scale evaluation on unconstrained FR (described in the Sect. 3.2.3: old and new protocol on LFW) in order to get good results. In the next section we review some of the benchmark competitions and evaluations done by independent organization and large research organizations.

3 Evaluation and Benchmark Competitions

The development of FR technology has started in 1993 and over of the period of time it has evolved to a very large extend including its applications from large scale nationwide deployment to ubiquitous wearable device computing. We have tried to summarize the entire evolution of FR benchmarks, competitions and algorithm evaluations in Fig. 11. The first FR technology test [65] took place in 1996 and this has lead to multiple FR vendor test conducted in 2000, 2002, 2006 and 2013. Links to all these FR vendor tests can be found in [66]. In between there are other competitions that took place, which include face recognition grand challenge (FRGC) 2005 [67], multiple biometric grand challenge (MBGC) 2009 [68], face and ocular challenge (FOCS) 2009 [69], good, bad and the ugly face challenge problem (GBU) 2009 [70] and multiple biometric evaluation (MBE) 2010 [71]. Furthermore, FR technology evaluation has been extended to mobile devices/environment like MOBIO in 2013 [72]. Generally, each of these competitions and evaluations takes place over 1–3 years time. Recently, due to the increasing popularity of social network and inexpensive "point-and-shoot" camera technology, people would just want to take pictures or videos, upload and recognize their friends, family and their acquaintances more-or-less automatically. This has spurred the point and shoot FR challenge (PaSC) in 2015 [73]. Going into details of each of them is beyond the scope of this chapter. In this section, we review some of the benchmark

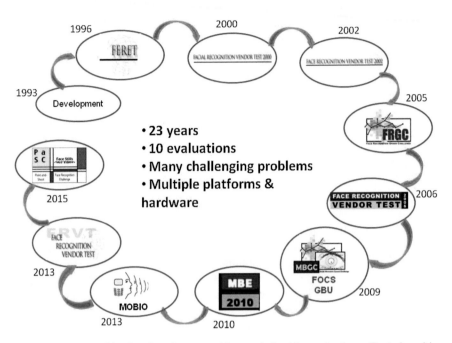

Fig. 11 Face recognition benchmarks, competitions and algorithm evaluations. (Best viewed in *color*)

evaluations and competitions that took place over the last few years, their protocols and summary of the evaluations. Later, we also discuss the emerging databases resulting from these competitions and evaluations.

3.1   FRVT 2013 Findings and Conclusions

The Face Recognition Vendor Test (FRVT) is a series of public evaluations for FR systems built by leading FR technology vendors. FRVT has been organized by the National Institute of Standards and Technology (NIST) in 2000, 2002, 2006 and 2013. FRVT succeeded the previous FERET evaluations held in 1994, 1995 and 1996 [15]. The latest one was FRVT 2013 [5], which started in middle 2012 and lasted until the mid of 2014. FRVT uses a large database to test both the accuracy and computational efficiency of various FR algorithms. The database consists of three parts. The first part is the law enforcement images (LEO) mugshot faces, which comprises about 86 % of the LEO database. The remaining 14 % images of the database were recorded by a webcam, which is referred to as LEO webcam. In addition, a smaller set of visa images consisting of well controlled frontal photographs of adults and children is also used. Besides the three types of face images, some sketch images based on the FERET dataset were also collected to support research in face sketch synthesis and recognition [74, 75]. For the competition, there are five tracks for the participants to participate:

1. Class A: Compare one-to-one verification (determine if two samples originate from the same person or not) accuracy.
2. Class B: Compare one-to-one verification accuracy but with an enrollment database present. This track was discontinued after the 2010 evaluation. Accuracy gains over class A are available.
3. Class C: Compare one-to-many identification (search to determine either that the person is not enrolled, or to determine the identity of the person). The FRVT test only evaluates on "open-set" identification algorithms because real-world applications are usually "open-set". Here, the "open-set" refers to the situation where a test face image might not be enrolled. The various partitions with numbers of enrolled individuals are 20,000, 160,000, 640,000 and 1,600,000.
4. Class D: Compare accuracy of determining the sex or age of a person in one or more input images. This separate class D track tests on determining whether the face in an image is frontal or non-frontal.
5. Class F: Find effectiveness of the algorithms that take one or more non-frontal images of a person as inputs and outputs one or more frontal images of the same person.
6. Class V: Find effectiveness of the algorithms that execute one-to-many identification of persons with frames extracted from surveillance video sequences.

The error measures used in FR evaluations such as for Class C are usually false alarms (search data from a person who has never been seen before is incorrectly associated with one or more enrollees' data) and misses (a search of an enrolled person's biometric does not return the correct identity).

From the results of algorithms submitted to the FRVT 2013 [5] for evaluation from various commercial vendors (NEC, Cognitec, etc.), the following points are observed:

(1) The age of the subjects strongly affects the identification accuracy. For all the algorithms evaluated, the older the person, the easier they are to be recognized. For children, both false alarm rate and miss rate are higher than other age groups. And infants are very difficult to identify.
(2) Sketch images are also used to match face photographs. For the most accurate algorithms, the rates of face not being among the top 50 candidates are quite high with 73.3 % for 3M/Cogent, 73.8 % for NEC, 78.5 % for Toshiba, 80.3 % for Morpho and 81.5 % for Neurotechnology.
(3) The image quality improvement is the largest contributor to the increase in recognition accuracy. The results show that there is a fourfold reduction in miss rate using high-quality mugshots vs. low-quality webcam images.
(4) The 2010 NIST FR evaluation showed that retention and use of all historical images increase accuracy considerably [76].

The FRVT 2013 provides independent evaluations of commercially available FR systems. These evaluations are aimed at helping the U.S. government agencies best evaluate and determine the scenarios where these technologies can be deployed. It also helps the FR research community to identify the limitations of current FR technologies and future research directions for improvement. As for the limitation of the dataset provided in FRVT 2013, neither the mughots nor the visa images have ideal properties. The mugshot images have too much pose variation while the visa images are degraded by the acquisition process and the JPEG compression.

3.2 Emerging Databases

3.2.1 The Good Bad and the Ugly (GBU) Datasets

In the past four decades, performance of FR on frontal still faces in controlled environment has improved significantly and achieved near perfect performance. However, frontal faces taken with uncontrolled environment (illumination) and expression remain challenging. As part of the Face and Ocular Challenge Series (FOCS) [69, 76], the Good, the Bad, and the Ugly (GBU) dataset tries to encourage algorithms that work well on matching "hard" face pairs but not at the expense of the performance on "easy" face pairs [70]. The GBU dataset consists of three partitions of frontal still face images, "Good", "Bad" and "Ugly". The three partitions were of different "difficulty" levels with the "Good" being the easiest partition, "Bad" being the average difficult partition and "Ugly" being the most challenging partition based on the analysis of results of the FRVT 2006 challenge [77, 78]. Some sample images are shown in Fig. 12.

Each of the three partitions has two sets, the target set and the query set. Each of the target and query sets in the three partitions contains 1085 images for 437 distinct people, 117 people with one image, 122 people with two images, 68 people with three images and 130 people with four images. The fusion algorithm based on the fusion of the top three performers of FRVT 2006 [78] were used to evaluate the similarity of face pairs and construct the three partitions. For FR, many factors contribute to the recognition performance with the big four factors being subject aging, pose variation, illumination and expression. The GBU dataset controls for subject aging, pose variation and the major factors that affect recognition are illumination and expression, as shown in Fig. 12. In order to avoid over-fitting on the data, the protocol of GBU does not allow training on images of subjects in the GBU dataset. A baseline algorithm, Local Region PCA (LRPCA) [70] is presented and evaluated to illustrate the training and evaluation protocol and provide a baseline performance for comparison.

Besides the original goal of stimulating research on "hard" FR problems, the GBU dataset can also be used to study other factors that could contribute to improving FR performance such as in [79]. In [79], the GBU dataset has been used to study the demographic effects on estimates of automatic FR performance. Based on their findings, the measures of FR performance rely both on the distribution of faces of matched identity as well as mismatched identities. They showed that the

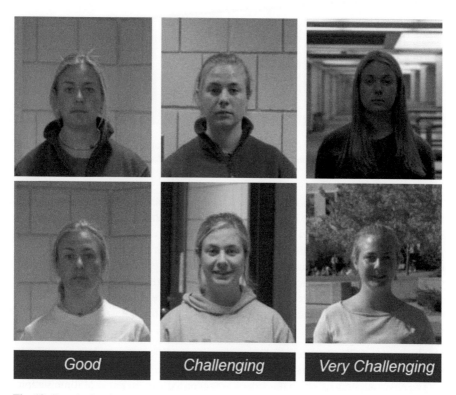

Fig. 12 Sample face images of 1 person from the GBU dataset, from Sinha et al. [4, 69]. The Good pair is referred to "Good", challenging pair as "Bad" and very challenging pair as "Ugly" (Best viewed in *color*)

demographic diversity differences in the non-matching distribution can radically change the estimates of FR algorithm performance. Thus, it poses a new challenge to find a method for tuning algorithm performance to the changing demographic environments where these FR systems will be used reliably.

3.2.2   FR in Mobile Environment

The MOBIO database provides the FR community a bi-modal (audio and video) dataset recorded in a less controlled environment by mobile phones. It also comes with an evaluation protocol together with a baseline algorithm to compare different algorithms developed by the participants of the FR competition in mobile environments hosted at the 2013 International Conference on Biometrics [80, 81]. The goal of the dataset is to stimulate research in the field of multi-modal recognition in a mobile environment. For first-person-view (FPV) or egocentric views face images, Mandal et al. [27] reported a database comprising of face images captured using wearable devices like Google Glass and head mounted web cameras.

Fig. 13 Sample face images of two persons from the MOBIO dataset. There are large variations in pose, illumination, makeups and hair style. (Best viewed in *color*)

MOBIO database was mostly collected by mobile phones with subjects speaking to a handhold mobile phone by answering a set of predefined questions as described in [80]. In total, the dataset consists of 61 h of audio-visual data recorded over a period of one and a half years. The participants consist of 100 males and 52 females, each of whom has 192 unique audio-video samples. For each participant, two phases were recorded, each of which contains six sessions of recordings, and the sessions are separated by several weeks. Some sample images are shown in Fig. 13.

Since the videos are recorded by mobile phones, the dataset has created the following challenges:

- The pose and illumination conditions vary across different samples,
- The quality of the speech recorded varies and
- The environments in which the videos are recorded vary in terms of illumination, background and acoustics.

For evaluation, the dataset is split into three non-overlapping partitions for training, development and evaluation. The training set is used to train the models, e.g. the project matrices for PCA. These images can be used as negative examples in a classification system for some systems. They can also be used for score normalization in training and testing. The development set is used to tune some meta-parameters of the models, e.g. the dimension of the PCA projection matrices. The evaluation set is used to test the models with data that haven's been seen in the training and tuning steps. As the goal of the dataset is to evaluate FR rather than face detection, the eye locations of some selected frames in each video are hand-labeled and provided to the participants.

The organizer provided a baseline algorithm for both speaker recognition and FR, and an algorithm based on fusion of the two modalities (video and audio) is also provided [81]. The baseline algorithm can process 15 frames per second and is suitable for running on mobile devices. For the competition, eight institutions participated and most of the algorithms submitted relied on one or more features of: local binary patterns, Gabor wavelet responses (especially Gabor phases) and color information. With score fusion, the University of Ljubljana and Alpineon Ltd. (UNILJ-ALP) performed best, achieving an equal error rate (ERR) of 2.751 % and 1.707 % on females and males respectively. Among those without fusion algorithms, the University of Campinas and Harvard University (UC-HU) team achieved the best performance of 4.709 % and 3.492 % on females and males, respectively, without relying on handcrafted features, but learned features with a convolutional neural network [82] instead.

The contribution of the dataset is threefold: first, it provides a challenging FR dataset with uncontrolled face videos; second, the dataset provides both audio and video for fusing the two modalities to improve the identity authentication performance; third, the whole dataset is recorded in mobile phones, and the evaluation requires a trade-off between performance and hardware requirement, which encourages algorithms designed for mobile devices.

The main drawback of the dataset is that in the FR evaluation, only one facial image was extracted from each video with the eye positions labeled manually. A more interesting problem is to look at how dynamics of the faces in the video can help improve the accuracy of FR. Although the algorithms from the participating institution were evaluated by the organizer, most of the datasets/partition information are not available online for reproducibility of the results.

3.2.3 "The Famous" Labeled Faces in the Wild, Its Old and New Protocols

There exist a large number of benchmark databases for evaluating FR algorithms, like FERET [15], AR [19] and ORL [83] just to name a few. A comprehensive list can be found in [84]. Most of the these databases are collected in controlled (studio) environment for studying certain aspects of FR (like expression and/or illumination) which are posed or unnatural. Under these controlled conditions, FR algorithms can achieve performance comparable to human beings. However, these algorithms cannot generalize well to data collected under different natural or spontaneous conditions. The LFW dataset [35] provides the FR community with uncontrolled face images from the web for pairwise matching/unmatching problem. The LFW dataset exhibits variability in lighting, pose, subject age, expression, race, gender and so on. The goals of the dataset are:

- Provide a large database of real world face images for the unseen pair matching problem of FR,
- Fit neatly into the detection-alignment-recognition pipeline, and
- Allow careful and easy comparison of FR algorithms.

The original LFW dataset contains 13,233 images of 5749 people, among which 1680 people have more than 1 image per person. The images were collected from online internet news articles and processed using Viola-Jones face detector [85] for detecting faces.

The Old Protocol

This dataset contains 300 pairs of genuine matches and 300 pairs of imposter matches for tenfold of cross validation leading to 3000 genuine and imposter matches each. The dataset is organized in 2 "views": view 1 is used for development training/testing purposes, where the training/test partitions are generated randomly and independently of the splits for tenfold of cross validation. This view is used for model selection and/or validation purposes. View 2 is used for performance testing and final evaluation of the algorithms to minimize fitting to the test data. View 2 is divided in ten subgroups such that the face pairs are mutually exclusive for tenfold of cross-validation, whose results are averaged to get the final performance of the model selected with view 1 data [86].

As running the Viola-Jones face detection algorithm [85] generated the face images, it fits well in the three-step detection-alignment-recognition pipeline for FR, (as explained in Sect. 2.2) and indeed, the latest LFW dataset includes four different sets of LFW images, the original and three different "aligned" images. The aligned versions include, (1) the "funneled images" (LFW-a) by Huang et al. [87], (2) for second version, an unpublished method is used for alignment of LFW-a [86] and (3) "deep funneled" images again by Huang et al. [88]. The last two funneled images produce superior results for most FV algorithms over the first two sets of images. From the evaluation of various algorithms, it is evident that the use of training data outside of LFW can have a significant impact on recognition performance. Numerous benchmarking results can be found in [46].

In conclusion, the LFW dataset provides the research community with a less controlled face dataset for FV system development. It has stimulated researchers to work on more "natural" and unconstrained FR problems that would generalize to data outside the existing dataset. However, as the face images were collected from news articles on the web, they are affected by the photographers' and editors' choice, so there were not many images under extreme lighting conditions. Since the faces are detected using the Viola-Jones detector, there are a limited number of faces with side views and views from above and below.

The New Challenging Protocol

If we think of a very common real-world scenario where 500,000 visitors visit an amusement park per day using facial biometrics, certain CVR at 0.1 % FAR implies that 500 people can falsely (with fake or shared identity) enter the park per day. This can be a big concern and loss to the park owner. Old LFW benchmark protocol contains 3000 pairs of genuine matches and 3000 pairs of imposter matches in total which are very limited to evaluate the large scale performance. Using old protocol, performance evaluation at false acceptance rate (FAR) of 0.1 % is not statistically significant as it requires to count only three imposters matching scores. A vast majority of researchers has been following this old protocol, that uses partial data of this database to evaluate their algorithms (for details see LFW results website [46]). So there is a need to enhance the LFW benchmark protocol and exploit all the available data.

Liao et al. [6] designed a division of the LFW dataset into development-set that contains a set of training and testing data to tune the parameters. Also, an evaluation-set is designed to evaluate the performance of FR with 85,341 genuine matches, 6,122,185 imposter matches in training; 156,915 genuine matches and 46,960,863 imposter matches in testing. The new protocol takes into account large number of genuine and imposter matches both in the training and testing datasets and hence, it can evaluate very low FARs (e.g. <0.1 %), which are statistically significant.

Liao et al. [6] implemented seven learning techniques: PCA, LDA, large margin nearest neighbor (LMNN), information theoretical metric learning (ITML), keep it simple and straightforward metric learning (KISSME), locally-adaptive decision functions (LADF) and joint Bayesian formulation using three features namely, hand-crafted feature LBP, a learning based descriptor local embedding (LE) and high dimensional LBP (HighDimLBP) feature. FAR and open-set identification rates are measured as performance indicators. The best results are obtained using joint Bayesian approach with HighDimLBP features [89], where the CVR achieved is 41.66 % rates at FAR = 0.1 % and open-set identification rate as 18.07 % at rank 1 and FAR = 1 %. Therefore, it is evident from this recent benchmark study of large-scale unconstrained FR [6] that the newer protocol is very challenging and more practical as compared to the previously evaluated results [46].

Although this work added some improvements to the LFW benchmark study by increasing the number of correct and false matches obtained by the data, the CVR is too low to be considered in real-world FR applications. The performance is still far from satisfactory as the verification and identification rates are very poor under the large-scale unconstrained FR setting.

3.2.4 YouTube Video Database

LFW is a database used for evaluating FR algorithms with still face images recorded in uncontrolled conditions. As for videos, there exist several methods that have performed well in video FR tasks by exploiting the fact that a single face might appear in a video in consecutive frames [90, 91]. But the datasets used for developing those algorithms are primarily collected in highly controlled lighting and shooting conditions with high quality storage. In contrast, the YouTube face Dataset (YTF) [14] complements the LFW by providing a database of face videos designed for studying the problem of FR in videos with uncontrolled lighting, shooting condition and video quality. The videos were downloaded from YouTube with identities from the LFW dataset. Each video in YTF comes with a label indicating the identities of a person appearing in that video.

The dataset contains 3425 video clips of 1595 different people. The duration of these video clips ranges from 48 frames to 6070 frames with an average length of 181.3 frames per person. Because the videos were downloaded from the YouTube using automatic tools, this dataset is highly uncontrolled in terms of lighting, shooting condition, video quality etc. Following the LFW protocol [35], the evaluation of algorithms on this dataset is a standard tenfold cross validation, pair-matching test. In the evaluation phase, 5000 video pairs are randomly selected form the dataset, in which half are matched pairs (same person) and half are unmatching pairs (different person). These pairs were divided into ten subgroups, each of which contains 250 matched pairs and 250 unmatched pairs. Each algorithm is trained on nine subsets and tested on the left 1 for 10 times with each of the subsets being the testing set once. The average performance is reported.

All video frames are encoded by several well-established image descriptors including LBP, center-symmetric LBP (CSLBP) and four-patch LBP. With these encodings, several types of methods have been evaluated with the YTF database. Because each video contains multiple frames and each frame can be encoded as a vector, the problem of matching the faces in a pair of videos becomes matching two sets of vectors. Three major groups of methods have been considered. The first group employs comparisons between pairs of face images from each of the two videos. The second group uses algebraic methods, which compare vector sets. A third group including the pyramid match kernel and the locality-constrained linear coding methods were effective in comparing sets of image descriptors. In total, the author of the dataset evaluated five groups of methods with three types of face image encoding and the results are shown in [14]. However, the best performance is reported using the DeepFace recording an accuracy of 91.4 %.

The contributions of the YouTube dataset and the evaluations include the following:

- A comprehensive dataset of labeled face videos in uncontrolled environment was presented together with benchmarks and pair-matching tests,
- The benchmark was used to compare a variety of existing video face matching methods and
- Stimulate further research in video FR in challenging and uncontrolled conditions.

3.3 Summary of the Emerging Databases

Four databases for FR have been discussed in the above subsections. These four databases together with six other emerging databases are summarized in Table 1. The FRVT 2013 provides with three main types of images for testing typical identity verification which could be deployed for detection of duplicates in databases, detection of fraudulent applications for credentials such as passports, criminal investigation, surveillance, and forensic clustering. The mugshot set and webcam set vary in their image quality and they can help study the effects of image quality on recognition performance. The evaluation also found that age of the people shown in the images also contribute to the performance of nearly all FR algorithms evaluated.

Table 1 Comparison of the emerging face recognition databases

Databases	Data source	Modality	Scenario	Data size	Subject size	Publicly available
LFW 2007 [6, 35]	Images from web news article	Still images	One-to-one FV & open-set FI	13,233	5749	Yes
GBU 2010 [69, 76]	Uncontrolled frontal images	Still images	One-to-one FV	1085 images in target & query sets × 3	437	No
YouTube 2011 [14]	Videos from YouTube	Videos	One-to-one FV	3425 clips	1595	Yes
MOBIO 2012 [80]	Collected with mobile phones	Videos with audio	One-to-many FI	61 h	152	Yes
Makeup database 2012 [92, 93]	Three categories from YouTube video makeup	Still female images	Makeup detection & one-to-one FV	604 + 204 + 154	151 + 51 + 124	Yes
FVRT 2013 [5]	Mugshot, webcam & visa images	Still images	Five tracks: one-to-one FV & FI, one-to-many FI, etc.	–	–	No
Indian movie face database 2013 [94]	Face images from movies	Still images	FI	34,512	100	Yes
McGillFaces database 2013 [95]	Indoor/outdoor uncontrolled face videos	Videos	Pose/gender/ face hair analysis	60 clips	60	Yes
Labeled wikipedia faces (LWF) 2014 [96]	Wikipedia biographic entries	Still images	One-to-one FV & FI	8500	1500	Yes
FaceScrub 2014 [97]	Public figures from searched queries	Still images	Detect target face from searched queries	107,818	530	Yes

Usually the older the people, the easier it is for the FR algorithm to recognize. Sketch faces based on FERET dataset was also used in the evaluation to support research in face sketch synthesis and recognition. The FRVT has provided a platform to test the commercial FR systems that have the potential to be deployed in different places by the US government and it also identities the future research directions for the FR research community.

GBU dataset provides three partitions of face images, each of different level of difficulty. The images were collected in a partially controlled environment where the pose and age are controlled but the expression, lighting are not. Because all faces are frontal faces, the only reason causing different recognition results is the representation of the faces in each image. FR algorithms can achieve better performance than humans in fully controlled condition [4]. While in fully uncontrolled conditions, no significant progress could be made. Thus the GBU dataset stimulates the development of robust frontal FR algorithms that could make progress in more challenging, partially controlled tasks without sacrificing its performance in easier ones.

The MOBIO face database consists of more than 61 h of audio-video bi-modal faces (also summarized in Table 1). The videos are recorded by handhold mobile phones, recording people speaking to the phone camera while answering a set of predefined questions. MOBIO provides the research community with a bi-modal dataset that could be used to evaluate speaker recognition, FR as well as their fusion. Since the videos are recorded by amateurs using mobile phones, there is large variability in pose, illumination, background environment as well as the audio-video quality. This nature of the dataset makes it challenging and encourages research to combine both modalities to improve the performance. However, in the evaluation stage, only individual frames containing faces were used to perform FR. A video based FR system should give better performance by exploiting the dynamics of the recorded faces. Another contribution of this dataset is to encourage the researchers to focus on the trade-off between performance and hardware requirement. Since the dataset is intended to stimulate development of algorithms that could find its applications in mobile devices, an important aspect of the evaluation to consider is the execution time and memory requirements.

Both LFW and YTF databases provide a large collection of faces recorded in uncontrolled conditions from the internet. For LFW, the face images are from online articles and each face comes with a label of the person's name. The YTF database takes a similar approach and the videos are downloaded from YouTube and also come with identity labels of the people. These two databases offer the FR community a good playground for developing and evaluating algorithms targeting at more natural and less controlled settings. For the LFW database, although the face images are more natural than those taken in fully controlled conditions, the images are often taken with good lighting and lack non-frontal faces.

In summary, from the benchmark databases presented above, we can see the following trend in FR research, benchmark database and protocol design:

1. As FR in controlled environment is considered a "solved" problem with some algorithms outperforming humans, the frontier of FR research is shifting to uncontrolled and more natural settings.
2. Coupled with powerful computing machines, improved algorithms for deep learning are able to discover patterns in large dataset. Hence larger labeled databases are desired in the FR community to develop large-scale and robust FR algorithms.
3. Nowadays, almost everyone cannot live without a mobile phone. FR systems on mobile phone and wearable devices would find its application in our everyday life. Thus robust FR algorithms running on mobile devices in natural settings will be of great value to the consumers.
4. As more and more algorithms are being developed for FR in videos, the dynamics of moving faces in videos should be further exploited to build more robust and accurate next generation FR systems.

From such papers, evaluations and benchmark competition results, it is apparent that unconstrained FR with large or small scale scenarios is largely an unsolved problem and should receive further attention. Human beings are amazing for FR under unconditioned settings. Even after years or with diverse makeups/appearences, human beings hardly fail to recognize an individual. Hence, it is imperative that we derive psychophysical and/or biological motivations from human beings on aspects that have made them experts in FR over centuries.

4   Human Recognition of Faces

The human face is perhaps the most important class of objects that we are interested to interact with. Our response to human faces is distinct from that to other classes of objects: there seemed to be a selective preference to human faces as we age. A study by Michael et al. [98] on 3-, 6-, 9-month old infants and adult groups revealed a greater percentage of gaze dwell time on faces with age. This selective attention of the human visual system towards other human faces might stem from having a default network in the brain that drives a series of involuntary cognitive processes: us thinking about recent events and speculating future ones that are founded on social interactions and involve the theory of mind [99] during periods of inactivity. Evidence from neuroimaging studies of brain diseases such as Alzheimer's, autism, schizophrenia depression etc., seemed to target and cripple this default network; therefore leading to the impairment of social cognitive abilities on varying degrees for patients with the aforementioned diseases.

The (hypothesized) existence of such network, one that attunes to social interactions and theory of mind, supports the fact that we gravitate towards connecting and understanding people above all others. This in turn, explains our selective preference to human faces and motivates the need to study how we perform the two main types of face recognition: face verification (for unfamiliar faces where the individual only has a sense of familiarity of having seen the face before e.g. acquaintances) [100] and face identification (for familiar faces where the individual has both a sense of familiarity of having seen the face before and is able to identify him/her by name) as the foundation to successfully navigate the social world.

That being said, it will be beyond the scope of this chapter to involve all aspects of neuroscience, neuroimaging and psychological studies to explain the neuroanatomy of the default network and social cognition. Henceforth, the coverage of this part of the review will be dedicated to psychophysical and neuroimaging discoveries about the FR capabilities in humans. The sections that follow are the introduction to the two main hypotheses on the motion advantage in recognizing faces, with four other subsections on the current most difficult conditions pertaining to human performance in FR (the Big Four! [4])—illumination-, facial expression-, view perspective-, and age-invariant recognition. These four sections are distinct from the challenges (scale, occlusion and motion blur) in machine recognition of faces as discussed in Sect. 2. Finally, we would like to offer a preview of our integrated experimental approach that might be feasible in transcending FR across the four difficult conditions to achieve performance inspired by humans in an unconstrained and naturalistic setting.

4.1 Temporal Cues That Aid Face Recognition: Two Hypotheses to Explain Motion Advantage

Motion brings not only a face, but also the personality of its owner, to life. We are inherently dependent on the dynamics of motion to infer the mental states of those whom we are interacting in numerous social contexts. Visual inputs of the changes in head movements, varying degrees of facial expressions, eye gaze directions etc. bombard our senses in a myriad of signals before being integrated into a general, yet uncannily accurate, perception of how the present moment of interaction feels like. Such visual cues are essential in guiding our predictions of the 'appropriate' actions to take within a particular social context. If your counterpart is speaking to you while his eye gaze kept darting towards the nearest exit or his watch, you would probably have inferred that he is in a hurry to leave and that you should quickly wrap things up and end the conversation.

Similarly, motion in dynamic faces gives rise to a plethora of information that elevates both face verification and identification as compared to when static faces are presented. Hence, research on human face processing is now delving into dynamic faces to not only simulate a more realistic context for face recognition and processing, it is also an attempt at dissecting and comprehending how the presence of motion leads to improved face verification and identification performances. Consequently, two hypotheses were formulated in an effort to explain the benefits that dynamic faces impart on human recognition of faces: the supplemental information hypothesis and representation enhancement hypothesis [7].

The supplemental information hypothesis embodies an idea that the ventral temporal cortex, which includes the lateral fusiform gyrus, occipital face area and other associated structures in the human brain, is responsible for processing both invariant face information and the idiosyncratic facial motions of individuals. Inevitably, this confines the realization of the supplemental information hypothesis in FR to recognizing familiar faces only [7]. Work by Lander and Chuang [51] provided evidence that non-rigid facial motion (movement of internal facial features such as blinking of the eyes and chewing movements of the mouth), more than rigid motion (global movement of head including pan, tilt, yaw and other head translations), improves FI of familiar individuals. "Distinctive" facial motion (a separate entity from distinctive facial features) as well as "naturalness" of facial motion (not artificially designed motion) suggested in [101], proved facilitative to facial identification of familiar faces as well.

On the other hand, the representation enhancement hypothesis posits that recognition performance of a novel face after a learning phase has a higher accuracy than that learnt from static faces. This hypothesis is founded on a perspective-oriented learning of unfamiliar faces; also known as the "structure-from-motion" learning [102], where the advantage relies on the fact that knowledge about the three-dimensional structure of an individual's face can be gathered from motion prior to subsequent recognition. Such a learning process is said to confer humans the ability of recognizing unfamiliar faces. In an experiment by Pilz et al. [103], subjects were primed with unfamiliar faces with emotions of either a frown or a smile in non-frontal viewing perspectives and asked to do FV with a target face in the frontal view with an opposite emotion from that of the primed face. Results revealed that subjects generally responded faster when primed faces were in non-rigid motion as compared to static ones.

These are some of the psychophysical experiments that are seemingly representative of the two different hypotheses mentioned. They can be considered seminal works, which inspire later research to further refine FR experiments for the sake of allowing a better understanding of how this is done so effortlessly in humans. The following sections will discuss interesting findings for different facets of FR spanning from the various fields of study (psychophysics and neuroimaging) for cross validation and inspiration.

4.2   How Do Human Beings Handle the Big 4?

Similar to the challenges faced by computational models described in Sect. 2.1, we human beings also face difficulties in recognizing individuals across various conditions. In the following four subsections, we review some of the popular human centered experiments and protocols so as to understand innovative strategies, prior learning, biases at various levels and how exactly human beings overcome these four big problems.

4.2.1 Face Recognition Across Different Illumination Conditions

Astonishingly, there is little work done in terms of psychophysics experiments conducted to investigate how humans do FR across illumination variations. Instead, a plethora of work mostly centralized around improving or developing new pipelines for computer vision under this category.

Nonetheless, Tarr et al. [104] have shown that recognition performance for human is dependent on the difference between the degree of facial illumination presented to subjects during the training and testing phases. Subjects were first allowed to study a sheet of ten different frontal face images with their corresponding names (e.g. Allen, Laura) printed. The faces were shown illuminated from the front, normal to the face. For each face, the lighting space was sampled in 15° increments in both the horizontal and vertical axes to the right of the camera axis. In each of the five experiments conducted, subjects were shown different subsets of illuminated faces in the training phase (an illustration is shown in Fig. 14) before proceeding to the full set in the testing phase with large illumination variations (similar to the images shown in Fig. 1).

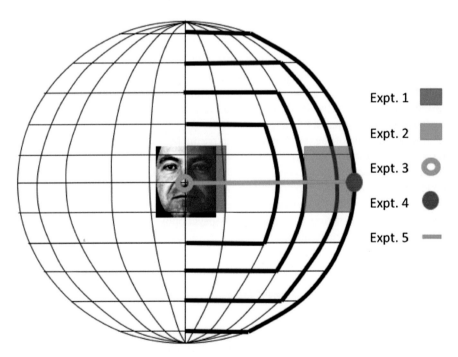

Fig. 14 Training sets for illumination variation experiments. Experiment 1 contains illuminations within 15° of the camera axis; Experiment 2 is a mirror of Experiment 1 with extreme lighting directions. Experiments 3 and 4 have one illumination condition each, (0°, 0°) and (75°, 0°) respectively. Experiment 5 contains illuminations along the horizontal meridian of the illumination space; from (0°, 0°) to (75°, 0°) (Best viewed in *color*)

Their results show that, in general, increasing the distance (i.e. the extent of difference) between illumination coordinates from the training and testing phases will decrease the FI performance of the subjects. Intuitively, we would expect performance to be worst for Experiments 2 and 4 (refer to Fig. 14), where subjects were trained with extreme illumination conditions ranging from 45° to 75° away from the normal. Yet, interestingly, the most prominent drop in performance was seen in Experiments 1, 3 and 5, where the face images were mostly illuminated from the frontal or near frontal coordinates. The authors reconciled this observation by explaining that because subjects were trained with extreme illumination conditions in Experiments 2 and 4, they were able to identify the faces with greater accuracies by using generic knowledge about geometry of faces as a class to infer their appearances under novel illumination circumstances. The ability of prediction can be attributed to the neural mechanisms of the posterior superior temporal sulcus (pSTS), which is responsible for processing changeable information in faces; where in this case it is the information on shape and surface orientations that is processed [105]. This could prove as evidence of the hypothesis that the dorsal stream pSTS identity representation might include a representation of facial shape that is independent of signature motions [102]. This experiment has shown that the humans are sensitive to the degree of face illumination conditions, and that learning from extreme degrees of illumination, albeit counter-intuitively, facilitates recognition of novel face configurations.

However, the experiment is still considered limited in terms of understanding how the human visual system actually compensate for dynamic variations in lighting. What the human visual system encodes is a continuum of illumination changes as the coordinates of the light source changes temporally, as opposed to the discreet increment of illumination changes in the experiment. There is, therefore, much to gather in terms of how the shape and geometry of an individual's face changes with illumination along the temporal dimension are encoded in the human visual system to confer us the high accuracy in FR under novel illumination conditions.

Sinha [106] revealed human psychophysical studies on a subset of the illumination spectrum of faces: contrast negation. It shows when concluding whether an image is a face, there is significant drop in performance with contrast negation.

As seen in Fig. 15, the patches in (a) and (b) have different overall brightness, but the images can still be discerned as illustrating the same object—a face. However, when comparing the patches in (a) and (c), where both have the same overall brightness, the object depicted may be perceived differently. It was concluded that the direction of brightness contrast, or otherwise known as contrast polarity, plays an important role in object perception and recognition.

Another study by Wallis et al. [107] using 3D face images confirmed the deduction that temporal cues in the context of varying illumination in motion functions like a 'perceptual glue' in human visual perception. Subjects showed the tendency to assume that they were viewing a single face sample when it was actually morphed to a different identity during the transition of varying illumination (refer to Fig. 16). The degree of this effect is influenced by the presence of a

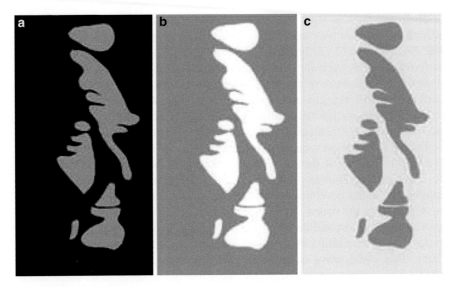

Fig. 15 Direction of contrast brightness affecting face detection, from Sinha [106]

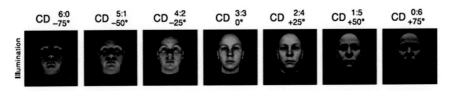

Fig. 16 Illumination varies during the morph transition from sample C to D in gradual ratio proportions, from Wallis et al. [107]

training phase where 'unlearning' the encoded visual representations of the sample is difficult, thereby supporting the representation enhancement hypothesis of the plausible "structure-from-motion" learning ability.

The context of any surface has an enormous effect on the color we see, e.g. illumination from the sun looks red in the evening, but yellow at noon and the recent 2015 debate on the color of a bodycon dress being blue-and-black or white-and-gold [108]. This is analogous to the scenario shown in Fig. 15. When there is a difference in contrast brightness and direction from the image's context to the object (i.e. both dark to light contrast seen in (a) and (b), but reversed for (c)), we see the conflation of illumination, reflectance and transmittance giving rise to the inverse optics problem; therefore leading to erroneous perception of the object [109, 110]. The peculiar way we see color and contrast, and hence, the way we perceive objects (especially faces) remains to be explained. Perhaps the answer to this problem is the way in which the human visual system copes with the inverse optics problem—a problem that could plausibly be simplified by investigating, on a frame-by-frame basis, how humans carry out FR across illumination variation leveraging on temporal dynamics in videos.

4.2.2 Face Recognition Across Different Facial Expressions

One of the most important skillsets for successful navigation in the social world is to accurately infer the mental and emotional states of others; hence our possession of a visual system that is attuned to human muscular motions [111]. One of the many such muscular motions is facial emotional expression—the main type of visual social cue that we infer from for information (such as the mental state of others, the intention behind their actions etc.) [112] on how appropriate we are to behave in a particular social context. Not only are facial expressions socially relevant, they are important in facilitating FI. The advantage for this aspect of FR is more profound when the facial expressions are presented in motion. Hereon, the main aspect of discussion will focus on the effects of emotional expressions on FR.

Firstly, there is the factor of idiosyncratic facial motion, which includes that of expressive (i.e. non-rigid motion) faces that promote higher FI accuracy. This is evident especially in the identification tasks involving faces that are familiar to the subjects during the experiment. The seminar work of [51] has shown that the accuracy for FI of familiar faces is the highest for expressing faces and that they outperform rigidly-moving (69.9 %, SD 14.5), talking (82.4 %, SD 11.7), and static faces (56.5 % SD 22.0) at 89.5 % (SD 6.8) identification rate. Further investigation in a separate experiment showed 77.3 % (SD 12.2) of expressing faces possess distinctive facial movements during the course of expression and hence the high recognition rate for expressing faces in motion is obtained. This particular finding concurs with the supplemental information hypothesis and that there is a very strong motion advantage for identifying familiar faces based on a set of signature non-rigid facial movements for every individual. In other words, it can be argued that the processes for face expression and those for FI are integrated from plausibly different neural mechanisms in a manner that facilitates better performance in FR.

Other interesting results from two of their experiments is that there is a higher identification rate for familiar samples when the dynamics of non-rigid facial motion is natural; not artificially created or modified, and that the speed of the facial motion during expression is naturally fluid; not sped up or slowed intentionally [101]. An explanation for these behavioral phenomena can be found in recent neuroimaging data utilizing whole-brain analysis to show that the STS is the region with the greatest BOLD response under the influence of increased information in dynamic faces [113, 114] and fluidity of its motion [114]. What is especially interesting is that it reinforces the idea of distinct processing mechanisms devoted to facial identity and expression respectively. Majority of the ventral temporal face-sensitive regions of the brain (i.e. bilateral fusiform face area (FFA), occipital face area (OFA), right inferior occipital gyrus (IOG) and the right fusiform gyrus (FG)) seemed to be sensitive to the increased amount of frame information in dynamic faces, while a separate processing area is dedicated to the fluidity of that motion—STS [115]. Giese et al.'s [116] computational model of biological motion recognition has specified both a motion pathway and a form pathway in which neurons in the middle temporal (MT) area, the middle superior temporal (MST) area and the kinetic occipital (KO) area are attuned to discern optic flow localities before sending its flow pattern to the STS for classification and identification in a feed-forward manner. This series of form detectors posits to supplement information from surface deformation of the face with the invariant face form learnt by the ventral temporal brain regions from multiple frames. Once again, this could potentially support the notion that enhanced FR from viewing the dynamics of natural facial expressions is not only dependent on the increased information presented in the form of increased number of frames, but relies on a disparate encoding process of an individual's non-rigid motion signature as well.

Another set of study by Rigby et al. [117] tested subjects on face processing where they have to make speeded expression (or identity) judgments of static and dynamic faces while identity (or expression) were held constant or varied. By showing that there was significant interference when processing static faces compared to dynamic faces, they provide evidence to support the idea that dynamic cues arising from the motion of facial expression, do facilitate a more efficient FI process. This dynamic advantage, however, was more obvious with the expression task as compared to the identity task. A plausible rationale behind such an observation could be that expressions causes global descriptors, which are crucial for holistic face processing, to be superseded by feature-based processing [28, 118]. Since facial expressions are considered socially salient and relevant [111], it is not surprising that humans will attend to and become adept at judging the types of expression.

Experiments on the composite effect of face processing, whereby feature-based face processing is dominant over holistic face processing, can serve as added evidence of having separate mechanisms for identity and expression recognition proficiency in humans. Underpinning this partial differentiation of the neuronal domains for identity and expression are human fMRI adaptation studies [119], as well as studies on prosopagnosic patients who possess the capability of recognizing facial expressions despite their disability in recognizing face identities [120, 121].

Concurring with this tenable anatomical discrimination of encoding for identity and expression, feature-based visual processing exuded in humans is demonstrated by Xiao et al. [122], who showed that subjects learning novel faces in non-rigid motion will have their feature-based FR less affected by irrelevant information in composite faces. They were more competent at verifying if the top and bottom halves of a face belonged to the same person after learning them in motion than with static faces. Such results are accounted for by the representation enhancement hypothesis—a 'structure-from-motion' type of learning. Perhaps, the motion information from the STS and other similarly committed brain regions is mapped in a piecewise manner to specific sites of the face according to the observed surface deformation in order for the brain to learn a set of signature facial movements for individuation. Using results from a classic experiment by Patterson and Baddeley [131] as an illustration, a simultaneous shift in both a face's viewing perspective and emotional expression between the learning and testing phases did not induce a significant drop in FI performance [123]. On the contrary, a sole change in viewing perspective will severely compromise FI performance during the testing phase. This suggests room for leveraging on the advantage of using dynamic facial expressions to extrapolate recognition from neutral or a set of orthogonal expressions. After all, the act of smiling not only stretches one's mouth such that it takes up a larger area relative to a neutral face; yet it allows the viewer to visually encode the unique shape and trajectory of that smile to aid face discrimination. It, therefore, will be worthwhile to investigate the exact mechanism of mapping such a dynamic learning process to invariant recognition of faces.

4.2.3   Face Recognition Across Different Viewing Perspectives

Being immersed in a social world, we interact with people under unconstrained conditions on a daily basis. One of such conditions is the constant change in viewing perspective of a face. Be it listening to a presentation at a conference or talking to a group of friends, we all succumb to viewing faces in a range of different angles relative to the horizontal and vertical axes from the normal of the frontal view. Hence, it is relevant to understand how such rigid motion (e.g. yaw or pitch) contributes useful input to the human visual system for robust FR.

Intuitively, presentation of a face moving across different perspectives to a subject will provide him with more information of the overall 3D structure of the individual; but which aspects of a dynamic face allows for better FR? It was argued that perhaps the human visual system is evolved to achieve a representational structure that includes object information across both temporal and spatial dimensions.

In a FV task across different view perspectives, Pilz et al. [103] established evidence that learning a novel face in motion will lead to heightened FV performance, along with a shorter response time, as compared to that when learning a static face. This observation was obtained despite the fact that the target face to be matched was presented in a different viewing perspective from that of the learned dynamic face. Souza et al. [124] discovered a heterogeneous distribution of view-selective face neurons in the anterior STS (aSTS) that might be able to explain how humans learn face identity from different view angles. They found that in the caudal region of the aSTS, majority of the face-sensitive neurons elicited responses to the right and left views of a face. On the other hand, face-sensitive neurons in the rostral region showed a peak in response to a single oblique view only. This could imply that the processing of a face's different perspectives is conducted by having different populations of neurons represent specific sets of view angle information. Therefore, when testing a novel face's identity is conducted after learning from a dynamic face, a faster response time with higher recognition accuracy can plausibly be explained with the integration of view angle information gathered by neurons in the different regions of the aSTS, along with the identity information from the FG.

A lot of the image-based recognition of objects (including faces) is carried out at the level of fine abstract features [107]. Neurons learning the invariant properties of a feature not only capture information on the object's transformations, they might also generalize such learning to a diversity of objects that might contain the same feature. Such a theory may offer an explanation as to why humans are competent in identifying objects from novel viewing angles. It also functions as evidence that temporal cues extracted from dynamic faces influence neural representations of objects by serving as the 'perceptual glue' to gel learnt concepts for subsequent recognition. However, the exact neural computation for such an abstract, generalized learning mechanism remains elusive to this day.

4.2.4   Face Recognition Across Age Differences

As humans age, the sands of time will etch a gradual, conspicuous trail of changes to the skin surface. Being a complex process, facial aging affects both the shape and texture of a face. Its manifestations vary among different age groups and ethnicities [125], with extrinsic factors like individual lifestyles and environmental conditions affecting the rate and extent of observable aging.

Understanding FR across time lapses in age is crucial especially in applications such as forensic art, electronic customer relationship management, security control and surveillance monitoring, biometrics, entertainment (e.g. accelerate actors' age in movies as required) and cosmetology [33]. Notwithstanding the relevance of FR in this aspect, there is very little work done using human psychophysics to study age invariant recognition in contrast to the vast amount of literature on computer vision techniques in this aspect.

Perhaps the computer can outperform human in terms of fine changes in facial such as the identification of craniofacial growth (i.e. changes in face shape) and the relative surface area and protrusion of facial features such as eyes, nose, ears, mouth, cheeks and chin (the cranium grows to cause sloping and shrinking of the forehead by releasing more space on the cranium's surface for those features) [126]. Skin texture will also be expected to change as collagen breaks down and the skin sags to form wrinkles due to its inability to maintain its former elasticity. At the same time, implications from previous exposure to the sun and age-related health problems like liver failure will begin to show in the form of hyper pigmentation patches and a yellowish complexion (e.g. jaundice) respectively on the skin's epidermal layers [127]. All these and more might not be as obvious to the human eye as it can be to a computer. However, given the premise that dynamic face information is omnipresent in reality, humans might be able to verify/identify a face with a much shorter processing time than machines. This age-invariant FR with motion is, unfortunately, not tested in any recent psychophysics or behavioral study for our evaluation. The closest we can get to human studies in this area of work is partially demonstrated by Suo et al. [128] where they did a simple human study after synthesizing new faces across different ages using a dynamic face aging model with multi-resolution and multi-layer image representations.

Given that their experiment is computational in nature (refer to Fig. 17 as an illustration) for an outline of their pipeline, the purpose of the included human study was to validate that their dynamic face aging model approximates to human perception. They did so in two separate experiments: one required the subjects to give estimates for the age of the face seen in original images and those generated by their model. The other task required subjects to match synthesized images of aged

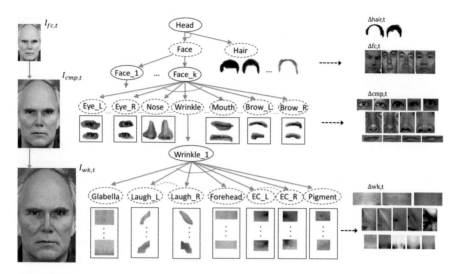

Fig. 17 Brief pipeline for dynamic face aging model

faces to their corresponding 'younger' face images. While subjects' estimates in the first task were consistently precise with the age of synthesized images, they did not perform as well in the second face-matching task. It was shown that identification performance decreases as the age difference between synthesized and real images increases.

Predicting how a person will look like in future as he/she ages is a hard task since everyone does so at different rates and that the signs of aging will differ from person to person. Moreover, using static images will be reducing FV to a picture-matching task that does not reflect the practical circumstances in which humans do so naturally—seeing faces in motion. It is plausible that certain facial features, given learning with motion, could serve as cues for invariant recognition across age (e.g. unique face surface deformations when smiling or frowning).

4.3 Transcending the Big Four: Evaluating Human Performance in Dynamic Perspective Invariant Face Recognition

We understand that prior work suggested that dynamic motion provides additional information, given an increase in number of frames to the identity of a face than static images, for an efficient FR system that allows human to navigate successfully in a social world. What remains unknown thus far is the type of additional information that can be gleaned from motion to support this inherent capability of mankind. In addition, the design of one or more FR tasks do not mimic the complex ones which we face in the natural world: recognizing a friend in a mall or an unfamiliar keynote speaker at a conference meeting when he/she is conversing with another (i.e. combination of changing view perspective with expressive motion), recognizing a relative whom you've not seen in years as he/she darts into a sheltered building on a sunny mid-afternoon (i.e. combination of view perspective illumination and age variations) etc. Hence, there exist the knowledge gap as to how humans tackle such challenging recognition circumstances (situations where several conditions are confounded) seamlessly and effortlessly.

With well-established work done with dynamic FR, we question if there exist a generic strategy for each type and/or combination of conditions employed by humans given unconstrained viewing of faces in motion. Some might reckon that machine FR performance could very well have surpassed that of human's [2], with a 97 % correction recognition rate by the standard LFW protocol. However, it entails a 3 % FAR (False Acceptance Rate) that is unsatisfactory for practical applications. Even at 0.1 % FAR, the algorithm cannot be implemented for large-scale recognition as discussed in Sect. 3.2.3, e.g. airport security which handles hundreds of thousands of people daily (large number of genuine and imposter matches) [6].

Therefore, we design our experiment to investigate the plausible facial features and eye-gaze strategies of previously unfamiliar faces to be learnt in dynamic motion for subsequent recognition tasks [129]. This psychophysical experiment to be conducted aims to obtain inspirations from highly-competent human subjects to determine if generic eye gaze scan path strategies, as well as crucial facial features, can be used to explain FV for the different realistic scenarios occurring around us on an everyday basis.

Subjects will be presented with pairs of dynamic face samples recorded in an array of different unconstrained settings and they will be asked to identify if the two faces belonged to the same person (i.e. FV). Key features from the tests can then be evaluated so that we may emulate the competence of the human recognition system to push the boundaries of machine FR.

5   Summary and Future Trends

Rigorous and huge amount of research efforts from diverse fields of studies like computer, cognitive and biological sciences, are aiming to tame the challenging problems of FR. As we have seen over the period of years, for constrained and well-conditioned limited cases, the field of FR has reached a certain level of maturity. However, a vast majority of unconstrained FR cases require further attention and new directions in their investigations. FR using videos is going to play a much bigger role in the years to come. As explained in Sect. 2.3.2, with hardware devices for computing, recording and storing the relevant data are becoming cheaper and more readily available, people will be able to perform their vision based tasks (such as FR) in video-to-video scenarios. The rich temporal information available in such modalities (captured under scenarios described in Sect. 2.1) makes it very appealing and attractive to many researchers. So it should be really exciting and challenging for researchers to find new methodologies for video based FR involving very high, large scale face voluminous data.

In recent times, DNN involved in deep learning architecture based methodologies using gigantic amount of training face images and hundreds of millions of parameters have shown surprisingly outstanding results. Results shown by DeepFace, DeepID and few others, are really impressive and they outperform most of the handcrafted features obtained using traditional machine learning approaches. However, their evaluations on large scale unconstrained FR problem as described in Sect. 3.2.3 are yet to be done. Moreover, their training images, architecture and learning frameworks are proprietary, thereby leaving very limited scopes for further research using large scale training images. One important factor that researchers can look into is how to develop DNN framework using lower number of training samples and how biologically inspired networks can be incorporated in DNN framework.

On the human FR aspect, researchers in cognitive science are moving away from how humans recognize still face images to recognition of faces in videos [130]. As explained in Sect. 4.1, the formulation of two hypotheses by O'Toole et al., is an attempt to explain the benefits that dynamic faces impart on human recognition of faces. The supplemental information hypothesis asserts FR depends on the representation of features and/or motion that is unique to an individual—his facial identity signature. Most experiments using familiar faces as stimuli fall under this category. Lander and Chuang showed that facial motion, specifically non-rigid motion, improves identification of faces when the dynamics are labeled 'distinctive', possess 'naturalness' (i.e. no artificial animations) in the motion and are viewed at naturalistic speeds.

On the other hand, the representation enhancement hypothesis posits that recognition performance of a novel face after a learning phase has a higher accuracy than that learnt from static faces. Lander and Bruce experiments have shown a heightened recognition performance after learning an unfamiliar face in motion. Although some argued that multiple static images of a face in different perspectives might be able to account for such learning advantage, studies by Pike et al. suggests that performance worsened when faces are learnt using a series of static images viewed in random order. These, and a few other works, showed that the dynamics of faces provide the viewer with a 3D structure that cannot be derived from multiple static views alone; hence making the study of faces in motion attractive.

The emerging directions discussed in Sect. 2.3 shed some light on how researchers are making fresh efforts in alleviating FR problems. One of the areas that has received attention is the biologically motivated approaches for FR. Understanding the invariance identity-preserving transformation theory may help to extract features that are invariant to certain transformations (may not be all). This would help to completely eradicate pre-processing stages in FR pipelines when processing raw images; thereby increasing their computational efficiencies and reducing the error rates at various stages.

Understanding how humans perform FR via behavioral studies can provide the first peek as to how the human brain gleans information from the external world in which we interact. This is the motivation driving our experiment: the first step to emulate naturalistic FR. However, the research sphere for FV is relatively nucleated as compared to FI; with the latter involving information association and retrieval from the human memory. As such, we propound the notion of a two-prong approach to investigate the human memory and human performance in FR to plausibly leapfrog the long standing hurdles of machine recognition of faces.

References

1. M. Turk, A. Pentland, Eigenfaces for recognition. J. Cogn. Neurosci. **3**(1), 71–86 (1991)
2. Y. Taigman, M. Yang, M. Ranzato, L. Wolf, Deepface: closing the gap to human-level performance in face verification, in *Proceedings of the IEEE Computer Society Conference on Computer Vision and Pattern Recognition*, Columbus, OH, 2014, pp. 1701–1708
3. Y. Sun, X. Wang, X. Tang, Deep learning face representation from predicting 10,000 classes, in *Proceedings of the IEEE Computer Society Conference on Computer Vision and Pattern Recognition*, Columbus, OH, 2014, pp. 1891–1898
4. P.J. Phillips, Face & Ocular Challenges. Presentation (2010), http://www.cse.nd.edu/BTAS_10/BTAS_Jonathon_Phillips_Sep_2010_FINAL.pdf
5. P. Grother, M. Ngan, Face Recognition Vendor Test (FRVT 2013) performance of face identification algorithms. Technical Report (2013), http://www.biometrics.nist.gov/cs_links/face/frvt/frvt2013/NIST_8009.pdf
6. S. Liao, Z. Lei, D. Yi, S.Z. Li, A benchmark study of large-scale unconstrained face recognition, in *IEEE International Joint Conference on Biometrics*, Clearwater, FL, 2014, pp. 1–8
7. A.J. O'Toole, D.A. Roark, H. Abdi, Recognizing moving faces: a psychological and neural synthesis. Trends Cogn. Sci. **6**(6), 261–266 (2002)
8. W.A. Bainbridge, P. Isola, A. Oliva, The intrinsic memorability of face photographs. J. Exp. Psychol. Gen. **4**(142), 1323–1334 (2013)
9. T.A. Busey, Formal models of familiarity and memorability in face recognition, in *Computational, Geometric, and Process Perspectives on Facial Cognition: Contexts and Challenges*, ed. by M.J. Wenger, J.T. Townsend (Lawrence Erlbaum Associates Publishers, Mahwah, 2001)
10. S. Georghiades, P.N. Belhumeur, D. Kriegman, From few to many: illumination cone models for face recognition under variable lighting and pose. IEEE Trans. Pattern Anal. Mach. Intell. **23**(6), 643–660 (2001)
11. L. Zhang, D. Samaras, Face recognition from a single training image under arbitrary unknown lighting using spherical harmonics. IEEE Trans. Pattern Anal. Mach. Intell. **28**(3), 351–363 (2006)
12. S. Vural, Y. Mae, H. Uvet, T. Arai, Illumination normalization for outdoor face recognition by using ayofa-filters. J. Pattern Recognit. Res. **6**(1), 1–18 (2011)
13. X. Zhao, S.K. Shah, I.A. Kakadiaris, Illumination alignment using lighting ratio: application to 3D-2D face recognition, in *Proceedings of International Conference on Automatic Face Gesture Recognition*, Shanghai, 2013, pp. 1–6
14. L. Wolf, T. Hassner, I. Maoz, Face recognition in unconstrained video with matched background similarity, in *Proceedings of the IEEE Computer Society Conference on Computer Vision and Pattern Recognition*, Colorado Springs, 2011, pp. 529–534
15. P.J. Phillips, H. Moon, S. Rizvi, P. Rauss, The FERET evaluation methodology for face recognition algorithms. IEEE Trans. Pattern Anal. Mach. Intell. **22**(10), 1090–1104 (2000)
16. The Face Recognition Technology (FERET) Normalization (2005), http://www.cs.colostate.edu/evalfacerec/data/normalization.html

17. C. Ding, D. Tao, A comprehensive survey on pose-invariant face recognition. CoRR abs/1502.04383 (2015), http://www.arxiv.org/abs/1502.04383
18. X. Zhang, Y. Gao, Face recognition across pose: a review. Pattern Recogn. **42**, 2876–2896 (2009)
19. A.M. Martinez, Recognizing imprecisely localized, partially occluded, and expression variant faces from a single sample per class. IEEE Trans. Pattern Anal. Mach. Intell. **24**(6), 748–763 (2002)
20. B. Mandal, X.D. Jiang, A. Kot, Verification of human faces using predicted eigenvalues, in *19th International Conference on Pattern Recognition*, Tempa, FL, 2008, pp. 1–4
21. J. Leibo, Q. Liao, T. Poggio, Subtasks of unconstrained face recognition, in *International Joint Conference on Computer Vision, Imaging and Computer Graphics*, Lisbon, vol. 2, 2014, pp. 113–121
22. P.J. Phillips, P.J. Flynn, T. Scruggs, K.W. Bowyer, J. Chang, K. Hoffman, J. Marques, J. Min, W. Worek, Overview of the face recognition grand challenge, in *Proceedings of the IEEE Computer Society Conference on Computer Vision and Pattern Recognition*, San Diego, CA, 2005, pp. 947–954
23. B. Mandal, W. Zhikai, L. Li, A. Kassim, Evaluation of descriptors and distance measures on benchmarks and first-person-view videos for face identification, in *International Workshop on Robust Local Descriptors for Computer Vision*, Singapore, 2014, pp. 585–599
24. X.D. Jiang, B. Mandal, A. Kot, Eigenfeature regularization and extraction in face recognition. IEEE Trans. Pattern Anal. Mach. Intell. **30**(3), 383–394 (2008)
25. M. Kawulok, J. Szymanek, Precise multi-level face detector for advanced analysis of facial images. IET Image Process. **6**(2), 95–103 (2012)
26. C. Zhang, Z. Zhang, A survey of recent advances in face detection. Technical Report MSR-TR-2010-66, http://www.research.microsoft.com/pubs/132077/facedetsurvey.pdf
27. B. Mandal, S. Ching, L. Li, V. Chandrasekha, C. Tan, J.-H. Lim, A wearable face recognition system on Google glass for assisting social interactions, in *Third International Workshop on Intelligent Mobile and Egocentric Vision*, Singapore, 2014, pp. 419–433
28. W. Zhao, R. Chellappa, P.J. Phillips, A. Rosenfeld, Face recognition: a literature survey. ACM Comput. Surv. **35**(4), 399–458 (2003)
29. B. Klare, A. Jain, On a taxonomy of facial features, in *Proceedings of International Conference on Biometrics: Theory, Applications and Systems* (2010), pp. 1–8
30. H. Zhou, A. Mian, L. Wei, D. Creighton, M. Hossny, S. Nahavandi, Recent advances on singlemodal and multimodal face recognition: a survey. IEEE Trans. Hum. Mach. Syst. **44**(6), 701–716 (2014)
31. P. Belhumeur, Ongoing challenges in face recognition, in *Frontiers of Engineering: Reports on Leading-Edge Engineering* (2006), pp. 5–14
32. X. Zou, J. Kittler, K. Messer, Illumination invariant face recognition: a survey, in *Proceedings of International Conference on Biometrics: Theory, Applications, and Systems* (2007), pp. 1–8
33. Y. Fu, G. Guo, T.S. Huang, Age synthesis and estimation via faces: a survey. IEEE Trans. Pattern Anal. Mach. Intell. **32**(11), 1955–1976 (2010)
34. A. Jain, B. Klare, U. Park, Face recognition: some challenges in forensics, in *Proceedings of International Conference on Automatic Face Gesture Recognition and Workshops* (2011), pp. 726–733
35. G. Huang, M. Ramesh, T. Berg, E.L. Miller, Labeled faces in the wild: a database for studying face recognition in unconstrained environments. Technical Report 07-49 (University of Massachusetts, Amherst, 2007)
36. T. Poggio, J. Mutch, F. Anselmi, J. Leibo, L. Rosasco, A. Tacchetti, The computational magic of the ventral stream: sketch of a theory (and why some deep architectures work). MIT-CSAIL-TR-2012-035
37. Q. Liao, J. Leibo, T. Poggio, Learning invariant representations and applications to face verification, in *Neural Information Processing Systems Foundation, Inc.*, Harrahs and Harveys, Lake Tahoe, USA (2013), pp. 1–9
38. S.Z. Li, A.K. Jain (eds.), *Handbook of Face Recognition*, 2nd edn. (Springer, Berlin, 2011)

39. J. Barr, K. Bowyer, P. Flynn, S. Biswas, Face recognition from video: a review. Int. J. Pattern Recognit. Artif. Intell. **26**(5), (2012), DOI: 10.1142/S0218001412660024
40. F. Anselmi, J. Leibo, L. Rosasco, J. Mutch, A. Tacchetti, T. Poggio, Unsupervised learning of invariant representations in hierarchical, architectures. arXiv preprint arXiv:1311.4158 (2013)
41. M. Riesenhuber, T. Poggio, Hierarchical models of object recognition in cortex. Nat. Neurosci. **2**(11), 1019–1025 (1999)
42. T. Serre, L. Wolf, S. Bileschi, M. Riesenhuber, T. Poggio, Robust object recognition with cortex-like mechanisms. IEEE Trans. Pattern Anal. Mach. Intell. **29**(3), 411–426 (2007)
43. Y. LeCun, Y. Bengio (eds.), Convolutional networks for images, speech, and time series, in *The Handbook of Brain Theory and Neural Networks*, ACM Digital Library (1995)
44. Q. Liao, J.Z. Leibo, Y. Mroueh, T. Poggio, Can a biologically-plausible hierarchy effectively replace face detection, alignment and recognition pipelines? arXiv:1311.4082v3 [cs.CV], no. 003 (2013)
45. E. Hubel, T. Wiesel, Receptive fields, binocular interaction and functional architecture in the cat's visual cortex. J. Physiol. **160**(1), 106 (1962)
46. Labeled faces in the wild (LFW) results (2015), http://www.vis-www.cs.umass.edu/lfw/results.html
47. L. Rowden, B. Klare, J. Klontz, A.K. Jain, Video-to-video face matching: establishing a baseline for unconstrained face recognition, in *IEEE 6th International Conference on Biometrics: Theory, Applications and Systems*, Washington, DC, 2013, pp. 1–8
48. Instagram, Online social network through images (2015), https://www.instagram.com/
49. Facebook, Online social network (2015), https://www.facebook.com/
50. A. Ishai, L.G. Ungerleider, A. Martin, J.L. Schouten, J.V. Haxby, Distributed representations of objects in the human ventral visual pathway. Proc. Natl. Acad. Sci. **96**, 9379–9384 (1999)
51. K. Lander, L. Chuang, Why are moving faces easier to recognize? Vis. Cogn. **12**(3), 429–442 (2005)
52. W.A. Bainbridge, P. Isola, I. Blank, A. Oliva, Establishing a database for studying human face photograph memory, in *Proceedings of the 34th Annual Conference of the Cognitive Science Society*, Austin, TX, 2012, pp. 1302–1307
53. Face Recognition Homepage Databases (2015), http://www.face-rec.org/databases/
54. U. Park, A.K. Jain, A. Ross, Face recognition in video: adaptive fusion of multiple matchers, in *IEEE Computer Workshop on Biometrics*, Minneapolis, 2007, pp. 1–8
55. H. Li, G. Hua, X. Shen, Z. Lin, J. Brandt, Eigen-PEP for video face recognition, in *Asian Conference on Computer Vision*, Singapore, 2014, pp. 17–33
56. Y. Chen, V. Patel, S. Shekhar, R. Chellappa, P. Phillips, Video-based face recognition via joint sparse representation, in *10th IEEE International Conference and Workshops on Automatic Face and Gesture Recognition*, Shanghai, 2013, pp. 1–8
57. R. Chellappa, J. Ni, V. M. Patel, Remote identification of faces: problems, prospects, and progress. Pattern Recogn. Lett. **33**(15), 1849–1859 (2012)
58. B. Mandal, X.D. Jiang, A. Kot, Dimensionality reduction in subspace face recognition, in *Sixth IEEE International Conference on Information, Communications & Signal Processing*, Singapore, 2007, pp. 1–5
59. X.D. Jiang, B. Mandal, A. Kot, Complete discriminant evaluation and feature extraction in kernel space for face recognition. Mach. Vis. Appl. (Springer) **20**(1), 35–46 (2009)
60. B. Mandal, H. Eng, Regularized discriminant analysis for holistic human activity recognition. IEEE Intell. Syst. **27**(1), 21–31 (2012)
61. X.D. Jiang, B. Mandal, A. Kot, Enhanced maximum likelihood face recognition. IEE Electron. Lett. **42**(19), 1089–1090 (2006)
62. Z. Zhu, P. Luo, X. Wang, X. Tang, Deep learning identity preserving face space, in *International Conference on Computer Vision*, Washington, DC, 2013, pp. 113–120
63. Y. Sun, X. Wang, X. Tang, Hybrid deep learning for face verification, in *IEEE International Conference on Computer Vision*, Sydney, 2013, pp. 1489–1496
64. T. Ahonen, A. Hadid, M. Pietikainen, Face description with local binary patterns: application to face recognition. IEEE Trans. Pattern Anal. Mach. Intell. **28**(12), 2037–2041 (2006)

65. Face Recognition Technology (FERET) (1996), http://www.nist.gov/itl/iad/ig/feret.cfm
66. Face Recognition Vendor Test (FRVT) (2015), http://www.nist.gov/itl/iad/ig/frvt-home.cfm
67. Face Recognition Grand Challenge (FRGC) (2005), http://www.nist.gov/itl/iad/ig/frgc.cfm
68. Multiple Biometric Grand Challenge (MBGC) (2009), http://www.nist.gov/itl/iad/ig/mbgc.cfm
69. Face and Ocular Challenge Series (FOCS): Good, Bad and the Ugly Database (2015), http://www.nist.gov/itl/iad/ig/focs.cfm
70. P. Phillips, J. Beveridge, B. Draper, G. Givens, A. O'Toole, D. Bolme, J. Dunlop, Y.M. Lui, H. Sahibzada, S. Weimer, An introduction to the good, the bad and the ugly face recognition challenge problem, in *IEEE International Conference on Automatic Face Gesture Recognition and Workshops* (2011), pp. 346–353
71. Multiple Biometrics Evaluation (MBE) (2010), http://www.nist.gov/itl/iad/ig/mbe.cfm
72. Competition on Face Recognition in Mobile Environment (MOBIO) (2013), https://www.idiap.ch/dataset/mobio
73. Point and Shoot Face Recognition Challenge (PaSC) (2015), http://www.nist.gov/itl/iad/ig/pasc.cfm
74. X. Wang, X. Tang, Face photo-sketch synthesis and recognition. IEEE Trans. Pattern Anal. Mach. Intell. **31**(11), 1955–1967 (2009)
75. W. Zhang, X. Wang, X. Tang, Coupled information-theoretic encoding for face photo-sketch recognition, in *Proceedings of the IEEE Computer Society Conference on Computer Vision and Pattern Recognition*, Colorado Springs, 2011, pp. 513–520
76. P. Grother, G.W. Quinn, P.J. Phillips, MBE 2010: report on the evaluation of 2D still-image face recognition algorithms. National Institute of Standards and Technology, NISTIR 7709
77. P.J. Phillips, W.T. Scruggs, A.J. OToole, P.J. Flynn, K.W. Bowyer, C.L. Schott, M. Sharpe, FRVT 2006 and ICE 2006 large-scale results. IEEE Trans. Pattern Anal. Mach. Intell. **32**(5), 831–846 (2010)
78. Face Recognition Vendor Test (FRVT2006) (2006), http://www.nist.gov/itl/iad/ig/frvt-2006.cfm
79. A. O'Toole, P. Phillips, X. An, J. Dunlop, Demographic effects on estimates of automatic face recognition performance, in *IEEE International Conference on Automatic Face Gesture Recognition and Workshops*, Santa Barbara, CA 2011, pp. 83–90
80. C. McCool, S. Marcel, A. Hadid, M. Pietikainen, P. Matejka, J. Cernocky, N. Poh, J. Kittler, A. Larcher, C. Levy, D. Matrouf, J.-F. Bonastre, P. Tresadern, T. Cootes, Bi-modal person recognition on a mobile phone: using mobile phone data, in *IEEE International Conference on Multimedia and Expo Workshops*, 2012, pp. 635–640
81. M. Gunther et al., Face recognition evaluation in mobile environment, in *International Conference on Biometrics*, Madrid, 2013, pp. 1–7
82. N. Pinto, D. Cox, Beyond simple features: a large-scale feature search approach to unconstrained face recognition, in *IEEE Automatic Face and Gesture Recognition*, Santa Barbara, CA, 2011, pp. 8–15
83. F. Samaria, A. Harter, Parameterization of a stochastic model for human face identification, in *Proceedings of 2nd IEEE Workshop on Applications of Computer Vision*, Sarasota, FL, 1994, pp. 138–142
84. Face Recognition Databases (2015), http://www.face-rec.org/databases/
85. P. Viola, M. Jones, Rapid object detection using a boosted cascade of simple features, in *Proceedings of the IEEE Computer Society Conference on Computer Vision and Pattern Recognition*, vol. 1 (2001), pp. 511–518
86. Labeled Faces in the Wild (LFW) (2015), http://www.vis-www.cs.umass.edu/lfw/
87. G. Huang, V. Jain, Unsupervised joint alignment of complex images, in *IEEE International Conference on Computer Vision*, Rio de Janeiro, 2007, pp. 1–8
88. G. Huang, M. Mattar, H. Lee, E.G. Learned-Miller, Learning to align from scratch, in *Advances in Neural Information Processing Systems*, vol. 25 (2012), pp. 764–772
89. D. Chen, X. Cao, L. Wang, F. Wen, J. Sun, Bayesian face revisited: a joint formulation, in *European Conference on Computer Vision*, Florence, 2012, pp. 566–579

90. M. Everingham, J. Sivic, A. Zisserman, Taking the bite out of automated naming of characters in tv video. Image Vis. Comput. **27**(5), 545–559 (2009)
91. D. Ramanan, S. Baker, S. Kakade, Leveraging archival video for building face datasets, in *IEEE International Conference on Computer Vision* (2007), pp. 1–8
92. A. Dantcheva, C. Chen, A. Ross, Can facial cosmetics affect the matching accuracy of face recognition systems? in *2012 IEEE 5th International Conference on Biometrics: Theory, Applications and Systems (BTAS)* (2012), pp. 391–398
93. C. Chen, A. Dantcheva, A. Ross, Automatic facial makeup detection with application in face recognition, in *International Conference on Biometrics* (IEEE, Madrid, 2013), pp. 1–8
94. S. Setty et al., Indian movie face database: a benchmark for face recognition under wide variations, in *National Conference on Computer Vision, Pattern Recognition, Image Processing and Graphics*, Jodhpur, 2013, pp. 726–733
95. M. Demirkus, J.J. Clark, T. Arbel, Robust semi-automatic head pose labeling for real-world face video sequences. Multimedia Tools Appl., **70**(1), 495–523 (2014)
96. M.K. Hasan, C.J. Pal, Experiments on visual information extraction with the faces of wikipedia, in *Proceedings of the 28th AAAI Conference on Artificial Intelligence*, 27–31 July 2014, Québec City, QC, 2014, pp. 51–58
97. H.-W. Ng, S. Winkler, A data-driven approach to cleaning large face datasets, in *2014 IEEE International Conference on Image Processing (ICIP)*, 2014, pp. 343–347
98. M.C. Frank, E. Vul, S.P. Johnson, Development of infants' attention to faces during the first year. Cognition **110**(2), 160–170 (2009)
99. R.L. Buckner, The serendipitous discovery of the brain's default network. NeuroImage **62**(2), 1137–1145 (2012)
100. B. Mandal, X. Jiang, H. Eng, A. Kot, Prediction of eigenvalues and regularization of eigenfeatures for human face verification. Pattern Recogn. Lett. **31**(8), 717–724 (2010)
101. K. Lander, L. Chuang, L. Wickham, Recognizing face identity from natural and morphed smiles. Q. J. Exp. Psychol. **59**(5), 801–808 (2006)
102. A. O'Toole, D. Roark, *Dynamic Faces: Memory for Moving Faces* (The MIT Press, Cambridge, 2011)
103. K.S. Pilz, I.M. Thornton, H.H. Bülthoff, A search advantage for faces learned in motion. Exp. Brain Res. **171**(4), 436–447 (2005)
104. M.J. Tarr, A.S. Georghiades, C.D. Jackson, Identifying faces across variations in lighting: psychophysics and computation. ACM Trans. Appl. Percept. **5**(2), 10:1–10:25 (2008)
105. J.V. Haxby, E.A. Hoffman, M.I. Gobbini, The distributed human neural system for face perception. Trends Cogn. Sci. **4**(6), 223–233 (2000)
106. P. Sinha, *Qualitative Representations for Recognition* (Springer, Berlin, 2002), pp. 249–262
107. G. Wallis, B.T. Backus, M. Langer, Learning illumination- and orientation-invariant representations of objects through temporal associations. J. Vis. **9**(7), 1–8 (2009)
108. The Science of Why No One Agrees on the Color of This Dress (2015), http://www.wired.com/2015/02/science-one-agrees-color-dress
109. D. Purves, *Brains: How They Seem to Work* (Pearson, Financial Times Press, New York, 2010)
110. D. Purves, R. Cabeza, S.A. Huettel, K.S. LaBar, M.L. Platt, M.G. Woldorff, *Principles of Cognitive Neuroscience*, 2nd edn., Sunderland, MA 01375-0407, USA, (Sinauer Associates, 2012)
111. R. Blake, M. Shiffrar, Perception of human motion. Annu. Rev. Psychol. **58**, 47–73 (2007)
112. M. Kamachi, V. Bruce, S. Mukaida, J. Gyoba, S. Yoshikawa, S. Akamatsu, Dynamic properties influence the perception of facial expressions. Perception **30**(7), 875–887 (2001)
113. D. Pitcher, D.D. Dilks, R.R. Saxe, C. Triantafyllou, N. Kanwisher, Differential selectivity for dynamic versus static information in face-selective cortical regions. NeuroImage **56**(4), 2356–2363 (2011)
114. J. Schultz, M. Brockhaus, H.H. Bülthoff, K.S. Pilz, What the human brain likes about facial motion. Cereb. Cortex **23**, 1167–1178 (2012)
115. C.P. Said, J.V. Haxby, A. Todorov, Brain systems for assessing the affective value of faces. Philos. Trans. R. Soc. B **366**, 1660–1670 (2011)

116. M.A. Giese, T. Poggio, Neural mechanisms for the recognition of biological movements. Nat. Rev. Neurosci. **4**, 179–192 (2003)
117. S. Rigby, B. Stoesz, L. Jakobson, How dynamic facial cues, stimulus orientation and processing biases influence identity and expression interference. J. Vis. **13**(9), 413–418 (2013)
118. X.D. Jiang, B. Mandal, A. Kot, Face recognition based on discriminant evaluation in the whole space, in *IEEE 32nd International Conference on Acoustics, Speech and Signal Processing*, Honolulu, Hawaii, 2007, pp. 245–248
119. J.S. Winston, R. Henson, M.R. Fine-Goulden, R.J. Dolan, fMRI-adaptation reveals dissociable neural representations of identity and expression in face perception. J. Neurophysiol. **92**(3), 1830–1839 (2004)
120. S. Bentin, J.M. DeGutis, M. D'Esposito, L.C. Robertson, Too many trees to see the forest: Performance, event-related potential, and functional magnetic resonance imaging manifestations of integrative congenital prosopagnosia. J. Cogn. Neurosci. **19**(1), 132–146 (2007)
121. B.C. Duchaine, H. Parker, K. Nakayama, Normal recognition of emotion in a prosopagnosic. Perception **32**, 827–838 (2003)
122. N.G. Xiao, P.C. Quinn, L. Ge, K. Lee, Elastic facial movement influences part-based but not holistic processing. J. Exp. Psychol. Hum. Percept. Perform. **39**(5), 1457–1467 (2013)
123. M.T. Posamentier, H. Abdi, Processing faces and facial expressions. Neuropsychol. Rev. **13**(3), 113–143 (2003)
124. W.C.D. Souza, S. Eifuku, R. Tamura, H. Nishijo, T. Ono, Differential characteristics of face neuron responses within the anterior superior temporal sulcus of macaques. J. Neurophysiol. **94**, 1252–1266 (2005)
125. U. Park, Y. Tong, A.K. Jain, Age-invariant face recognition. IEEE Trans. Pattern Anal. Mach. Intell. **32**(5), 947–954 (2010)
126. B. Mandal, X.D. Jiang, A. Kot, Multi-scale feature extraction for face recognition, in *IEEE International Conference on Industrial Electronics and Applications*, Singapore, 2006, pp. 1–6
127. T. Igarashi, K. Nishino, S.K. Nayar, The appearance of human skin. Technical Report, Columbia University, New York (2005)
128. J. Suo, F. Min, S. Zhu, S. Shan, X. Chen, A multi-resolution dynamic model for face aging simulation, in *IEEE Computer Vision and Pattern Recognition*, Minneapolis, MN, 2007, pp. 1–8
129. R. Lim, M.R. Sayed, B. Mandal, K.T. Ma, L. Li, J.H. Lim, Evaluating human performance in dynamic perspective invariant face recognition (accepted), in *11th Asia-Pacific Conference on Vision*, Singapore, 2015
130. G. Kreiman, Kreiman's lab (2015), http://www.klab.tch.harvard.edu/publications/publications.html
131. K. Patterson, A.D. Baddeley, When face recognition fails. J. Exp. Psychol. Hum. Learn. Mem. **3**(4), 406–417 (1977)

Labeled Faces in the Wild: A Survey

Erik Learned-Miller, Gary B. Huang, Aruni RoyChowdhury, Haoxiang Li, and Gang Hua

Abstract In 2007, Labeled Faces in the Wild was released in an effort to spur research in face recognition, specifically for the problem of face verification with unconstrained images. Since that time, more than 50 papers have been published that improve upon this benchmark in some respect. A remarkably wide variety of innovative methods have been developed to overcome the challenges presented in this database. As performance on some aspects of the benchmark approaches 100 % accuracy, it seems appropriate to review this progress, derive what general principles we can from these works, and identify key future challenges in face recognition. In this survey, we review the contributions to LFW for which the authors have provided results to the curators (results found on the LFW results web page). We also review the cross cutting topic of alignment and how it is used in various methods. We end with a brief discussion of recent databases designed to challenge the next generation of face recognition algorithms.

1 Introduction

Face recognition is a core problem and popular research topic in computer vision for several reasons. First, it is easy and natural to formulate well-posed problems, since individuals come with their own label, their name. Second, despite its well-posed nature, it is a striking example of *fine-grained classification*—the variation of two images within a class (images of a single person) can often exceed the variation between images of different classes (images of two different people). Yet human observers have a remarkably easy time ignoring nuisance variables such as pose and

E. Learned-Miller (✉) • A. RoyChowdhury
University of Massachusetts, Amherst, MA, USA
e-mail: elm@cs.umass.edu; arunirc@cs.umass.edu

G.B. Huang
Howard Hughes Medical Institute, Janelia Research Campus, Ashburn, VA, USA
e-mail: gbhuang@cs.umass.edu

H. Li • G. Hua
Stevens Institute of Technology, Hoboken, NJ, USA
e-mail: hli18@stevens.edu; ganghua@gmail.com

© Springer International Publishing Switzerland 2016   189
M. Kawulok et al. (eds.), *Advances in Face Detection and Facial Image Analysis*,
DOI 10.1007/978-3-319-25958-1_8

expression and focusing on the features that matter for identification. Finally, face recognition is of tremendous societal importance. In addition to the basic ability to identify, the ability of people to assess the emotional state, the focus of attention, and the intent of others are critical capabilities for successful social interactions. For all these reasons, face recognition has become an area of intense focus for the vision community.

This chapter reviews research progress on a specific face database, Labeled Faces in the Wild (LFW), that was introduced to stimulate research in face recognition for images taken in common, everyday settings. In the remainder of the introduction, we review some basic face recognition terminology, provide the historical setting in which this database was introduced, and enumerate some of the specific motivations for introducing the database. In Sect. 2, we discuss the papers for which the curators have been provided with results. We group these papers by the protocols for which they have reported results. In Sect. 3, we discuss alignment as a cross-cutting issue that affects almost all of the methods included in this survey. We conclude by discussing future directions of face recognition research, including new databases and new paradigms designed to push face recognition to the next level.

1.1 Verification and Identification

In this chapter, we will refer to two widely used paradigms of face recognition: *identification* and *verification*. In identification, information about a specific set of individuals to be recognized (the *gallery*) is gathered. At test time, a new image or group of images is presented (the *probe*). The task of the system is to decide which of the gallery identities, if any, is represented by the probe. If the system is guaranteed that the probe is indeed one of the gallery identities, this is known as *closed set* identification. Otherwise, it is *open set* identification, and the system is expected to identify when an image does not belong to the gallery.

In contrast, the problem of *verification* is to analyze two face images and decide whether they represent the *same* person or two *different* people. It is usually assumed that neither of the photos shows a person from any previous training set.

Many of the early face recognition databases and protocols focused on the problem of identification. As discussed below, the difficulty of the identification problem was so great that researchers were motivated to simplify the problem by controlling the number of image parameters that were allowed to vary simultaneously. One of the salient aspects of LFW is that it focused on the problem of verification exclusively, although it was certainly not the first to do so.[1] While the use of the images in LFW originally grew out of a motivation to study learning from one

---

[1] Other well-known benchmarks had previously used verification. See, for example, this benchmark [80].

example and fine-grained recognition, a side effect was to render the problem of face recognition in real-world settings significantly easier—easier enough to attract the attention of a wide range of researchers.

1.2 Background

In the early days of face recognition by computer, the problem was so daunting that it was logical to consider a divide-and-conquer approach. What is the best way to handle recognition in the presence of lighting variation? Pose variation? Occlusions? Expression variation? Databases were built to consider each of these issues using carefully controlled images and experiments.[2] One of the most comprehensive efforts in this direction is the CMU Multi-PIE[3] database, which systematically varies multiple parameters over an enormous database of more than 750,000 images [38].

Studying individual sources of variation in images has led to some intriguing insights. For example, in their efforts to characterize the structure of the space of images of an object under different lighting conditions, Belhumeur and Kriegman [15] showed that the space of faces under different lighting conditions (with other factors such as expression and pose held constant) forms a convex cone. They propose doing lighting invariant recognition by examining the distance of an image to the convex cones defined for each individual.

Despite the development of methods that could successfully recognize faces in databases with well-controlled variation, there was still a gap in the early 2000s between the performance of face recognition on these controlled databases and results on real face recognition tasks, for at least two reasons:

- Even with two methods, call them *A* and *B*, that can successfully model two types of variation separately, it is not always clear how to combine these methods to produce a method that can address both sources of variation. For example, a method that can handle significant occlusions may rely on the precise registration of two face images for the parts that are not occluded. This might render the method ineffective for faces that exhibit both occlusions and pose changes. As another example, the method cited above to handle lighting variations [15] relies on all of the other parameters of variation being fixed.
- There is a significant difference between handling *controlled* variations of a parameter, and handling *random* or *arbitrary* values of a parameter. For example, a method that can address five specific poses may not generalize well to arbitrary poses. Many previously existing databases studied fixed variations of parameters such as pose, lighting, and decorations. While useful, this does not guarantee the

---

[2]For a list of databases that were compiled before LFW, see the original LFW technical report [49].
[3]The abbreviation PIE stands for Pose, Illumination, and Expression.

handling of more general cases of these parameters. Furthermore, there are too many sources of variation to effectively cover the set of possible observations in a controlled database. Some databases, such as the ones used in the 2005 Face Recognition Grand Challenge [77], used certain "uncontrolled settings" such as an office, a hallway, or outdoor environments. However, the fact that these databases were built manually (rather than mining previously existing photos) naturally limited the number of settings that could be included. Hence, while the settings were uncontrolled in that they were not carefully specified, they were drawn from a small fixed set of empirical settings that were available to the database curators. Algorithms tuned for such evaluations are not required to deal with a large amount of previously unseen variability.

In 2006, while results on some databases were saturating, there was still poor performance on problems with real-world variation.

1.3 Variations on Traditional Supervised Learning and the Relationship to Face Recognition

In parallel to the work in the early 2000s on face identification, there was a growing interest in the machine learning community in variations of the standard supervised learning problem with large training sets. These variations included:

- learning from small training sets [35, 68],
- transfer learning—that is, sharing parameters from certain classes or distributions to other classes or distributions that may have less training data available [76], and
- semi-supervised learning, in which some training examples have no associated labels (e.g. [73]).

Several researchers chose face verification as a domain in which to study these new issues [21, 30, 36]. In particular, since face verification is about deciding whether two face images match (without any previous examples of those identities), it can be viewed as an instance of learning from a single training example. That is, letting the two images presented be I and J, I can be viewed as a single training example for the identity of a particular person. Then the problem can be framed as a binary classification problem in which the goal is to decide whether image J is in the same class as image I or not.

In addition, face verification is an ideal domain for the investigation of transfer learning, since learning the forms of variation for one person is important information that can be transferred to the understanding of how images of another person vary.

One interesting paper in this vein was the work of Chopra et al. from CVPR 2005 [30]. In this paper, a convolutional neural network (CNN) was used to learn a metric between face images. The authors specifically discuss the structure of the face recognition problem as a problem with a large number of classes and small numbers of training examples per class. In this work, the authors reported results on the relatively difficult AR database [66]. This paper was a harbinger of the recent highly successful application of CNNs to face verification.

1.3.1 Faces in the Wild and Labeled Faces in the Wild

Continuing the work on fine-grained recognition and recognition from a small number of examples, Ferencz et al. [36, 57] developed a method in 2005 for deciding whether two images represented the same object. They presented this work on data sets of cars and faces, and hence were also addressing the face verification problem. To make the problem challenging for faces, they used a set of news photos collected as part of the Berkeley "Faces in the Wild" project [18, 19] started by Tamara Berg and David Forsyth. These were news photos taken from typical news articles, representing people in a wide variety of settings, poses, expressions, and lighting. These photos proved to be very popular for research, but they were not suited to be a face recognition benchmark since (a) the images were only noisily labeled (more than 10 % were labeled incorrectly), and (b) there were large numbers of duplicates. Eventually, there was enough demand that the data were relabeled by hand, duplicates were removed, and protocols for use were written. The data were released as "Labeled Faces in the Wild" in conjunction with the original LFW technical report [49].

There were several goals behind the introduction of LFW. These included

- stimulating research on face recognition in unconstrained images;
- providing an easy-to-use database, with low barriers to entry, easy browsing, and multiple parallel versions to lower pre-processing burdens;
- providing consistent and precise protocols for the use of the database to encourage fair and meaningful comparisons;
- curating results to allow easy comparison, and easy replication of results in new research papers.

In the following section, we take a detailed look at many of the papers that have been published using LFW. We do not review all of the papers. Rather we review papers for which the authors have provided results to the curators, and which are documented on the LFW results web page.[4] We now turn to describing results published on the LFW benchmark.

---

[4]http://vis-www.cs.umass.edu/lfw/results.html.

2 Algorithms and Methods

In this section, we discuss the progression of results on LFW from the time of its release until the present. LFW comes with specific sets of image pairs that can be used in training. These pairs are labeled as "same" or "different" depending upon whether the images are of the same person. The specification of exactly how these training pairs are used is described by various protocols.

2.1 The LFW Protocols

Originally, there were two distinct protocols described for LFW, the *image-restricted* and the *unrestricted* protocols. The unrestricted protocol allows the creation of additional training pairs by combining other pairs in certain ways. (For details, see the original LFW technical report [49].)

As many researchers started using additional training data from outside LFW to improve performance, new protocols were developed to maintain fair comparisons among methods. These protocols were described in a second technical report [47].

The current six protocols are:

1. Unsupervised.
2. Image-restricted with no outside data.
3. Unrestricted with no outside data.
4. Image-restricted with label-free outside data.
5. Unrestricted with label-free outside data.
6. Unrestricted with labeled outside data.

In order to make comparisons more meaningful, we discuss the various protocols in three groups.

In particular, we start with the two protocols allowing no outside data. We then discuss protocols that allow outside data not related to identity, and then outside data with identity labels. We do not address the unsupervised protocol in this review.

2.1.1 Why Study Restricted Data Protocols?

Before starting on this task, it is worth asking the following question: Why might one wish to study methods that do not use outside data when their performance is clearly inferior to those that do use additional data? There are several possible answers to this question.

Utility of methods for other tasks. One reason to consider methods which use limited training data is that they can be used in other settings in which training data are limited. That is, it may be the case that in recognition problems other than face recognition, there may not be available the hundreds of thousands or millions of images that are available to train face recognizers. Thus, a method that uses less training data is more transportable to other domains.

Statistical efficiency versus asymptotic optimality. It has been known since the mid-seventies [87] that many methods, such as K-nearest neighbors (K-NN), continue to increase in accuracy with increasing training data until they reach optimal performance (also known as the *Bayes error rate*). In other words, if one only cares about accuracy with unlimited training data and unlimited computation time, there is no method better than K-NN.

Thus, we know not only that many methods will continue to improve as more training data is added, but that many methods, including some of the simplest methods, will achieve optimal performance. This makes the question of *statistical efficiency* a primary one. The question is not *whether* we can achieve optimal accuracy (the Bayes error rate), but rather, how fast (in terms of training set size) we get there. Of course, a closely related question is which method performs best with a fixed training set size.

At the same time, using equivalent data sets for training removes the question that plagues papers trained on huge, proprietary data sets: how much of their performance is due to algorithmic innovation, and how much is simply due to the specifics of the training data?

Despite our interest in fixed training set protocols, at the same time, the practical issues of how to collect large data sets, and find methods that can benefit from them the most, make it interesting to push performance as high as possible with no ceiling on the data set size. The protocols of LFW consider all of these questions.

Human learning and statistical efficiency. Closely related to the previous point is to note that humans solve many problems with very limited training data. While some argue that there is no particular need to mimic the way that humans solve problems, it is certainly interesting to try to discover the *principles* which allow them to learn from small numbers of examples. It seems likely that these principles will improve our ability to design efficient learning algorithms.

2.1.2 Order of Discussion

Within each protocol, we primarily discuss algorithms in the order with which we received the results. Note that this order does not always correspond to the official publication order.[5] We make every effort to remark on the first authors to use a particular technique, and also to refer to prior work in other areas or on other databases that may have used similar techniques previously. We apologize for any oversights in advance. Note that some methods, especially some of the commercial

---

[5]Some authors have sent results to the curators before papers have been accepted at peer-reviewed venues. In these cases, as described on the LFW web pages, we highlight the result in our results table in red, indicating that it has not yet been published at a peer-reviewed venue. In most cases, the status of such results are updated once the work has been accepted at a peer-reviewed venue. However, we maintain the original order in which we received the results.

Table 1 This table summarizes the new LFW protocols

Allowed information → / Protocol ↓	Same/different labels for LFW training pairs allowed?	Identity info for LFW training images allowed?	Annotations for LFW training data allowed?	Non-LFW images allowed?	Non-LFW annotations allowed?	Same/different labels for non-LFW pairs allowed?	Identity info for non-LFW images allowed?
Unsupervised	No	No	Yes	Yes	Yes	No	No
Image-restricted, no outside data	Yes	No	No	No	No	No	No
Unrestricted, no outside data	Yes	Yes	No	No	No	No	No
Image-restricted, label-free outside data	Yes	No	Yes	Yes	Yes	No	No
Unrestricted, label-free outside data	Yes	Yes	Yes	Yes	Yes	No	No
Unrestricted, labeled outside data	Yes	Yes	Yes	Yes	Yes	Yes	Yes

There are six protocols altogether, shown in the left column. The allowability for each category of data is shown to the right. The second LFW technical report gives additional details about these protocols [47]

Table 2 CNN top results: as some of the highest results on LFW have been from using supervised convolutional neural networks (CNNs), we compare the details of the top-performing CNN methods in a separate table

Method	Net. loss	Outside data	# models	Aligned	Verif. metric	Layers	Accu.
DeepFace [97]	Ident.	4M	4	3D	Wt. chi-sq.	8	97.35 ± 0.25
Canon. view CNN [114]	Ident.	203K	60	2D	Jt. Bayes	7	96.45 ± 0.25
DeepID [92]	Ident.	203K	60	2D	Jt. Bayes	7	97.45 ± 0.26
DeepID2 [88]	Ident. + verif.	203K	25	2D	Jt. Bayes	7	99.15 ± 0.13
DeepID2+ [93]	Ident. + verif.	290K	25	2D	Jt. Bayes	7	99.47 ± 0.12
DeepID3 [89]	Ident. + verif.	290K	25	2D	Jt. Bayes	10–15	99.53 ± 0.10
Face++ [113]	Ident.	5M	1	2D	L2	10	99.50 ± 0.36
FaceNet [82]	Verif. (triplet)	260M	1	No	L2	22	99.60 ± 0.09
Tencent [8]	–	1M	20	Yes	Jt. Bayes	12	99.65 ± 0.25

N.B.—unknown parameters that were not mentioned in the corresponding papers are denoted with a "–"

ones, do not give much detail about their implementations. Rather than devoting an entire section to methods for which we have little detail, we summarize them in Sect. 2.5. We now start with protocols incorporating labeled outside data.

2.2   Unrestricted with Labeled Outside Data

This protocol allows the use of same and different training pairs from outside of LFW. The only restriction is that such data sets should not include pictures of people whose identities appear in the test sets. The use of such outside data sets has dramatically improved performance in several cases.

2.2.1   Attribute and Simile Classifiers for Face Verification, 2009 [53]

Kumar et al. [53] present two main ideas in this paper. The first is to explore the use of describable attributes for face verification. For attribute classifiers 65 describable visual traits such as gender, age, race, and hair color are used. At least 1000 positive and 1000 negative pairs of each attribute were used for training each attribute classifier. The paper gives the accuracy of each individual attribute classifier. Note that the attribute classifier does *not* use labeled outside data, and thus, when not used in conjunction with the simile classifier, qualifies for the *unlabeled outside data* protocols.

The second idea develops what they call *simile* classifiers, in which various classifiers are trained to rate face parts as "similar" or "not similar" to the face parts of certain reference individuals. To train these "simile" classifiers, multiple images of the same individuals (from outside of the LFW training data) are used, and thus this method uses outside labeled data.

The original paper [53] gives an accuracy of $85.29 \pm 1.23\%$ for the hybrid system, and a follow-up journal paper [54] gives slightly higher results of $85.54 \pm 0.35\%$. These numbers should be adjusted downward slightly to 84.52 and 84.78 % since there was an error in how their accuracies were computed.[6]

This paper was also notable in that it gave results for human recognition on LFW (99.2 %). While humans had an unfair advantage on LFW since many of the LFW images were celebrities, and hence humans have seen prior images of many test subjects, which is not allowed under any of the protocols, these results have nevertheless been widely cited as a target for research. The authors also noted that

---

[6]The authors reported that their classifier failed to complete, due to a failed preprocessing step, in 53 out of 6000 cases. According to the footnote in their journal paper, they scored about 85 % of these cases as correct. However, according to the protocol, if an answer is not given, the test sample must be considered incorrect.

humans could do remarkably well using only close crops of the face (97.53 %), and even using only "inverse crops", including none of the face, but portions of the hair, body, and background of the image (94.27 %).

2.2.2 Face Recognition with Learning-Based Descriptor, 2010 [26]

Cao et al. [26] develop a visual dictionary based on unsupervised clustering. They explore K-means, principal components analysis (PCA) trees [37] and random projection trees [37] to build the dictionary. While this was a relatively early use of learned descriptors, they were not learned discriminatively, i.e. to optimize performance.

One of the other main innovative aspects of this paper was building verification classifiers for various combinations of poses such as frontal-frontal, or rightfacing-leftfacing, to optimize feature weights conditioned on the specific combination of poses. This was done by finding the nearest pose to training and test examples using the Multi-PIE data set [38]. Because the Multi-PIE data set uses multiple images of the same subject, this paper is put in the category with outside labeled data. However, it seems plausible that this method could be used on a subset of multi-PIE that did not have images of the same person, as long there was a full range of labeled poses. Such a method, if pursued would qualify these techniques for the category *image-restricted with label-free outside data.*

The highest accuracy reported for their method was 84.45 ± 0.46 %.

2.2.3 An Associate-Predict Model for Face Recognition, 2011 [110]

This paper was one of the first systems to use a large additional amount of outside labeled data, and was, perhaps not coincidentally, the first system to achieve over 90 % on the LFW benchmark.

The main idea in this paper (similar to some older work [13]) was to *associate* a face with one person in a standard reference set, and use this reference person to *predict* the appearance of the original face in new poses and lighting conditions.

Building on the previous work by one of the co-authors [26], this paper also uses different strategies depending upon the relative poses of the presented face pair. If the poses of the two faces are deemed sufficiently similar, then the faces are compared directly. Otherwise, the associate-predict method is used to try to map between the poses. The best accuracy of this system on the *unrestricted with labeled outside data* was 90.57 ± 0.56 %.

2.2.4 Leveraging Billions of Faces to Overcome Performance Barriers in Unconstrained Face Recognition, 2011 [95]

This proprietary method from `Face.com` uses 3D face frontalization and illumination handling along with a strong recognition pipeline and achieves $91.30 \pm 0.30\%$ accuracy on LFW. They report having amassed a huge database of almost 31 billion faces from over a billion persons.

They further discuss the contribution of effective 3D face alignment (or *frontalization*) to the task of face verification, as this is able to effectively take care of out-of-plane rotation, which 2D based alignment methods are not able to do. The 3D model is then used to render all images into a frontal view. Some details are given about the recognition engine—it uses non-parametric discriminative models by leveraging their large labeled data set as exemplars.

2.2.5 Tom-vs-Pete Classifiers and Identity-Preserving Alignment for Face Verification, 2012 [16]

This work presented two significant innovations. The first was to do a new type of non-affine warping of faces to improve correspondences while preserving as much information as possible about identity. While previous work had addressed the problem of non-linear pose-normalization (see, for example, the work by Asthana et al. [10, 11]), it had not been successfully used in the context of LFW.

In particular, as the authors note, simply warping two faces to maximize similarity may reduce the ability to perform verification by eliminating discriminative information between the two individuals. Instead, a warping should be done to maximize similarity while maintaining identity information. The authors achieve this identity-preserving warping by adjusting the warping algorithm so that parts with informative deviations in geometry (such as a wide nose) are preserved better (see the paper for additional details). This technique makes about a 2% (91.20–93.10%) improvement in performance relative to more standard alignment techniques.

This paper was also one of the first evaluated on LFW to use the approximate symmetry of the face to its advantage. Since using the above warping procedure tends to distort the side of the face further from the camera, the authors reflect the face, if necessary, such that the side closer to the camera is always on the right side of the photo. This results in the right side of the picture typically being more faithful to the appearance of the person. As a result, the learning algorithm which is subsequently applied to the flipped faces can learn to rely more on the more faithful side of the face. It should be noted, however, that the algorithm pays a price when the person's face is asymmetric to begin with, since it may need to match the left side of a person's face to their own right side. Still this use of facial symmetry improves the final results.

The second major innovation was the introduction of so-called *Tom-vs-Pete* classifiers as a new type of learned feature. These features were developed by using external labeled training sets (also labeled with part locations) to develop binary classifiers for pairs of identities, such as two individuals named Tom and Pete. For each of the $\binom{n}{2}$ pairs of identities in the external training set, k separate classifiers are built, each using SIFT features from a different region of the face. Thus, the total number of Tom-vs-Pete classifiers is $k \times \binom{n}{2}$. A subset of these were chosen by maximizing discriminability.

The highest accuracy of their system was $93.10 \pm 1.35\,\%$. However, they increased accuracy (and reduced the standard error) a bit further by adding attribute features based upon their previous work, to $93.30 \pm 1.28\,\%$.

2.2.6 Bayesian Face Revisited: A Joint Formulation, 2012 [28]

One of the most important aspects of face recognition in general, viewed as a classification problem, is that all of the classes (represented by individual identities) are highly similar. At the same time, within each class is a significant amount of variability due to pose, expression, and so on. To understand whether two images represent the same person, it can be argued that one should model both the distribution of identities, and also the distribution of variations within each identity.

This basic idea was originally proposed by Moghaddam et al. in their well-known paper "Bayesian face recognition" [69]. In that paper, the authors defined a difference between two images, estimated the distribution of these differences conditioned on whether the images were drawn from the same identity or not, and then evaluated the posterior probability that this difference was due to the two images coming from different identities.

Chen et al. [28] point out a potential shortcoming of the probabilistic method applied to image differences. They note that by forming the image difference, information available to distinguish between two classes (in this case the "same" versus "different" classes of the verification paradigm) may be thrown out. In particular, if $\mathbf{x}$ and $\mathbf{y}$ are two image vectors of length N, then the pair of images, considered as a concatenation of the two vectors, contains $2N$ components. Forming the difference image is a linear operator corresponding to a projection of the image pair back to N dimensions, hence removing some of the information that may be useful in deciding whether the pair is "same" or "different". This is illustrated in Fig. 1. To address this problem, Chen et al. focus on modeling the joint distribution of image pairs (of dimension $2N$) rather than the difference distribution (of dimension N). This is an elegant formulation that has had a significant impact on many of the follow-up papers on LFW.

Another appealing aspect of this paper is the analysis that shows the relationship between the joint Bayesian method and the reference-based methods, such as the simile classifier [53], the multiple one-shots method [96], and the associate-predict method [110]. The authors show that their method can be viewed as equivalent to a reference method in the case that there are an infinite number of references, and that the distributions of identities and within class variance are Gaussian.

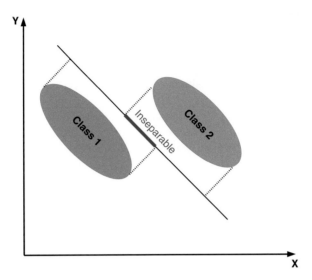

Fig. 1 When the information from two images is projected to a lower dimension by forming the difference, discriminative information may be lost. The joint Bayesian approach [28] strives to avoid this projection, thus preserving some of the discriminative information

The accuracy of this method while using outside data for training (*unrestricted with labeled outside data*) was 92.42 ± 1.08 %.

2.2.7 Blessing of Dimensionality: High-Dimensional Feature and Its Efficient Compression for Face Verification, 2013 [29]

This paper argues that high-dimensional descriptors are essential for high performance, and also describes a method for compression termed as *rotated sparse regression*. They construct the high-dimensional feature using local binary patterns (LBP), histograms of oriented gradients (HOG) and others, extracted at 27 facial landmarks and at five scales on 2D aligned images. They use principal components analysis (PCA) to first reduce this to 400 dimensions and use a supervised method such as linear discriminant analysis (LDA) or a joint Bayesian model [28] to find a discriminative projection. In a second step, they use L1-regularized regression to learn a sparse projection that directly maps the original high-dimensional feature into the lower-dimensional representation learned in the previous stage.

They report accuracies of 93.18 ± 1.07 % under the *unrestricted with label-free outside data* protocol and 95.17 ± 1.13 % using their WDRef (99,773 images of 2995 subjects) data set for training following the *unrestricted with labeled outside data* protocol.

2.2.8 A Practical Transfer Learning Algorithm for Face Verification, 2013 [25]

This paper applies transfer learning to extend the high performing joint Bayesian method [28] for face verification. In addition to the data likelihood of the target domain, they add the KL-divergence between the source and target domains as a regularizer to the objective function. The optimization is done via closed-form updates in an expectation-maximization framework. The source domain is the non-public WDRef data set used in their previous versions [28, 29] and the target is set to be LFW. They use the high-dimensional LBP features from [29], reducing its size from over 10,000 dimensions to 2000 by PCA.

They report $96.33 \pm 1.08\%$ accuracy on LFW in the *unrestricted with labeled outside data* protocol, which improves over the results from using joint Bayesian without the transfer learning on high dimensional LBP features [29].

2.2.9 Hybrid Deep Learning for Face Verification, 2013 [91]

This method [91] uses an elaborate hybrid network of convolutional neural networks (CNNs) and a Classification-RBM (restricted Boltzmann machine), trained directly for verification. A pair of 2D aligned face images are input to the network. At the lower part, there are 12 groups of CNNs, which take in images each covering a particular part of the face, some in colour and some in grayscale. Each group contains five CNNs that are trained using different bootstrap samples of the training data. A single CNN consists of four convolutional layers and a max-pooling layer. Similar to [48], they use local convolutions in the mid- and high-level layers of the CNNs. There can be eight possible "input modes" or combinations of horizontally flipping the input pair of images and each of these pairs are fed separately to the networks. The output from all these networks is in layer L0, having $8 * 5 * 12$ neurons. The next two layers average the outputs, first among the eight input modes and then the five networks in a group. The final layer is a classification RBM (models the joint distribution of class labels, binary input vectors and binary hidden units) with two outputs that indicate same or different class for the pairs, which is discriminatively trained by minimizing the negative log probability of the target class given the input, using gradient descent. The CNNs and the RBM are trained separately; then the whole model is jointly fine-tuned using back-propagation. Model averaging is done by training the RBM with five different random sets of training data and averaging the predictions. They create a new training data set, "CelebFaces", consisting of 87,628 images of 5436 celebrities collected from the web. They report $91.75 \pm 0.48\%$ accuracy on the LFW in the *unrestricted with label-free outside data* protocol and $92.52 \pm 0.38\%$ following the *unrestricted with labeled outside data* protocol.

2.2.10 POOF: Part-Based One-vs-One Features for Fine-Grained Categorization, Face Verification, and Attribute Estimation, 2013 [17]

When annotations of parts are provided, this method learns highly discriminative features between two classes based on the appearance at a particular landmark or part that has been provided. They formulate face verification as a fine-grained classification task, for which this descriptor is designed to be well suited.

For training a single "POOF" or Part-Based One-vs-One Feature, it is provided a pair of classes to distinguish and two part locations—one for alignment and the other for feature extraction. All the images of the two classes are aligned with respect to the two locations using similarity transforms with 64 pixels horizontal distance between them. A crop of 64×128 at the mid-point of the two locations is taken and grids of 8×8 and 16×16 are placed on it. Gradient direction histograms and color histograms are used as base features for each cell and concatenated. A linear support vector machine (SVM) is trained on these to separate the two classes. These SVM weights are used to find the most discriminative cell locations and a mask is obtained by thresholding these values. Starting from a given part location as a seed, its connected component is found in the thresholded mask. Base features from cells in this connected component are concatenated and another linear SVM is used to separate the two classes. The score from this SVM is the score of that part-based feature.

They learn a random subset of 10,000 POOFs using the database in [16], getting two 10,000-dimensional vectors for each LFW pair. They use both absolute difference ($|f(A) - f(B)|$) and product ($f(A).f(B)$) of these vectors to train a same-versus-different classifier on the LFW training set. They report $93.13 \pm 0.40\%$ accuracy on LFW following the *unrestricted with labeled outside data* protocol.

2.2.11 Learning Discriminant Face Descriptor for Face Recognition, 2014 [58]

This approach learned a "Discriminative Face Descriptor" (DFD) based upon improving the LBP feature (which are essentially differences in value of a particular pixel to its neighbours). They use the Fisher criterion for maximizing between class and minimizing within class scatter matrices to learn discriminative filters to extract features at the pixel level as well as find optimal weights for the contribution of neighbouring pixels in computing the descriptor. K-means clustering is used to find the most dominant clusters among these discriminant descriptors (typically of length 20). They reported best performance using $K = 1024$ or 2048.

They used the LFW-a images and cropped the images to 150×130. They further used a spatial grid to encode separate parts of the face separately into their DFD representation and also apply PCA whitening. The descriptors themselves were learned using the FERET data set (*unrestricted with labeled outside data*), however the authors note that the distribution of images in FERET is quite different from that of LFW—performance on LFW is an indicator of the generalizable power of their descriptor. They report an LFW accuracy of $84.02 \pm 0.44\%$.

2.2.12 Face++, 2014

We discuss two papers from the Face++/Megvii Inc. group here, both involving supervised deep learning on large labeled data sets. These, along with Facebook's DeepFace [97] and DeepID [92], exploited massive amounts of labeled outside data to train deep convolutional neural networks (CNNs) and reach very high performance on LFW.

In the first paper from the Face++ group, a new structure, which they term the pyramid CNN [34] is used. It conducts supervised training of a deep neural network one layer at a time, thus greatly reducing computation. A four-level Siamese network trained for verification was used. The network was applied on four face landmarks and the outputs were concatenated. They report an accuracy of 97.3% on the LFW *unrestricted with labeled outside data* protocol.

The Megvii Face Recognition System [113] was trained on a data set of five million labeled faces of around 20,000 identities. A ten-layer network was trained for identification on this data set. The second-to-last layer, followed by PCA, was used as the face representation. Face verification was done using the L2 norm score, achieving $99.50 \pm 0.36\%$ accuracy. With the massive training data set size, they argue that the advantages of using more sophisticated architectures and methods become less significant. They investigate the long tail effect of web-collected data (lots of persons with very few image samples) and find that after the first 10,000 most frequent individuals, including more persons with very few images into the training set does not help. They also show in a secondary experiment that high performance on LFW does not translate to equally high performance in a real-world security certification setting.

2.2.13 DeepFace: Closing the Gap to Human-Level Performance in Face Verification, 2014 [97]

This paper from Facebook [97] has two main novelties—a method for 3D face frontalization[7] and a deep neural net trained for classification. The neural network featured 120 million parameters, and was trained on 4000 identities having four million images (the non-public *SFC* data set). This paper was one of the first papers to achieve very high accuracies on LFW using CNNs. However, as mentioned above,

---

[7]See Sect. 3 for a discussion of previous work on 3D frontalization.

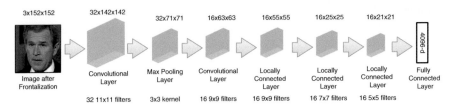

Fig. 2 The architecture of the DeepFace convolutional neural network [97]. This type of architecture, which has been widely used in other object recognition problems, has become a dominant presence in the face recognition literature

other papers that used deep networks for face recognition predated this by several years [48, 70]. Figure 2 shows the basic architecture of the DeepFace CNN, which is typical of deep architectures used on other non-face benchmarks such as ImageNet.

3-D frontalized RGB faces of size 152×152 are taken as input, followed by 32 11×11 convolution filters (C1), a max-pooling layer (2×2 size with stride of two pixels, M2) and another convolutional layer with 16 9×9 filters (C3). The next three layers (L4–6) are locally connected layers [48], followed by two fully connected layers (F7–8). The 4096-dimensional F7 layer output is used as the face descriptor. ReLU activation units are used as the non-linearity in the network and *dropout regularization* is applied to F7 layer. L_2-normalization is applied to the descriptor. Training the network for 15 epochs took three days. The weighted χ^2 distance is used as the verification metric. Three different input image types (3D aligned RGB, grayscale with gradient magnitude and orientation and 2-D aligned RGB) are used, and their scores are combined using a kernel support vector machine (SVM). Using the *restricted* protocol, this reaches 97.15 % accuracy. Under the *unrestricted* protocol, they train a Siamese network (initially using their own SFC data set, followed by two epochs on LFW pairs), reaching 97.25 % after combining the Siamese network with the above ensemble. Finally, adding four randomly-seeded DeepFace networks to the ensemble a final accuracy of 97.35 ± 0.25 % is reached on LFW following the *unrestricted with labeled outside data* protocol.

2.2.14 Recover Canonical-View Faces in the Wild with Deep Neural Networks, 2014 [114]

In this paper, the authors train a convolutional neural network to recover the canonical view of a face by training it on 2D images without any use of 3D information. They develop a formulation using symmetry and matrix-rank terms to automatically select the frontal face image for each person at training time. Then the deep network is used to learn the regression from face images in arbitrary view to the canonical (frontal) view.

After this canonical pose recovery is performed, they detect five landmarks from the aligned face and train a separate network for each patch at each landmark along with one network for the entire face. These small networks (two convolutional and two pooling layers) are connected at the fully connected layer and trained on the CelebFaces data set [91] with the cross-entropy loss to predict identity labels. Following this, a PCA reduction is done, and an SVM is used for the verification task, resulting in an accuracy of $96.45 \pm 0.25\%$ under the *unrestricted with labeled outside data* protocol.

2.2.15 Deep Learning Face Representation from Predicting 10,000 Classes, 2014 [92]

In this approach, called "DeepID" [92], the authors trained a network to recognize 10,000 face identities from the "CelebFaces" data set [91] (87,628 face images of 5436 celebrities, non-overlapping with LFW identities). The CNNs had four convolutional layers (with 20, 40, 60 and 80 feature maps), followed by max-pooling, a 160-dimensional fully-connected layer (DeepID-layer) and a softmax layer for the identities. The higher convolutional layers had locally shared weights. The fully-connected layer was connected to both the third and fourth convolutional layers in order to see multi-scale features, referred to as a "skipping" layer. Faces were globally aligned based on five landmarks. The input to a network was one out of 60 patches, which were square or rectangular and could be both colour or grayscale. Sixty CNNs were trained on flipped patches, yielding a $160 * 2 * 60$ dimensional descriptor of a single face. PCA reduction to 150 dimensions was done before learning the joint Bayesian model, reaching an accuracy of 96.05%. Expanding the data set (CelebFaces+ [88]) and using the joint Bayesian model for verification gives them a final accuracy of $97.45 \pm 0.26\%$ under the *unrestricted with labeled outside data* protocol.

2.2.16 Surpassing Human-Level Face Verification Performance on LFW with GaussianFace, 2014 [65]

This method uses multi-task learning and the discriminative Gaussian process latent variable model (DGP-LVM) [55, 100] to be one of the top performers on LFW. The DGP-LVM [100] maps a high-dimensional data representation to a lower-dimensional latent space using a discriminative prior on the latent variables while maximizing the likelihood of the latent variables in the Gaussian process (GP) framework for classification. GPs themselves have been observed to be able to make accurate predictions given small amounts of data [55] and are also robust to situations when the training and test data distributions are not identical. The authors were motivated to use DGP-LVM over the more usual GPs for classifications as the former, by virtue of its discriminative prior, is a more powerful predictor.

The DGP-LVM is reformulated using a kernelized linear discriminant analysis to learn the discriminative prior on latent variables and multiple source domains are used to train for the target domain task of verification on LFW. They detail two uses of their *Gaussian Face* model—as a binary classifier and as a feature extractor. For the feature extraction, they use clustering based on GPs [51] on the joint vectors of two faces. They compute first and second order statistics for input joint feature vectors and their latent representations and concatenate them to form the final feature. These GP-extracted features are used in the GP-classifier in their final model.

Using 200,000 training pairs, the "GaussianFace" model reached $98.52 \pm 0.66\%$ accuracy on LFW under the *unrestricted with labeled outside data* protocol, surpassing the recorded human performance on close-cropped faces (97.53%).

2.2.17   Deep Learning Face Representation by Joint Identification-Verification, 2014 [88]

Building on the previous model, DeepID [92], "DeepID2" [88] used both an *identification signal* (cross-entropy loss) and a *verification signal* (L2 norm verification loss between DeepID2 pairs) in the objective function for training the network, and expanded the CelebFaces data set to "CelebFaces+", which has 202,599 face images of 10,177 celebrities from the web. 400 aligned face crops were taken to train a network for each patch and a greedy selection algorithm was used to select the best 25 of these. A final 4000 ($25 * 160$) dimensional face representation was obtained, followed by PCA reduction to 180-dimensions and joint Bayesian verification, achieving 98.97% accuracy.

The network had four convolutional layers and max-pooling layers were used after the first three convolutional layers. The third convolutional layer was locally connected, sharing weights in 2×2 local regions. As mentioned before, the loss function was a combined loss from identification and verification signals. The rationale behind this was to encourage features that can discriminate identity, and also reduce intra-personal variations by using the verification signal. They show that using either of the losses alone to train the network is sub-optimal and the appropriate loss function is a weighted combination of the two.

A total of seven networks are trained using different sets of selected patches for training. The joint Bayesian scores are combined using an SVM, achieving $99.15 \pm 0.13\%$ accuracy under the *unrestricted with labeled outside data* protocol.

2.2.18   Deeply Learned Face Representations are Sparse, Selective and Robust, 2014 [93]

Following on with the DeepID "family" of models, "DeepID2+" [93] increased the number of feature maps to 128 in the four convolutional layers, the DeepID size to 512 dimensions and expanded their training set to around 290,000 face images from 12,000 identities by merging the CelebFaces+ [88] and WDRef [29] data sets. Another interesting novelty of this method was the use of a loss function at multiple layers of the network, instead of the standard supervisory signal (loss function) in the top layer. They branched out 512-dimensional fully-connected layers at each of the four convolutional layers (after the max-pooling step) and added the loss function (a joint identification-verification loss) after the fully-connected layer for additional supervision at the early layers. They show that removal of the added supervision lowers their performance, as well as some interesting analysis on the sparsity of the neural activations. They report that only about half the neurons get activated for an image, and each neuron activates for about half the images. Moreover they found a difference of less than 1 % when using a binary representation by thresholding, which led them to state that the fact that a neuron is activated or not is more important than the actual value of that activation.

This report an accuracy of $99.47 \pm 0.12\%$ (*unrestricted with labeled outside data*) using the joint Bayesian model trained on 2000 people in their training set and combining the features from 25 networks trained on the same patches as DeepID2 [88].

2.2.19 DeepID3: Face Recognition with Very Deep Neural Networks, 2015 [89]

"DeepID3" uses a deeper network (10–15 feature extraction layers) with Inception layers [94] and stacked convolution layers (successive convolutional layers without any pooling layer in between) on a similar overall pipeline to DeepID2+ [93]. Similar to DeepID2+, they include unshared weights in later convolutional layers, max-pooling in early layers and the addition of joint identification-verification loss functions to branched-out fully connected layers from each pooling layer in the network.

They train two networks, one using the stacked convolution and the other using the recently-proposed Inception layer used in the GoogLeNet architecture, which was a top-performer in the ImageNet challenge in 2015 [94]. The two networks reduce the error rate of DeepID2+ by 0.81 % and 0.26 %, respectively.

The features from both the networks on 25 patches is combined into a vector of about 30,000 dimensions. It is PCA reduced to 300 dimensions, followed by learning a joint Bayesian model. It achieved $99.53 \pm 0.10\%$ verification accuracy on LFW (*unrestricted with labeled outside data*).

2.2.20 FaceNet: A Unified Embedding for Face Recognition and Clustering, 2015 [82]

This model from Google, called the FaceNet [82], uses 128-dimensional representations from very deep networks, trained on a 260-million image data set using a *triplet loss* at the final layer—the loss separates a positive pair from a negative pair by a margin. An online hard negative exemplar mining strategy within each mini-batch is used in training the network. This loss directly optimizes for the verification task and so a simple L2 distance between the face descriptors is sufficient.

They use two variants of networks. In NN1, they add $1 \times 1 \times d$ convolutional layers between the standard Zeiler&Fergus CNN [112] resulting in 22 layers. In NN2, they use the recently proposed Inception modules from GoogLeNet [94] which is more efficient and has 20 times lesser parameters. The L2-distance threshold for verification is estimated from the LFW training data. They report results, following the *unrestricted with labeled outside data* protocol, on central crops of LFW ($98.87 \pm 0.15\,\%$) and when using a proprietary face detector ($99.6 \pm 0.09\,\%$) using the NN1 model, which is the highest score on LFW in the *unrestricted with labeled outside data* protocol. The scores from using the NN2 model were reported to be statistically in the same range.

2.2.21 Tencent-BestImage, 2015 [8]

This commercial system followed the *unrestricted with labeled outside data* protocol and built their system combining an alignment system, a deep convolutional neural network with 12 convolution layers, and the joint Bayesian method for verification. The whole system was trained on their data set—"BestImage Celebrities Face" (BCF), which contains about 20,000 individuals and one million face images and is identity-disjoint with respect to LFW. They divided the BCF data into two subsets for training and validation. The network was trained on the BCF training set with 20 face patches. The features from each patch were concatenated, followed by PCA and the joint Bayesian model learned on BCF validation set. They report an accuracy of $99.65 \pm 0.25\,\%$ on LFW under the *unrestricted with labeled outside data* protocol.

2.3 Label-Free Outside Data Protocols

In this section, we discuss two of the LFW protocols together— *image-restricted with label-free outside data* and *unrestricted with label-free outside data*. While these results are curated separately for fairness on the LFW page, conceptually they are highly similar, and are not worth discussing separately.

These protocols allow the use of outside data such as additional faces, landmark annotations, part labels, and pose labels, as long as this additional information does not contain any information that would allow making pairs of images labeled "same" or "different". For example, a set of images of a single person (even if the person were not labeled) or a video of a person would not be allowed under these protocols, since any pair of images from the set or from the video would allow the formation of a "same" pair.

Still, large amounts of information can be used by these methods to understand the general structure of the space of faces, to build supervised alignment methods, to build attribute classifiers, and so on. Thus, these methods would be expected to have a significant advantage over the "no outside data" protocols.

2.3.1   Face Recognition Using Boosted Local Features, 2003 [50]

One of the earliest methods applied to LFW was developed at Mitsubishi Electric Research Labs (MERL) by Michael Jones and Paul Viola [50]. This work built on the authors' earlier work in boosting for face detection [101], adapting it to learn a similarity measure between face images using a modified AdaBoost algorithm. They use filters that act on a pair of images as features, which are a set of linear functions that are a superset of the "rectangle" filters used in their face detection system. A threshold on the absolute difference of the scalar values returned by a filter applied on a pair of faces can be used to determine valid or invalid variation of a particular property or aspect of a face (the validity being with respect to whether the faces belong to the same identity).

The technical report was released before LFW, and so does not describe application to the database, but the group submitted results on LFW after publication, achieving $70.52 \pm 0.60\%$.

2.3.2   LFW Results Using a Combined Nowak Plus MERL Recognizer, 2008 [46]

This early system combined the method of Nowak and Jurie [74] with an unpublished method [46] from Mitsubishi Electric Research Laboratory (MERL), and thus technically counts as a method whose full details are not published. However, some details are given in a workshop paper [46].

The MERL system initially detects a face using a Viola-Jones frontal face detector, followed by alignment based on nine facial landmarks (also detected using a Viola-Jones detector). After alignment, some simple lightning normalization is done. The score of the MERL face recognition system [50] is then averaged with the score from the best-performing system of that time (2007), by Nowak and Jurie [74].

The accuracy of this system was $76.18 \pm 0.58\%$.

2.3.3   Is That You? Metric Learning Approaches for Face Identification, 2009 [39]

This paper presents two methods to learn robust distance measures for face verification, the logistic discriminant-based metric learning (LDML) and marginalized K-nearest neighbors (MkNN) classifier. The LDML learns a Mahalanobis distance between two images to make the distances between positive pairs smaller than the distances between negative pairs and obtain a probability that a pair is positive in a standard linear logistic discriminant model. The MkNN classifies an image pair belongs to the same class with the marginal probability that both of them are assigned to the same class using a K-nearest neighbor classifier. In the experiments, they represent the images as stacked multi-scale local descriptors extracted at nine facial landmarks. The facial landmarks detector is trained with outside data. Without using the identify labels for the LFW training data, the LDML achieves $79.27 \pm 0.6\%$ under the *image-restricted with label-free outside data* protocol. They obtain this accuracy by fusing the scores from eight local features including LBP, TPLBP, FPLBP, SIFT and their element-wise square root variants with a linear combination. This multiple feature fusion method is shown to be effective in a number of literatures. Under the *unrestricted with label-free outside data* protocol, they show that the performance of LDML is significantly improved with more training pairs formed using the identity labels. And they obtain their best performance $87.50 \pm 0.4\%$ accuracy by linearly combining the 24 scores with the three methods LDML, large margin nearest neighbor (LMNN) [102] and MkNN over the eight local features.

2.3.4 Multiple One-Shots for Utilizing Class Label Information, 2009 [96]

The authors extend the one-shot similarity (OSS) introduced in [104] which we will describe under the *image-restricted, no outside data* protocol. In brief, the OSS for an image pair is obtained by training a binary classifier with one image in the pair as the positive sample and a set of pre-defined negative samples to classify the other image in the pair. This paper extends the OSS to be multiple one-shots similarity vector by producing OSS scores with different negative sample sets. Each set reflecting either a different subject or a different pose. In their face verification system, the faces are firstly aligned with a commercial face alignment system. The aligned faces are published as the "aligned" LFW data set or LFW-a data set. The face descriptors are then constructed by stacking local descriptors extracted densely over the face images. The information theoretic metric learning (ITML) method is adopted to obtain a Mahalanobis matrix to transform the face descriptors and a linear SVM classifies a pair of faces to be matched or not based on the multiple OSS scores. They achieve their best result $89.50 \pm 0.51\%$ accuracy by combining 16 multiple OSS scores including eight descriptors (SIFT, LBP, TPLBP, FPLBP and their square root variants) under two settings of the multiple OSS scores (the subject-based negative sample sets and pose-based negative sample sets).

2.3.5 Attribute and Simile Classifiers for Face Verification, 2009 [53]

We discussed the attribute and simile classifiers for face verification [53] under the *unrestricted with labeled outside data* protocol. The authors' result with the attribute classifier qualifies for the *unrestricted with label-free outside data* protocol. They reported their result with the attribute classifier on LFW as $85.25 \pm 0.60\%$ accuracy in their follow-up journal paper [54].

2.3.6 Similarity Scores Based on Background Samples, 2010 [105]

In this paper, Wolf et al. [105] extend the one-shot similarity (OSS) introduced in [104] to the two-shot similarity (TSS). The TSS score is obtained by training a classifier to classify the two face images in a pair against a background face set. Although the TSS score by itself is not discriminative for face verification, they show that the performance is improved by combining the TSS scores with OSS scores and other similarity scores. They extend the OSS and TSS framework to use linear discriminant analysis instead of an SVM as the online trained classifier. In addition to OSS and TSS, they propose to represent each image in the pair with its rank vector obtained by retrieving similar images from the background face set. The correlation between the two rank vectors provides another dimensionality of the similarity measure of the face pair. In their experiments, they use the LFW-a data set to handle alignment. Combining the similarities introduced above with eight variants of local descriptors, they obtain an accuracy of 86.83 ± 0.34 % under the *image-restricted with label-free outside data* protocol.

2.3.7 Rectified Linear Units Improve Restricted Boltzmann Machines, 2010 [70]

Restricted Boltzmann machines (RBMs) are often formulated as having binary-valued units for the hidden layer and Gaussian units for the real-valued input layer. Nair and Hinton [70] modify the hidden units to be "noisy rectified linear units" (NReLUs), where the value of a hidden unit is given by the rectified output of the activation and some added noise, i.e. $max(0, x + N(0, V))$, where x is the activation of the hidden unit given an input, and $N(0, V)$ is the Gaussian noise. RBMs with 4000 NReLU units in the hidden layer are first pre-trained generatively, then discriminatively trained as a feed-forward fully-connected network using back-propagation (in the latter case the Gaussian noise term is dropped in the rectification).

In order to model face pairs, they use a "Siamese" network architecture, where the same network is applied to both faces and the cosine distance is the symmetric function that combines the two outputs of the network. They show that that NReLUs are *translation equivariant* and *scale equivariant*(the network outputs change in the same way as the input), and combined with the *scale invariance* of cosine distance the model is analytically invariant to the rescaling of its inputs. It is not translation invariant. LFW images are center-cropped to 144×144, aligned based on the eye location and sub-sampled to 32×32 3-channel images. Image intensities are normalized to be zero-mean and unit-variance. They report an accuracy of $80.73 \pm 1.34\%$ (*image-restricted with label-free outside data*). It should be noted that because the authors use manual correction of alignment errors, this paper does not conform to the LFW protocols, and thus need not be used as a comparison against fully automatic methods.

2.3.8   Face Recognition with Learning-based Descriptor, 2010 [26]

This paper was discussed under the *unrestricted with labeled outside data* protocol. With the holistic face as the only component, the method qualifies for the *image-restricted with label-free outside data* protocol, under which the authors obtain an accuracy of 81.22 ± 0.35 %.

2.3.9   Cosine Similarity Metric Learning for Face Verification, 2011 [72]

This paper proposes cosine similarity metric learning (CSML) to learn a transformation matrix to project faces into a subspace in which cosine similarity performs well for verification. They define the objective function to maximize the margin between the cosine similarity scores of positive pairs and cosine similarity scores of negative pairs while regularizing the learned matrix by a predefined transformation matrix. They empirically demonstrate that this straightforward idea works well on LFW and that by combining scores from six different feature descriptors their method achieves an accuracy of 88.00 ± 0.38 % under the *image-restricted with label-free outside data* protocol. Subsequent communication with the authors revealed an error in the use of the protocol. Had the protocol been followed properly, our experiments suggest that the results would be about three percent lower, i.e., about 85 %. Still, this method has played an important role in subsequent research as a popular choice for the comparison of feature vectors.

2.3.10   Beyond Simple Features: A Large-Scale Feature Search Approach to Unconstrained Face Recognition, 2011 [31]

This method [31] uses the biologically-inspired V1-like features that are designed to approximate the initial stage of the visual cortex of primates. It is essentially a cascade of linear and non-linear functions. These are stacked into two and three layer architectures, HT-L2 and HT-L3 respectively. These models take in 100×100 and 200×200 grayscale images as inputs. A linear SVM is trained on a variety of vector comparison functions between two face descriptors. Model selection is done on 5915 HT-L2 and 6917 HT-L3 models before the best five were selected. Multiple kernels were used to combine data augmentations (rescaled crops of 250×250, 150×150 and 125×75), blend the top five models within each "HT class", and also blend models across HT classes. The HT-L3 gives an accuracy of 87.8 % while combining all of the models gives a final accuracy of 88.13 ± 0.58 % following the *image-restricted with label-free outside data* protocol.

2.3.11   Face Verification Using the LARK Representation, 2011 [84]

This work extends previous work [83] in which two images are represented as two local feature sets and the matrix cosine similarity (MCS) is used to separate faces from backgrounds. All kinds of visual variations are addressed implicitly in the MCS which is the weighted sum of the cosine similarities of the local features. In this work, the authors present the locally adaptive regression kernel (LARK) local descriptor for face verification. LARK is defined as the self-similarity between a center pixel and its surroundings. In particular, the distance between two pixels is the geodesic distance. They consider an image as a 3D space which includes the 2D coordinates and the gray-scale value at each pixel. The geodesic distance is then the shortest path on the image surface. PCA is then adopted to reduce the dimensionality of the local features. They further apply an element-wise logistic function to generate a binary-like representation to remove the dominance of large relative weights to increase the discriminative power of the local features. They conduct experiments on LFW under both the unsupervised protocol and the image restricted protocol.

In the unsupervised setting, they compute LARKs of size 7×7 from each face image. They evaluate various combinations of different local descriptors and similarity measures and report that the LBP with Chi-square distance achieves the best 69.54 % accuracy among the baseline methods. Their method achieves 72.23 % accuracy. Under the *image-restricted with label-free outside data* protocol, they adopt the OSS with LDA for face verification and achieve an accuracy of 85.10 ± 0.59 % by fusing scores from 14 combinations of local descriptors (SIFT, LBP, TPLBP and pcaLARK) and similarity measures (OSS, OSS with logistic function, MCS and MCS with logistic function).

2.3.12   Probabilistic Models for Inference About Identity, 2012 [62]

This paper presents a probabilistic face recognition method. Instead of representing each face as a feature vector and measuring the distances between faces in the feature space, they propose to construct a model in which identity is a hidden variable in a generative description of the image data. Other variations in pose, illumination and etc., is described as noise. The face recognition is then framed as a model comparison task.

More concretely, they present a probabilistic latent discriminant analysis (PLDA) model to describe the data generation. In PLDA, the data generation depends on the latent identity variable and an intra-class variation variable. This design helps factorize the identity subspace and within-individual subspace. The model is learned by expectation-maximization (EM) and the face verification is conducted by looking at the likelihood ratio of an image pair generated by a similar pair model over a dissimilar pair model. The PLDA model is further extended to be a mixture of

PLDA models to describe the potential non-linearity of the face manifold. Extensive experiments are conducted to evaluate the PLDA and its variants in face analysis. Their face verification result on LFW is $90.07 \pm 0.51\,\%$ under the *unrestricted with label-free outside data* protocol.

2.3.13   Large Scale Strongly Supervised Ensemble Metric Learning, with Applications to Face Verification and Retrieval, 2012 [43]

High-dimensional overcomplete representations of data are usually informative but can be computationally expensive. This paper proposes a two-step metric learning method to enforce sparsity and to avoid features with little discriminability and improve computational efficiency. The two-step design is motivated by the fact that straightforwardly applying the group lasso with row-wise and column-wise L_1 regularization is very expensive in high-dimensional feature spaces. In the first step, they iteratively select μ groups of features. In each iteration, the feature group which gives the largest partial derivative of the loss function is chosen and the Mahalanobis matrix of a weak metric for the selected feature group is learned and assembled into a sparse block diagonal matrix $A_\dagger$. With an eigenvalue decomposition, they obtain a transformation matrix to reduce the feature dimensionality. After that, in the second step, another Mahalanobis matrix is learned to exploit the correlations between the selected feature groups in the lower-dimensional subspace. They adopt the projected gradient descent method to iteratively learn the Mahalanobis matrix.

In their experiments, they use the LFW-a data set and center crop the face images to 110×150. By concatenating two types of features (covariance matrix descriptors and soft local binary pattern histograms) after the first step, they achieve $92.58 \pm 1.36\,\%$ accuracy under the *image-restricted with label-free outside data* protocol.

2.3.14   Distance Metric Learning with Eigenvalue Optimization, 2012 [111]

In this paper, the authors present an eigenvalue optimization framework for learning a Mahalanobis metric. They learn the metric by maximizing the minimal squared distances between dissimilar pairs while maintaining an upper bound for the sum of squared distances between similar pairs. They further show that this is equivalent to an eigenvalue optimization problem. Similarly, the previous metric learning method LMNN can also be formulated as a general eigenvalue decomposition problem.

They further develop an efficient algorithm to solve this optimization problem, which will only involve the computation of the largest eigenvector of a matrix. In the experiments, they show that the proposed method is more efficient than other metric learning methods such as LMNN and ITML. On LFW, they evaluate this method with both the LFW *funneled* data set and the LFW-a data set. They use

SIFT features computed at the fiducial points for faces on the *funneled* LFW data set and achieve 81.27 ± 2.30 % accuracy. On the "aligned" LFW data set, they evaluate three types of features including concatenated raw intensity values, LBP and TPLBP. Combining the scores from the four different features with a linear SVM, they achieve 85.65 ± 0.56 % accuracy under the *image-restricted with label-free outside data* protocol.

2.3.15   Learning Hierarchical Representations for Face Verification with Convolutional Deep Belief Networks, 2012 [48]

In this work [48], a local convolutional deep belief network is used to generatively model the distribution of faces. Then, a discriminatively learned metric (ITML) is used for the verification task. The shared weights of convolutional filters (10×10 in size) in the CRBM (convolutional RBM) makes it possible to use high-resolution images as input. Probabilistic max-pooling is used in the CRBM to have local translation invariance and still allow top-down and bottom-up inference in the model.

The authors argue that in images like faces, that exhibit clear spatial structure, the weights of a hidden unit being shared across the locations in the whole image is not desirable. On the other hand, using a layer with fully-connected weights may not be computationally tractable without either subsampling the input image or first applying several pooling layers. In order to exploit this structure, the image is divided into overlapping regions and the weight-sharing in the CRBM is restricted to be local. Contrastive divergence is used to train the local CRBM. Two layers of these CRBMs are stacked to form a deep belief network (DBN). The *local CRBM* is used in the second layer of their network. In addition to using raw pixels, the uniform LBP descriptor is also used as input to the DBN. The two features are combined at the score level by using a linear SVM. The LFW-a face images are used as input, with three croppings at sizes 150×150, 125×75, 100×100, resized to the same size before input to the DBN. The deep learned features give competitive performance (86.88 ± 0.62 %) to hand-crafted features (87.18 ± 0.49 %), while combining the two gives the highest of 87.77 ± 0.62 % (*image-restricted with label-free outside data*).

2.3.16   Bayesian Face Revisited: A Joint Formulation, 2012 [28]

We discussed this paper under the *unrestricted with labeled outside data* protocol. The authors also present their result under the *unrestricted with label-free outside data* protocol. Combining scores of four descriptors (SIFT, LBP, TPLBP and FPLBP), they achieve an accuracy of 90.90 ± 1.48 % on LFW.

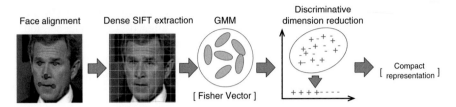

Fig. 3 The Fisher vector face encoding work-flow [85]

2.3.17 Blessing of Dimensionality: High-dimensional Feature and Its Efficient Compression for Face Verification, 2013 [29]

We discussed this paper under the *unrestricted with labeled outside data* protocol. Without using the WDRef data set for training, they report an accuracy of 93.18 ± 1.07 % under the *unrestricted with label-free outside data* protocol.

2.3.18 Fisher Vector Faces in the Wild, 2013 [85]

In this paper, Simonyan et al. [85] adopt the Fisher vector (FV) for face verification. The FV encoding had been shown to be effective for general object recognition. This paper demonstrates that this encoding is also effective for face recognition. To address the potential high computational expense due to the high dimensionality of the Fisher vectors, the authors propose a discriminative dimensionality reduction to project the vectors into a low dimensional subspace with a linear projection.

To encode a face image with FV, it is first processed into a set of densely extracted local features. In this paper, the dense local feature of an image patch is the PCA-SIFT descriptor augmented by the normalized image patch location in the image. They train a Gaussian mixture model (GMM) with diagonal covariance over all the training features. As shown in Fig. 3, to encode a face image with FV, the face image is first aligned with respect to the fiducial points. The Fisher vector is then the stacked, average first and second order differences of the image features over each GMM component center. To construct a compact and discriminative face representation, the authors propose to adopt a large-margin dimensionality reduction step after the Fisher vector encoding.

In their experiments, they report their best result as 93.03 ± 1.05 % accuracy on LFW under the *unrestricted with label-free outside data* protocol.

2.3.19 Fusing Robust Face Region Descriptors via Multiple Metric Learning for Face Recognition in the Wild, 2013 [32]

In this paper, the authors present a region-based face representation. They divide each face image into spatial blocks and sample image patches from a fixed grid of positions. The patches are then represented by nonnegative sparse codes and sum pooled to construct the representation for the block. PCA whitening is then applied to reduce its dimensionality. After processing each image into a sequence of block representations, the distance between two images are the fusion of pairwise block-to-block distances. They further propose a metric learning method to jointly learn the sequence of Mahalanobis matrices for discriminative block-wise distances. Their best result on LFW is $89.35 \pm 0.50\%$ (*image-restricted with label-free outside data*) fusing 8 distances from two different scales of face images and four different spatial partitions of blocks.

2.3.20   Towards Pose Robust Face Recognition, 2013 [108]

This paper presents a pose adaptive framework to handle pose variations in face recognition. Given an image with landmarks, they present a fitting algorithm to fit a 3D shape of the given face. The 3D shape is used to project the pre-defined 3D feature points to the 2D image to reliably locate facial feature points. They then extract descriptors around the feature points with Gabor filtering and concatenate local descriptors to represent the face. In their method, an additional technique to address self-occlusion is to use descriptors from the less-occluded half face for matching. In their experiments, they show that this pose adaptive framework can handle pose variations well in unconstrained face recognition. They obtain $87.77 \pm 0.51\%$ accuracy on LFW (*image-restricted with label-free outside data*).

2.3.21   Similarity Metric Learning for Face Recognition, 2013 [23]

This paper presents a framework to learn a similarity metric for unconstrained face recognition. The learned metric is expected to be robust to the large intra-personal variation and discriminative in order to differentiate similar image pairs from dissimilar image pairs. The robustness is introduced by projecting the face representations into the intra-personal subspace, which is spanned by the top eigenvectors of the intra-personal covariance matrix after the whitening process. After mapping the images to the intra-personal subspace, the discrimination is incorporated in learning the similarity metric. The similarity metric is defined as the difference of the image pair similarity against the distance measure, parameterized by two matrices respectively. The matrices are learned by minimizing the hinge loss and regularizing the two matrices to identity matrices. In their experiments, they use LBP and TBLBP descriptors on the LFW-a data set and SIFT descriptors on the LFW *funneled* data set computed at nine facial key points. Under the *image-restricted with label-free outside data* protocol, combining six scores from the three descriptors and their square roots variants they achieve $89.73 \pm 0.38\%$ accuracy. Under the *unrestricted with label-free outside data* protocol, they generate more training pairs with the identity labels and improve the accuracy to $90.75 \pm 0.64\%$.

2.3.22   Fast High Dimensional Vector Multiplication Face Recognition, 2013 [12]

In this method, the authors propose the over-complete LBP (OCLBP) descriptor, which is the concatenation of LBP descriptors extracted with different block and radius sizes. The OCLBP based face descriptor is then processed by Whiten-PCA and LDA. They further introduce a non-linear dimensionality reduction technique Diffusion Maps (DM) with the proposed framework. Extensive experiments are conducted with different local features and dimensionality reduction methods combinations. They report $91.10 \pm 0.59\%$ accuracy under the *image-restricted with label-free outside data* protocol and $92.05 \pm 0.45\%$ under the *unrestricted with label-free outside data* protocol.

2.3.23   Discriminative Deep Metric Learning for Face Verification in the Wild, 2014 [41]

In this method, referred to as DDML, a verification loss between pairs of faces is directly incorporated into a deep neural network, resulting in a non-linear distance metric that can be trained end-to-end using the back-propagation algorithm. The rationale for the verification loss is that the squared Euclidean distance between positive pairs is smaller than that between negative pairs, formulated as a large margin metric learning problem. The network is initialized randomly with three layers and *tan h* as the nonlinear activation function. They use 80×150 crops of the *LFW-a* (aligned) data set and extract Dense SIFT (45 SIFT descriptors from 16×16 non-overlapping patches, resulting in a 5760-dimensional vector), LBP features (10×10 non-overlapping blocks to get a 7080-dimensional vector) and Sparse SIFT (SIFT computed on nine fixed landmarks at three scales on the *funneled* LFW images, resulting in a 3456-dimensional vector). These features are projected down to 500 dimensions using PCA Whitening. Multiple features are fused at the score level by averaging. Their final accuracy is $90.68 \pm 1.41\%$ (*image-restricted with label-free outside data*).

2.3.24   Large Margin Multi-Metric Learning for Face and Kinship Verification in the Wild, 2014 [42]

In this paper, a large margin multi-metric learning (LM^3L) method is proposed to exploit discriminative information from multiple features of the same face image. Extracting multiple features from the face images, the distance between two face images is the weighted sum of Mahalanobis distances in each image feature. LM^3L jointly learn the distance metrics in different features and the weights of the features by optimizing each distance metric to be discriminative while minimizing the difference of distances in different features of the same image pair. They evaluate the method on the LFW-a data set with SIFT, LBP and Sparse SIFT features. With all the features, they achieve an accuracy of 89.57 ± 1.53 % under the *image-restricted with label-free outside data* protocol.

2.3.25 Effective Face Frontalization in Unconstrained Images, 2014 [40]

To show the importance of 3D frontalization to the task of face verification, in this paper Hassner et al. [40] evaluate their *alignment technique* using an earlier face recognition method [104], so that the impact of 3D alignment is not subsumed by the representation power of a more powerful model like the deep network.

Prior work in 3D frontalization of faces [97] would try to reconstruct the 3D surface of a face and then use this 3D model to general views, usually of a canonical pose. This paper explores the alternative of using a single 3D reference surface, without trying to modify the 3D head model to fit every query face's appearance. Although the exact head shape of a query face would be containing discriminative information regarding identity, the final 3D shape fitted to the query face would be an approximation largely dependent upon the accuracy of facial landmark localization. Solving the simpler problem by using an unmodified 3D shape model is shown to give qualitatively equivalent frontalization results, and performance improvement over 2D keypoint alignment methods is demonstrated on face verification and gender estimation tasks.

The frontalized faces of LFW, termed "LFW3D", provided a 3 % boost over the LFW-a aligned images. By combining multiple feature descriptors and models by stacking linear SVM scores, they reach an accuracy of 91.65 ± 1.04 % on the *image-restricted with label-free outside data* protocol of LFW.

2.3.26 Multi-Scale Multi-Descriptor Local Binary Features and Exponential Discriminant Analysis for Robust Face Authentication, 2014 [75]

In this paper, the authors represent an face image as the concatenation of region based descriptors which are stacked histograms of local descriptors over multiple scales. They further utilize the exponential discriminant analysis (EDA) to address the small-sample-size problem in LDA to learn a discriminative subspace for the face image feature. And they adopt the within class covariance normalization to project the feature after EDA into a subspace, in which the directions contribute to large intra-class distances have lower weights. They obtain their best result 93.03 ± 0.82 % on LFW (*image-restricted with label-free outside data*) by fusing scores from three different local features.

2.4   No Outside Data Protocols

The most restrictive LFW protocols are the "no outside data" protocols, including *image restricted with no outside data* and *unrestricted with no outside data*. We present these results together as well.

2.4.1   Face Recognition Using Eigenfaces, 1991 [99]

Turk and Pentland [99] introduce the eigenpicture method by Sirovich and Kirby [86] to face recognition. The eigenfaces approach they developed is a very important face recognition method in early years. The eigenfaces are the eigenvectors spanning the PCA subspace of a set of training faces. To recognize the unseen face, it is projected to a low-dimensional subspace with the eigenfaces and compared to the average face of each person.

As an early work on face recognition, it is mainly for recognizing frontal faces. Because it assumes faces are well aligned, the PCA subspace keeps mostly variations related to the identity which spans a good "face space". And after projecting faces into the "face space", the representations are all low-dimensional weight vectors. As a result, they can build a near-real-time face recognition system with this eigenface approach for both face detection and recognition. This is an impressive progress considering the limited computational power in early years.

The eigenface approach is designed for well-aligned frontal faces. For the real-world faces in LFW, it achieves $60.02 \pm 0.79\%$ verification accuracy in the *image-restricted with no outside data* protocol.

2.4.2   Learning Visual Similarity Measures for Comparing Never Seen Objects, 2007 [74]

Nowak and Jurie [74] present a method to recognize general objects. Without having the class labels in training stage, they present a method to learn to differentiate if two images are for the same object from training image pairs with only "same" and "different" labels. This is a typical setting for the *image-restricted with no outside data* protocol on LFW.

In the proposed method, they first extract corresponded image patches from the image pair. Then the differences between the corresponded image patches are quantized with an ensemble of randomized binary trees to obtain a vectorized representation for the image pair. A binary linear SVM is applied to the vectorized representation to predict whether the image pair is the "same" or "different".

This method achieves $72.45 \pm 0.40\%$ accuracy on the original LFW data set and $73.93 \pm 0.49\%$ with the *funneled* LFW data set.

2.4.3   Unsupervised Joint Alignment of Complex Images, 2007 [45]

This is the method that generated the *funneled* LFW data set. In this paper, Huang et al. [45] present a method to align images unsupervisedly in the *image-restricted with no outside data* setting of LFW. It is observed that the face recognition accuracy is improved when the recognition method is applied after an alignment stage. The method extends the congealing-style [56] method to handle real-world images. Compared with other domain specific alignment algorithms, congealing does not require manual labeling of specific parts of the object in the training stage. In congealing, a distribution field is defined as the sequence of feature values at a pixel location across a sequence of images. The congealing process is to iteratively minimize the entropy of the distribution field by applying affine transformations to the images. In this work, they use soft quantized SIFT features in congealing.

It shows that with the images aligned by this proposed method, the verification accuracy is improved. For example, the method by Nowak and Jurie [74] achieves $72.45 \pm 0.40\%$ accuracy on the original LFW data set but is improved to $73.93 \pm 0.49\%$ after aligning images with the proposed method.

2.4.4   Descriptor Based Methods in the Wild, 2008 [104]

In this paper, Wolf et al. [104] evaluate the descriptor-based methods on LFW with LBP descriptor, Gabor filter and two variants of LBP descriptor named Three-Patch LBP (TPLBP) and Four-Patch LBP (FPLBP). The TPLBP and FPLBP are produced by comparing the values of three or four patches to produce a bit value in the code assigned to each pixel. For each descriptor, they use both the Euclidean distance and Hellinger distance to evaluate the similarity of a face pair. Then they train a linear SVM to fuse the eight kinds of prediction scores and achieve an improved performance after fusion.

Besides the evaluation of these descriptor-based methods, they also adopt the one-shot learning for face verification. In this method, a binary classifier is learned online using one face in the given face pair as positive example with a set of negative examples. The binary classifier then evaluates the other face image to obtain the one-shot similarity (OSS). The same process is applied for each face in the pair to obtain an average similarity score of the face pair. They evaluate this method on LFW with the four descriptors and their element-wise square root variants. Combining the 8 scores also improve the accuracy.

Their best result on LFW is $78.47 \pm 0.51\%$ by fusing all 16 scores with a linear SVM, under the *image-restricted with no outside data* protocol.

2.4.5 Multi-Region Probabilistic Histograms for Robust and Scalable Identity Inference, 2009 [81]

Sanderson and Lovell [81] present a region-based face representation. They divide each face into several fixed regions. 2D DCT (Discrete Cosine Transform) features are extract densely from each region. Then a soft quantized histogram is constructed for each region with a Gaussian mixture model as the visual dictionary. The distance between two faces are defined as the average L_1 distances of the corresponded region histograms.

They also propose a two-step method in constructing the soft histogram for acceleration. The Gaussian components are clustered into K clusters. The K Gaussian components nearest to the cluster center are evaluated first in the histogram construction to obtain K likelihoods Then the Gaussian components are evaluated cluster by cluster in the descending order with respect to the likelihoods until the total number of evaluated Gaussian components exceeds the threshold.

The above distance between two faces is normalized by dividing the average pairwise distance of the two faces and a set of cohort faces. They observe this distance normalization method is effective that it brings additional 2.57 % average accuracy. This work achieves $72.95 \pm 0.55 \%$ accuracy on LFW (*image-restricted with no outside data*).

2.4.6 How Far Can You Get with a Modern Face Recognition Test Set Using Only Simple Features? 2009 [78]

Pinto et al. [78] present that it is possible to achieve a good recognition performance on LFW by combining several low-level simple features. They extract 48 variants of V1-like features by varying parameters such as the size of Gabor filters and spatial frequencies. To combine the effectiveness of different features, they adopt the multiple kernel learning (MKL) to jointly learn a weighted linear combination of the 48 kernels and the parameters of the kernel SVM for classification. Their best result on LFW is $79.35 \pm 0.55 \%$ following the *image-restricted with no outside data* protocol.

2.4.7 Probabilistic Elastic Matching for Pose Variant Face Verification, 2013 [60]

Li et al. [60] present an elastic matching method to handle the pose variations in face verification, reporting results under the *image-restricted with no outside data* protocol of LFW. Without relying on a sophisticated face alignment system, they resort to identify the corresponded regions to compare with in matching two faces. As long as the selected corresponded regions are from a semantically consistent face part, the matching could be invariant to pose variations. In their method, a set of face part models as a Gaussian mixture model (GMM) is

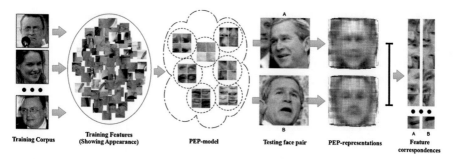

Training Corpus   **Training Features**   **PEP-model**   **Testing face pair**   **PEP-representations**   **Feature**
       (Showing Appearance)   **correspondences**

Fig. 4 The training and testing work-flow of the probabilistic elastic matching [60]

learned over all training features. The feature is densely extracted local descriptor augmented by the spatial locations of the image patch in the image. Incorporating the spatial information at the feature-level make each Gaussian component of the GMM capture the joint spatial-appearance distribution of certain face structure. With this GMM, a face can be represented as a sequence of features each of which induces the highest probability on a Gaussian component of the GMM.

In the experiments, they center crop the face images to 150×150 and densely extract local descriptors. Given an image, they concatenate the selected sequence of features to be its face representation. An image pair is then represented as the element-wise difference of the two face representations. A SVM is trained from matched and mismatched face pairs for face verification. In their following-up work, they name the GMM the Probabilistic Elastic Part (PEP) model and the face representation is named PEP-representation. The work-flow is illustrated in Fig. 4. The best result reported in the paper is $84.08 \pm 1.20\%$ on the *funneled* LFW fusing the prediction scores obtained with the SIFT and LBP features with a linear SVM.

2.4.8 Efficient Processing of MRFs for Unconstrained-Pose Face Recognition, 2013 [9]

Arashloo and Kittler [9] present a method to handle the pose variations via dense pixel matching across face images with MRFs. They propose to reduce the processing time of inference in MRF-based image matching by parallelizing the computation on GPU cores. The major contribution of this paper is how it parallelizes the computation on GPU cores. After adopting the dual decomposition for the MRF optimization, the original problem is decomposed into a set of subproblems. To efficiently solve the subproblems, they further propose several techniques such as incremental subgradient updates and multi-resolution analysis. After obtaining the image matching, multi-scale LBP descriptors are extracted from matched image regions. They stack the descriptors, apply PCA and use the cosine similarity score for face verification. Their best result on LFW is $79.08 \pm 0.14\%$ under the *image-restricted with no outside data* protocol.

2.4.9 Fisher Vector Faces in the Wild, 2013 [85]

We discussed this paper under the *unrestricted with label-free outside data* protocol. They also report their result under the restricted protocol, in which they obtain $87.47 \pm 1.49\%$ accuracy on LFW.

2.4.10 Eigen-PEP for Video Face Recognition, 2014 [61]

Li et al. [61] develop the Eigen-PEP method upon their early work [60]. With the probabilistic elastic matching, a face image or a face track can be represented as a set of face parts. Since the faces are implicitly aligned in a part-based representation, the similar idea from the eigenfaces [99] is adopted here to build a low-dimensional face representation. They use the joint Bayesian classifier [28] for verification. They construct a two-frame face track for each image by adding the mirrored face and achieve $88.97 \pm 1.32\%$ accuracy on the *funneled* LFW in this paper (*image-restricted with no outside data*).

2.4.11 Class-Specific Kernel Fusion of Multiple Descriptors for Face Verification Using Multiscale Binarised Statistical Image Features, 2014 [79]

In this paper, Arashloo et al. [79] address the pose variations via dense pixel matching with their prior work [9]. They then extract three kinds of descriptors, the multi-scale binarized statistical image feature, the multi-scale LBP and the multi-scale local phase quantization feature from the matched image regions. The image representations are embedded into a discriminative subspace with a class-specific kernel discriminant analysis approach. Their best result on the *funneled* LFW data set is $95.89 \pm 1.94\%$ achieved by combining the results of the three image representations (*image-restricted with no outside data*).

2.4.12 Hierarchical-PEP Model for Real-World Face Recognition, 2015 [59]

In this paper, Li and Hua [59] present a Hierarchical-PEP model to hierarchically apply the probabilistic elastic part (PEP) model combined with a PCANet [27] to achieve an improved face verification accuracy. They point out that the parts selected after the elastic matching could still present significant visual appearance variations due to the pose variations of the faces. Applying the PEP model to the parts could further introduce pose-invariance in the part representations. After that, the dimensionality of the part representation is discriminatively reduced by a net of

PCA and Linear Discriminant Embedding (LDE). They achieve $91.10 \pm 1.47\%$ accuracy in this paper on the *funneled* LFW data set under the *image-restricted with no outside data* protocol.

2.5 Other Methods

Here we include those methods for which details are too brief to merit a separate section. These are usually proprietary methods from commercial systems where in-depth detail is not available.

2.5.1 Colour & Imaging Technology (TCIT), 2014 [4]

TCIT calculates the average position of the facial area and judges the identical person or other person by face recognition using the facial area. Face Feature Positioning is applied to get the face data template which is used to verify different faces. They report an accuracy of $93.33 \pm 1.24\%$ *(unrestricted with labeled outside data)*.

2.5.2 betaface.com, 2014 [1]

They have used original LFW images, converted to grayscale, auto-aligned with their alignment system and followed unrestricted protocol with labeled outside data. LFW data was not used for training or fine-tuning. Their reported accuracy is $98.08 \pm 0.16\%$ *(unrestricted with labeled outside data)*.

2.5.3 insky.so, 2015 [2]

They used original LFW images to run the test procedure, without doing any training on the LFW images. They report $95.51 \pm 0.13\%$ accuracy *(unrestricted with labeled outside data)*.

2.5.4 Uni-Ubi, 2015 [5]

They used original LFW images, converted to grayscale, auto-aligned with their face detector and alignment system. LFW was not used for training or fine-tuning. They report $99.00 \pm 0.32\%$ accuracy *(unrestricted with labeled outside data)*.

2.5.5 VisionLabs ver. 1.0, 2013 [6]

The method makes use of metric learning and dense local image descriptors. External data is only used implicitly for face alignment. They report $92.90 \pm 0.31\,\%$ accuracy for the unrestricted training setup *(unrestricted with label-free outside data)*, using LFW-a aligned images.

2.5.6 Aurora Computer Services Ltd: Aurora-c-2014-1, 2014 [7]

The face recognition technology is comprised of Aurora's proprietary algorithms, machine learning and computer vision techniques. They report results using the *unrestricted with label-free outside data* training protocol, achieving $93.24 \pm 0.44\,\%$. The aligned and funneled sets and some external data were used solely for alignment purposes.

3 Pose and Alignment

One of the most significant issues in face verification is how to address variations in pose. For instance, consider the restricted case in which both faces are guaranteed to be from the same pose, but the pose may vary. The most informative features for comparison will likely change if presented with two profile faces versus two frontal faces. An ideal verification system would presumably account for these differences.

Even more vexing than the above case of how to select features conditioned on pose, however, is the more general problem of how to compare two images that exhibit significantly different poses. Many of the errors seen in the top systems show that these situations are among the most difficult to address (see Figs. 5 and 6). Because pose is a cross-cutting issue that virtually every verification system must address in some fashion, we treat it as a separate topic in this section.

LFW was designed to fit into what we call the Detection-Alignment-Recognition pipeline. In particular, by including in LFW only images from the OpenCV Viola-Jones face detector, the designers facilitated the building of end-to-end face recognition systems. Given a recognizer that works well on LFW, the practitioner can pair this with the Viola-Jones face detector to produce an end-to-end system with more predictable performance.

A consequence of the decision to use only faces detected by this specific detector, however, is that most LFW faces are within 20 degrees of frontal, and just a small percentage show some greater degree of yaw angle. This makes addressing pose in LFW a bit easier than it might be for databases with even greater pose variation, such as the recent IJB-A database [52]. Still, the techniques used on LFW to address pose encompass a wide range of strategies and can be expected to be incorporated into systems designed for new and more difficult benchmarks.

There are many approaches to addressing pose in verification problems. These include

1. aligning the input images, either by transforming both to a canonical pose, or by transforming one of them to the other;
2. building mappings that allow inference of what one view looks like given another view;
3. conditioning on pose, such as building separate classifiers for each category of pose pairs;
4. having no explicit mechanism for addressing pose, but rather providing a learning algorithm, and enough training data, so that a model can learn to compare images across pose.

In this section, we review some of the mechanisms that authors have used to address pose variation in LFW, and their relative successes and drawbacks. Tables 3 and 4 enumerate all of the alignment methods used in the papers reviewed in this survey. They are grouped by strategy of alignment (alignment type). The papers using a specific method are given in the rightmost column of the tables.

3.1 Alignment, Transformation, and Part Localization

Probably the most common way of addressing pose changes is to attempt to transform images to a canonical pose or position as a pre-processing step. Because LFW images are the results of detections of the Viola-Jones face detector [101], they are already roughly centered and scaled. However, it seems intuitive that improving the consistency of the head position in preprocessing should improve verification performance. Huang et al. [45] were the first authors to show that alignment improves verification performance on LFW, for at least two different alignment methods.

Landmark-based methods. One common way to align face images is to find landmarks, such as the corners of the eyes and the mouth. Once the landmarks have been detected, one can either transform the image such that the landmarks are placed into a standard position, or simply sample patches or features at the landmark locations. This approach has been taken by many authors. These methods are shown in Tables 3 and 4 under the alignment type of *Landmark* [1, 3–5, 8, 14, 26, 33, 46, 63, 90, 106, 107, 110].

In particular, the LFW-a alignment [106] was widely used by many verification systems. These images were produced by aligning seven fiducial points to fixed locations with a similarity transform. Subsequent methods explored improving the accuracy of the landmark detectors. For instance, Sun et al. [90] performed detection using a deep convolutional network cascade, which allowed for using larger context and implicit geometric constraints, leading to better performance in difficult conditions due to factors such as occlusion and extreme pose angles. Other methods have explored fitting a larger number of landmarks (generally more than

Table 3 Alignment techniques: part 1

Alignment type	Common name	Brief description	Method reference	Method usage in LFW
None	–	No pre-processing. Use of raw LFW images	–	Eigenfaces [99], Nowak [74], Multi-Region histograms [81], Learning-based descriptor [26] (no global alignment, but parts aligned), Associate-predict [110], FaceNet [82]
Manual alignment	–	Used Machine Perception Toolbox from MPLab, UCSD to detect eye location, manually corrected eye coordinates for worst 2000 detections, used coordinates to rotate and scale images	Nair et al. [70]	NReLU [70]
Congeal1	Funneling or SIFT-congealing	(1) GMM on SIFT features (2) Jointly align images to minimize entropy of SIFT cluster IDs.	Huang et al. [45]	Nowak [74], Funneling [45], Hybrid descriptor-based [104], HT Brain-inspired [78], LDML-MKNN [39], PEP [60], Eigen-PEP [61], POP-PEP [59]
Congeal2	Deep funneling or deep congealing	(1) Boltzmann machine model of unaligned face images (2) Adjust image to maximize its likelihood under model	Huang et al. [44]	–
MRF based	MRF-based alignment	MRF for matching is done using Daisy features and multi-scale LBP histograms. It starts by using images from LFW-a or LFW-funneled (Congeal1)	Arashloo et al. [9]	MRF-MLBP [9], MRF-fusion-CSKDA [79]
Landmark1	Buffy	(1) Build classifiers for each of K different landmarks (2) Similarity transform of detected landmarks	Everingham et al. [33]	SFRD+multiple-metric [32], Fisher Vector Faces [85]

Landmark2	MERL alignment	(1) Nine landmarks located using Viola-Jones detector (2) Similarity transform puts landmarks in standard position	Nowak-MERL [46]	Nowak-MERL [46]
Landmark3	LFW-a	This is a slightly modified version of "Buffy" used by face.com	Wolf et al. [106]	Wolf et al. [106] (Journal version of [96] and [105]), Cosine [72], Large scale feature search [31], LARK [84], DML-eigen [111], Conv-DBN [48], CMD (Ensemble metric) [43], LBP PLDA [62], MRF-MLBP [9], Pose-robust [108], Similarity metric [23], DFD [58], VisionLabs [6], DDML [41], Face and Kinship [42], Multi-scale LBP [75]
Landmark4	Component-based discriminative search	(1) Detect possible modes or positions of face components (2) "Direction classifiers" used to find best alignment direction between image patch and face component	Liang et al. [63]	Learning-based [26], Joint-Bayesian [28]
Landmark5	Explicit shape regression (5 landmark rectification)	Coarse to fine regression *Note:* the primary benefit of this method is not really to "align" the image but rather to find landmarks which are used as conditional feature locations	Cao et al. [24]	"Blessing" of dimensionality [29], TL-Joint Bayesian [25]

This table and Table 4 summarize the various alignment techniques used in conjunction with LFW

Table 4 Alignment techniques: part 2

Alignment type	Common name	Brief description	Method reference	Method usage in LFW
Landmark6	Associate-predict face alignment	Four landmarks are detected using a standard facial point detector and used to determine twelve facial components	Yin et al. [110]	Associate-Predict Model [110]
Landmark7	Consensus of exemplars	Performs the alignment based not on the part locations in the image itself, but on "generic" parts—where the parts would be for an average person (using 120 reference faces) with the same pose and expression as the test image. Avoids over-alignment which could distort identity information	Belhumeur et al. [14]	Tom-vs-Pete [16], POOF [17]
Landmark8	CNN for facial landmark	3-Level cascaded CNN, with the input as the face region from a face detector, and each level regressing to the 5 output keypoints	Sun et al. [90]	Hybrid CNN-RBM, 2013 [91], DeepID [92]
Landmark9	SDM (Intraface)	At training, SDM learns the sequence of descent directions that minimizes the mean of sampled NLS (non-linear least squares) functions. At test time, these learned directions are used instead of the Jacobean or Hessian, which makes it much faster computationally	Xiong et al. [107]	DeepID2 [88], DeepID2+ [93], DeepID3 [89]
Landmark10	TCIT	Commercial system for face alignment	TCIT [4]	TCIT [4]
Landmark11	betaface	Commercial system for face alignment	betaface.com [1]	betaface.com [1]

Landmark12	Uni-Ubi	Commercial system for face alignment	Uni-Ubi [5]	Uni-Ubi [5]
Landmark13	Tencent-BestImage	Commercial system for face alignment	Tencent-BestImage [8]	Tencent-BestImage [8]
Landmark14	OKAO Vision	Commercial system for face alignment	OMRON [3]	OMRON [3]
3D-1	3D pose normalization	Uses a 3D head model for normalizing pose. Very similar to the DeepFace approach	3D pose normalization, 2011 [11]	–
3D-2	DeepFace	(1) Landmark-based 2D alignment (6 landmarks) (2) Dense landmark identification (67 landmarks) (3) Iterative projection from 3-D mask to estimate pose (4) Reproject landmarks and image from frontal pose	Taigman et al. [97]	Billion Faces [95], DeepFace [97], Effective face frontalization [40]
3D-3	Fast 3D Model Fitting	(1) Detects landmarks using a three-view Active Shape Model (2) Solves for pose and shape by matching 34 landmarks to 3D vertex index on a deformable face model	Yi et al. [108]	Towards Pose Robust FR [108]

(continued)

Table 4 (continued)

			High-Fidelity Pose Normalization [115]
3D-4	3D Morphable Model	(1) Detects landmarks using SDM (Landmark11) on image (2) Does pose-adaptive filtering of the 3DMM to handle non-correspondences between 2D and 3D landmarks	Zhu et al. [115]
View	Recover canonical-view face	Recovers the canonical view of a face using a deep neural network (commercial system)	Zhu et al. [114] — CNN view-recovery [114]

This table and Table 3 summarize the various alignment techniques used in conjunction with LFW

50) to face images, using techniques such as boosted regression in Cao et al. [24], or through approximate second order optimization in Xiong and De la Torre [107].

As one moves from similarity transforms to more complex classes of transformations for producing alignments, a natural question is whether discriminative verification information may be lost in the alignment process. For instance, if an individual's face has a narrow nose, and landmarks are placed at the extremes of the width of the nose, then positioning these landmarks into a canonical position will remove this information.

Berg and Belhumeur [16] addressed this issue in the context of their piecewise affine alignment using 95 landmarks. In order to preserve identity information, they warped the image not based on the detected landmarks themselves, but rather by the inferred landmarks of a generic face in the same pose and expression as the test image to be aligned. This is accomplished by using a reference data set containing 120 individuals. For each individual, the image whose landmark positions most closely match the test image is found, and these landmark positions are then averaged across all the subjects to yield the generic face landmarks. By switching from a global affine alignment to a piecewise alignment, they increase the accuracy of their system from 90.47 to 91.20 %, and by additionally using their identity-preserving generic warp, they achieve a further increase in accuracy to 93.10 %.

Note that since all of these landmark-based methods rely on the training of landmark detectors, they require additional labeling beyond that provided by LFW, and hence require any verification methods which use them to abandon the category of *no outside data*. The unsupervised methods, discussed next, do not have this property.

Two-dimensional unsupervised joint alignment methods (congealing). In contrast to methods that rely on trained part localizers, other methods are unsupervised and attempt to align methods using image similarity. One group of such methods is known as *congealing* [44, 45, 68]. In congealing, a set of images are *jointly aligned* by transforming each image to maximize a measure of similarity to the other images. This can be viewed as maximizing the likelihood of each image with respect to all of the others, or alternatively, as minimizing the entropy of the full image set. Once a set of images has been aligned, it can be used to produce a "machine" that aligns new image samples efficiently. This new machine is called a *funnel*. Thus, images aligned with congealing are referred to as *funneled* images. Since congealing can be done using only the training set images for a particular test set, it relies on no additional annotations, and is compatible with the *no outside data* protocols.

The LFW web site provides a two additional versions of the original LFW images that have been aligned using congealing. The first is referred to as *funneled*. In this version of the database, each image was processed with the congealing method of Huang et al. [45]. This method was shown to improve classification rates over some of the early landmark-based alignment methods, but was not as effective as some of the later landmark methods, such as the one used in LFW-a.

An improved version of congealing was developed [44], and was used to produce another version of LFW, known as the *deep-funneled* version. This method used a feature representation learned from a multi-layer Boltzmann machine to align images under the congealing framework. This unsupervised method appears to be comparable to most of the landmark-based methods with respect to the final classification accuracy, and has the advantage of being unsupervised.

One other notable unsupervised method was presented by Arashloo et al. [9, 79] in two separate papers. They start from the *funneled* LFW images and use a Markov random field to further warp the images so that they are more similar.

Frontalization and other methods using 3D information. Another idea to handle differences in views is to attempt to transform views to a canonical frontal pose, sometimes known as *frontalization*. This is clearly beyond the abilities of methods which only perform affine or landmark-based alignment, since the process of transforming a profile face view to a frontal view requires an implicit understanding of the geometry of the head, occluded areas, and the way other features, such as hair, appear from different perspectives.

Early work along these lines was done at Mitsubishi [10, 11], although this did not result in state-of-the-art results on LFW. More recently, Taigman et al. [97] developed a frontalization method that contributed a modest improvement to accuracy on LFW, although most of their gains are attributable to their CNN architecture and the large training sets. Finally, two other methods are essentially landmark-based, but used 3D models to fit 3D landmark coordinates to 2D images [108, 115].

3.2   Conditioning on Pose Explicitly

Rather than transforming images so that they are all approximately frontal, another approach to dealing with pose variability is to apply strategies separately to different types of image pairs. For example, if one classifies each input image as left-facing (A), frontal (B), or right-facing (C), then we can define nine types of input pairs: AA, AB, AC, BA, BB, BC, CA, CB, CC. One approach is to train separate classifiers for each group of these images, focusing on the peculiarities of each group. By reflecting right-facing images (C), to be left-facing (B), we can reduce the total number of pair categories to just four: AA, AB, BA, BB, although doing this may eliminate information about asymmetric faces. As mentioned in Sect. 2.2.2, this approach was used by Cao et al. [26].

The associate-predict model proposed by Yin et al. [110] uses the above strategy to separate pairs of test images into those pairs that have similar pose (AA, BB, and CC), which the authors refer to as *comparable* images, and those that do not have similar pose. For the comparable images, the authors run a straightforward computation of an image distance. For the non-comparable images, the authors "associate" features of a face with the features of a set of reference faces, and "predict" the appearance of the feature from a new viewpoint by using the feature appearance from the closest matching reference person, in the desired view.

3.3   Learning Our Way Out of the Pose Problem

As discussed in Sect. 2, almost all of the current dominant methods use some CNN architecture and massive training sets. One of the original motivations for using convolutional neural networks was to introduce a certain amount of invariance to the position of the inputs. In addition, max-pooling operators, which take the maximum feature response over a neighborhood of filter responses, also introduce some invariance to position.

However, the invariance introduced by CNNs and max-pooling can also eliminate important positional information in many cases, and it may be difficult to analyze whether the subtle geometrical information required to discriminate among faces is preserved through these types of operations. While many deep learning approaches have shown excellent robustness to small misalignments, all that we are aware of continue to show modest improvements by starting with aligned images. Even the highest performing system (FaceNet [82]) improves from 98.87 % without explicit alignment to 99.63 % by using a trained alignment system. This seems to suggest that a system dedicated to alignment may relieve a significant burden on the discriminative system. Of course, given enough training data, such an advantage may dissolve, but at this point it still seems worthwhile to produce alignments as a separate step in the process.

4   The Future of Face Recognition

As this chapter is being written, the highest reported accuracy on LFW described by a peer-reviewed publication stands at $99.63 \pm 0.09 \%$, by Schroff et al. [82]. This method reported only 22 errors on the entire test set of 6000 image pairs. These errors are shown in Figs. 5 and 6. Furthermore, five of these 22 errors correspond to labeling errors in LFW, meaning that only 17 pairs represent real errors. Accounting for the five ground-truth errors in LFW, the highest accuracy should not go above $\frac{5995}{6000} \approx 99.9\%$, so the results for the protocol *unrestricted with labeled outside data* are very close to the maximum achievable by a perfect classifier.[8] With accuracy rates this high, it is time to ask the question "What next?" High accuracy on verification protocols does not necessarily imply high accuracy on other common face recognition protocols such as identification. In addition, some

---

[8]For a classifier to get more than 5995 of the 6000 test examples correct according to the benchmark, it must actually report the wrong answer on at least one of the five incorrectly labeled examples in LFW. Of course it is always possible that a classifier could get extremely lucky and "miss" just the right five examples that correspond to labeling errors in the database while getting all of the other examples, corresponding to correctly labeled test data, correct. However, a method that has a very low error rate overall, and at the same time "accidentally" reports the correct answers for the labeled errors, is likely to be fitting to the test data in some manner.

Fig. 5 All of the errors produced by the FaceNet verification algorithm [82] on matched pairs. The pairs surrounded in *red* are labeling errors in LFW. Thus, while these were flagged as errors, the FaceNet system actually gave the correct answer (correctly identifying these pairs as mismatches). The remainder of the pairs (without *red boxes*) were identified incorrectly as mismatches. They are in fact matches. It is interesting to note that the rightmost pair in the third row shows the actress Nicole Kidman, but in the rightmost image of this pair, she is wearing an artificial nose, in order to appear more like Virginia Woolf in the film *The Hours* [103]. Thus, this case represents an extreme variation of an individual that would not normally be encountered in everyday life, and it is not clear that one should train a system until this example is evaluated correctly

real-world applications of face recognition involve imaging that is significantly more challenging than LFW. Next, we explore some aspects of face recognition that still need to be addressed.

4.1 Verification Versus Identification

As discussed in Sect. 1, the LFW protocols are defined for the face verification problem. Even for such realistic images, the problem of verification, for some image pairs, can often be quite easy. It is not uncommon that two random individuals have

Fig. 6 All of the errors produced by the FaceNet verification algorithm [82] on mismatched pairs. These are the only pairs of mismatched images that were incorrectly reported as matched pairs by Schroff et al. [82]

large differences in appearance. In addition, given two images of the same person taken randomly from some distribution of "same" pairs, it is quite common that such images are highly similar. Thus, verification is, by its nature a problem in which many examples are easy.

For identification, on the other hand, the difficulty of identifying a person is directly related to the number of people in the gallery. With a small gallery, identification can be relatively easy. On the other hand, with a gallery of thousands or millions of people, identifying a probe image can be extremely difficult. The reason for this is simple and intuitive—the more people in a gallery, the greater the chance that there are two individuals that are highly similar in appearance.

It is for this reason that many standard biometric benchmarks use evaluation criteria that are independent of the gallery size, using a combination of the *true accept rate* (TAR) and *false accept rate* (FAR) for open set recognition. The true accept rate is defined to be the percentage of probes which, when compared to the matching gallery identity, are identified as matches. The false accept rate is the percentage of incorrect identities to which a probe is matched. Because it is defined as a percentage, it is independent of the gallery size. It is common to fix the FAR and report the TAR at this fixed FAR, as in "a TAR of 85% at a FAR of 0.1%".

To understand the relationship between accuracies on verification and identification, it is instructive to consider how a high-accuracy verification system might perform in a realistic identification scenario. In particular, consider a verification system that operates at 99.0 % accuracy. On average, for 100 matched pairs, and 100 mismatched pairs, we would expect it to make only two errors. Now consider such a system used in a closed set identification scenario with 901 gallery subjects. For example, this might represent a security system in a large office building.

In a typical case of identification under these parameters, in addition to matching the correct subject, we would expect 1 % of the 900 mismatched gallery identities to be rated as a "match" with the probe image by the verification classifier. That is, we would expect to have one correct identity and nine incorrect identities to be above the match threshold of our verification system. The job of the identification system would then be to sort these in the correct order by selecting the one true match as the "best match" from among the ten that were above threshold. This is quite difficult since by definition the ten selected images look like the probe identity. If we are successful at selecting the correct match from this set of ten similar identities 50 % of the time, which is already quite impressive, then the total identification rate is merely 50 %.

In larger galleries, the problem of course becomes even more difficult. In a pool of 9901 gallery subjects, achieving 50 % identification with a 99 % accurate identification system would require finding the correct identity from among 100 examples that looked similar to the probe. This informal analysis demonstrates why identification can be so much harder than verification. In addition, these examples describe closed set identification. Open set identification is even more difficult, as one must try to determine whether the probe is in the gallery at all.

4.2   New Databases and Benchmarks

In order to study the identification problem with a gallery and probe images, one needs a data set established for this purpose. Some authors have developed protocols from the images in LFW for this purpose, e.g. [20], sometimes by augmenting LFW images with other image sets [64]. Other authors have augmented the images in LFW to study image retrieval with large numbers of distractors [22]. However, the time is ripe for new databases and benchmarks designed specifically for new problems, especially identification problems. Several new databases aim to address these needs.

In this section, we discuss several new face recognition databases and benchmarks, and the new issues they allow researchers to address. We only include discussions of publicly available databases. These include the CASIA database of faces, the FaceScrub database, IJB-A database and benchmark from IARPA, and the MegaFace database from the University of Washington.

4.2.1   IJB-A Database and Benchmark

The recently announced IJB-A database [52] is designed to study the problems of open set identification, verification, and face detection. It includes both images and videos of each subject and establishes standard protocols.

The database includes images and videos from 500 subjects in unconstrained environments, and all media have creative commons licensing. In order to get a wider range of poses and other conditions than LFW, the images were identified and localized by hand, rather than using an automatic face detector (as with LFW) which is likely to be biased towards easier-to-detect faces.

One interesting element of the protocols provided with this database is that a distinction is made (for identification protocols) about whether a classifier was trained on gallery images or not. Another interesting aspect of this database is that probes are presented as media collections rather than single images. Thus, a probe may consist of a combination of individual images and video. Thus, this encourages exploration of how to best use multiple probe images at the same time to increase accuracy.

4.2.2 The FaceScrub and CASIA Data Sets

This section describes two distinct databases known as FaceScrub and CASIA. The FaceScrub data set [71] contains 107,818 images of celebrities automatically collected from the web, and verified using a semi-automated process. It contains 530 different individuals, with an average of approximately 200 images per person. As such it is an important example of a *deep* data set, rather than a *broad* data set, meaning that it has a large number of images per individual. The data set is released under a creative commons license, and the URLs, rather than the images themselves are released.

The CASIA-WebFace data set, or simply CASIA, consists of 494,414 images, and is similar in spirit to the FaceScrub data set. It is described here [109].

The automatic processing of the images in these databases has two important implications:

- First, because images that are outliers are automatically rejected, there is a limit to the degree of variability seen in the images. For example, heavily occluded images may be marked as outliers, even if they contain the appropriate subject.
- Second, it is difficult to know the percentage of correct labels in the database. While the authors could presumably estimate this fairly easily, they have not reported these numbers in either FaceScrub or CASIA.

Despite these drawbacks, these large and deep databases are two that are currently available to researchers to train large face recognition systems with large numbers of parameters, and because of that, they are valuable resources.

4.2.3 MegaFace

Another new database designed to study large scale face recognition is MegaFace [67], a database of one million face images derived from the Yahoo 100 Million Flickr creative commons data set [98]. This database, which contains one image each of one million *different* individuals, is designed to be used with other databases to allow the addition of large numbers of distractors.

In particular, the authors describe protocols that are used in conjunction with FaceScrub, described in the previous section. All of the images in MegaFace are first registered in a gallery, with one image each. Then, for each individual in FaceScrub, a single image of that person is also registered in the gallery, and the remaining images are used as test examples in an identification paradigm. That is, the goal is to identify the single matching individual from among the 1,000,001 identities in the gallery.

The paradigms discussed in this work are important in addressing the ability to identify individuals in very large galleries, or in the open set recognition problem. The authors show that several methods that perform well on the standard LFW benchmark quickly deteriorate as distractors are added. A notable exception is the FaceNet system [82], which shows remarkable robustness to distractors.

5 Conclusions

In this chapter, we have reviewed the progress on the Labeled Faces in the Wild database from the time it was released until the current slew of contributions, which are now coming close to the maximum possible performance on the database. We analyzed the role of alignment and noted that current algorithms can perform almost as well without any alignment after the initial face detection, although most algorithms do get a small benefit from alignment preprocessing. Finally, we examined new emerging databases that promise to take face recognition, including face detection and multimedia paradigms, to the next level.

Acknowledgements This research is based upon work supported in part by the Office of the Director of National Intelligence (ODNI), Intelligence Advanced Research Projects Activity (IARPA), via contract number 2014-14071600010. The views and conclusions contained herein are those of the authors and should not be interpreted as necessarily representing the official policies or endorsements, either expressed or implied, of ODNI, IARPA, or the U.S. Government. The U.S. Government is authorized to reproduce and distribute reprints for Governmental purpose notwithstanding any copyright annotation thereon.

Research reported in this publication was also partly supported by the National Institute Of Nursing Research of the National Institutes of Health under Award Number R01NR015371. The content is solely the responsibility of the authors and does not necessarily represent the official views of the National Institutes of Health. This work is also partly supported by US National Science Foundation Grant IIS 1350763 and GH's start-up funds from the Stevens Institute of Technology.

References

1. betaface.com, http://betaface.com (2015)
2. inksy.so, http://www.insky.so/ (2015)
3. Omron from Okao Vision, http://www.omron.com/technology/index.html (2009)

4. Taiwan Colour & Imaging Technology (TCIT), http://www.tcit-us.com/ (2014)
5. Uni-ubi, http://uni-ubi.com (2015)
6. Visionlabs ver. 1.0, http://www.visionlabs.ru/face-recognition (2014)
7. Aurora Computer Services Ltd: Aurora-c-2014-1, http://www.facerec.com/ (2014)
8. Tencent-Bestimage, http://bestimage.qq..com (2014)
9. S.R. Arashloo, J. Kittler, Efficient processing of MRFs for unconstrained-pose face recognition, in *2013 IEEE Sixth International Conference on Biometrics: Theory, Applications and Systems (BTAS)* (IEEE, Washington, 2013), pp. 1–8
10. A. Asthana, M.J. Jones, T.K. Marks, K.H. Tieu, R. Goecke, Pose normalization via learned 2D warping for fully automatic face recognition, in *The British Machine Vision Conference*, Citeseer, 2011
11. A. Asthana, T.K. Marks, M.J. Jones, K.H. Tieu, M. Rohith, Fully automatic pose-invariant face recognition via 3D pose normalization, in *2011 IEEE International Conference on Computer Vision (ICCV)* (IEEE, Washington, 2011), pp. 937–944
12. O. Barkan, J. Weill, L. Wolf, H. Aronowitz, Fast high dimensional vector multiplication face recognition, in *2013 IEEE International Conference on Computer Vision (ICCV)* (IEEE, Washington, 2013), pp. 1960–1967
13. E.Bart, S.Ullman, Class-based feature matching across unrestricted transformations. IEEE Trans. Pattern Anal. Mach. Intell. **30**(9), 1618–1631 (2008)
14. P.N. Belhumeur, D.W. Jacobs, D.J. Kriegman, N. Kumar, Localizing parts of faces using a consensus of exemplars. IEEE Trans. Pattern Anal. Mach. Intell. **35**(12), 2930–2940 (2013)
15. P.N. Belhumeur D.J. Kriegman, What is the set of images of an object under all possible illumination conditions? Int. J. Comput. Vis. **28**(3),245–260 (1998)
16. T. Berg, P.N. Belhumeur, Tom-vs-Pete classifiers and identity-preserving alignment for face verification, in *The British Machine Vision Conference*, vol. 2, Citeseer (2012), p. 7
17. T. Berg, P.N. Belhumeur, POOF: part-based one-vs.-one features for fine-grained categorization, face verification, and attribute estimation, in *2013 IEEE Conference on Computer Vision and Pattern Recognition (CVPR)* (IEEE, Washington, 2013), pp. 955–962
18. T.L. Berg, A.C. Berg, J. Edwards, D.A. Forsyth, Who's in the picture? in *Neural Information Processing Systems*, 2005
19. T.L. Berg, A.C. Berg, J. Edwards, M. Maire, R. White, Y.-W. Teh, E. Learned-Miller, D.A. Forsyth, Names and faces in the news, in *Proceedings of the 2004 IEEE Computer Society Conference on Computer Vision and Pattern Recognition, 2004 (CVPR 2004)*, vol. 2 (IEEE, Washington, 2004), p. II-848
20. L. Best-Rowden, H. Han, C. Otto, B.F. Klare, A.K. Jain, Unconstrained face recognition: identifying a person of interest from a media collection. IEEE Trans. Inf. Forensics Secur. **9**(12), 2144–2157 (2014)
21. D. Beymer, T. Poggio, Face recognition from one example view, in *Proceedings of the Fifth International Conference on Computer Vision, 1995* (IEEE, Washington, 1995), pp. 500–507
22. B. Bhattarai, G. Sharma, F. Jurie, P.Pérez, Some faces are more equal than others: hierarchical organization for accurate and efficient large-scale identity-based face retrieval, in *Computer Vision-ECCV 2014 Workshops* (Springer, Heidelberg, 2014), pp. 160–172
23. Q. Cao, Y. Ying, P. Li, Similarity metric learning for face recognition, in *2013 IEEE International Conference on Computer Vision (ICCV)* (IEEE, Washington, 2013), pp. 2408–2415
24. X. Cao, Y. Wei, F. Wen, J. Sun, Face alignment by explicit shape regression. Int. J. Comput. Vis. **107**(2), 177–190 (2014)
25. X. Cao, D. Wipf, F. Wen, G. Duan, J. Sun, A practical transfer learning algorithm for face verification, in *2013 IEEE International Conference on Computer Vision (ICCV)* (IEEE, Washington, 2013), pp. 3208–3215
26. Z. Cao, Q. Yin, X. Tang, J. Sun, Face recognition with learning-based descriptor, in *2010 IEEE Conference on Computer Vision and Pattern Recognition (CVPR)* (IEEE, Washington, 2010), pp. 2707–2714
27. T.-H. Chan, K. Jia, S. Gao, J. Lu, Z. Zeng, Y. Ma, PCANet: a simple deep learning baseline for image classification?. IEEE Tran. Image Process. **24**(12), 5017–5032 (2015)

28. D. Chen, X. Cao, L. Wang, F. Wen, J. Sun, Bayesian face revisited: a joint formulation, in *Computer Vision—ECCV 2012* (Springer, Heidelberg, 2012), pp. 566–579
29. D. Chen, X. Cao, F. Wen, J. Sun, Blessing of dimensionality: high-dimensional feature and its efficient compression for face verification, in *2013 IEEE Conference on Computer Vision and Pattern Recognition (CVPR)* (IEEE, Washington, 2013), pp. 3025–3032
30. S. Chopra, R. Hadsell, Y. LeCun, Learning a similarity metric discriminatively, with application to face verification, in *IEEE Computer Society Conference on Computer Vision and Pattern Recognition, 2005 (CVPR 2005)*, vol. 1 (IEEE, Washington, 2005), pp. 539–546
31. D. Cox, N. Pinto, Beyond simple features: a large-scale feature search approach to unconstrained face recognition, in *2011 IEEE International Conference on Automatic Face & Gesture Recognition and Workshops (FG 2011)* (IEEE, Washington, 2011), pp. 8–15
32. Z. Cui, W. Li, D. Xu, S. Shan, X. Chen, Fusing robust face region descriptors via multiple metric learning for face recognition in the wild, in *2013 IEEE Conference on Computer Vision and Pattern Recognition (CVPR)* (IEEE, Washington, 2013), pp. 3554–3561
33. M. Everingham, J. Sivic, A. Zisserman, "Hello! My name is…Buffy"—Automatic naming of characters in TV video, in *The British Machine Vision Conference*, vol. 2 (2006), p. 6
34. H. Fan, Z. Cao, Y. Jiang, Q. Yin, C. Doudou, Learning deep face representation (2014). ArXiv preprint arXiv:1403.2802
35. L. Fei-Fei, R. Fergus, P. Perona, One-shot learning of object categories. IEEE Trans. Pattern Anal. Mach. Intell. **28**(4), 594–611 (2006)
36. A. Ferencz, E.G. Learned-Miller, J. Malik, Building a classification cascade for visual identification from one example, in *Tenth IEEE International Conference on Computer Vision, 2005 (ICCV 2005)*, vol. 1 (IEEE, Washington, 2005), pp. 286–293
37. Y. Freund, S. Dasgupta, M. Kabra, N. Verma, Learning the structure of manifolds using random projections, in *Advances in Neural Information Processing Systems* (2007), pp. 473–480
38. R. Gross, I. Matthews, J. Cohn, T. Kanade, S. Baker, Multi-PIE. Image Vis. Comput. **28**(5), 807–813 (2010)
39. M. Guillaumin, J. Verbeek, C. Schmid, Is that you? Metric learning approaches for face identification, in *2009 IEEE 12th International Conference on Computer Vision* (IEEE, Washington, 2009), pp. 498–505
40. T. Hassner, S. Harel, E. Paz, R. Enbar, Effective face frontalization in unconstrained images (2014). ArXiv preprint arXiv:1411.7964
41. J. Hu, J. Lu, Y.-P. Tan, Discriminative deep metric learning for face verification in the wild, in *2014 IEEE Conference on Computer Vision and Pattern Recognition (CVPR)* (IEEE, Washington, 2014), pp. 1875–1882
42. J. Hu, J. Lu, J. Yuan, Y.-P. Tan, Large margin multimetric learning for face and kinship verification in the wild, in *Proc. ACCV*, 2014
43. C. Huang, S. Zhu, K. Yu, Large scale strongly supervised ensemble metric learning, with applications to face verification and retrieval (2012). ArXiv preprint arXiv:1212.6094
44. G. Huang, M. Mattar, H. Lee, E.G. Learned-Miller, Learning to align from scratch, in *Advances in Neural Information Processing Systems* (2012), pp. 764–772
45. G.B. Huang, V. Jain, E. Learned-Miller. Unsupervised joint alignment of complex images, in *IEEE International Conference on Computer Vision (ICCV)* (IEEE, Washington, 2007), pp. 1–8
46. G.B. Huang, M.J. Jones, E. Learned-Miller, LFW results using a combined Nowak plus MERL recognizer, in *Workshop on Faces in 'Real-Life' Images: Detection, Alignment, and Recognition*, 2008
47. G.B. Huang, E. Learned-Miller, Labeled faces in the wild: updates and new reporting procedures. Technical Report UM-CS-2014-003, University of Massachusetts, Amherst (May 2014)

48. G.B. Huang, H. Lee, E. Learned-Miller, Learning hierarchical representations for face verification with convolutional deep belief networks, in *2012 IEEE Conference on Computer Vision and Pattern Recognition (CVPR)* (IEEE, Washington, 2012), pp. 2518–2525
49. G.B. Huang, M. Ramesh, T. Berg, E. Learned-Miller, Labeled faces in the wild: a database for studying face recognition in unconstrained environments. Technical Report 07-49, University of Massachusetts, Amherst (October 2007)
50. M. Jones, P.A. Viola, Face recognition using boosted local features. Technical Report TR2003-25, Mitsubishi Electric Research Laboratory (2003)
51. H.-C. Kim, J. Lee, Clustering based on Gaussian processes. Neural Comput. **19**(11), 3088–3107 (2007)
52. B.F. Klare, B. Klein, E. Taborsky, A. Blanton, J. Cheney, K. Allen, P. Grother, A. Mah, A.K. Jain, Pushing the frontiers of unconstrained face detection and recognition: IARPA Janus benchmark A, in *2015 IEEE Conference on Computer Vision and Pattern Recognition (CVPR)* (IEEE, Washington, 2015), pp. 1931–1939
53. N. Kumar, A.C. Berg, P.N. Belhumeur, S.K. Nayar, Attribute and simile classifiers for face verification, in *2009 IEEE 12th International Conference on Computer Vision* (IEEE, Washington, 2009), pp. 365–372
54. N. Kumar, A.C. Berg, P.N. Belhumeur, S.K. Nayar, Describable visual attributes for face verification and image search. IEEE Trans. Pattern Anal. Mach. Intell. **33**(10), 1962–1977 (2011)
55. N.D. Lawrence, Gaussian process latent variable models for visualisation of high dimensional data. Adv. Neural Inf. Process. Syst. **16**(3), 329–336 (2004)
56. E.G. Learned-Miller, Data driven image models through continuous joint alignment. IEEE Trans. Pattern Anal. Mach. Intell. **28**(2), 236–250 (2006)
57. E.G. Learned-Miller, A. Ferencz, J. Malik, Learning hyper-features for visual identification, in *Advances in Neural Information Processing Systems 17: Proceedings of the 2004 Conference*, vol. 17 (MIT Press, Cambridge, 2005), p. 425
58. Z. Lei, M. Pietikainen, S.Z. Li, Learning discriminant face descriptor. IEEE Trans. Pattern Anal. Mach. Intell. **36**(2), 289–302 (2014)
59. H. Li, G. Hua, Hierarchical-PEP model for real-world face recognition, in *Proceedings of the IEEE Conference on Computer Vision and Pattern Recognition* (2015), pp. 4055–4064
60. H. Li, G. Hua, Z. Lin, J. Brandt, J. Yang, Probabilistic elastic matching for pose variant face verification, in *2013 IEEE Conference on Computer Vision and Pattern Recognition (CVPR)* (IEEE, Washington, 2013), pp. 3499–3506
61. H. Li, G. Hua, X. Shen, Z. Lin, J. Brandt, Eigen-PEP for video face recognition, in *Asian Conference on Computer Vision (ACCV)*, 2014
62. P. Li, Y. Fu, U. Mohammed, J.H. Elder, S.J. Prince, Probabilistic models for inference about identity. IEEE Trans. Pattern Anal. Mach. Intell. **34**(1), 144–157 (2012)
63. L. Liang, R. Xiao, F. Wen, J. Sun, Face alignment via component-based discriminative search, in *Computer Vision—ECCV 2008* (Springer, Heidelberg, 2008), pp. 72–85
64. S. Liao, Z. Lei, D. Yi, S.Z. Li. A benchmark study of large-scale unconstrained face recognition, in *2014 IEEE International Joint Conference on Biometrics (IJCB)* (IEEE, Washington, 2014), pp. 1–8
65. C. Lu, X. Tang, Surpassing human-level face verification performance on LFW with Gaussianface (2014). ArXiv preprint arXiv:1404.3840
66. A. Martinez, R. Benavente, The AR face database. Technical Report CVC Technical Report 24, Ohio State University (1998)
67. D. Miller, I. Kemelmacher-Shlizerman, S.M. Seitz, MegaFace: a million faces for recognition at scale (2015). ArXiv preprint arXiv:1505.02108
68. E.G. Miller, N.E. Matsakis, P. Viola, Learning from one example through shared densities on transforms, in *Proceedings of the IEEE Conference on Computer Vision and Pattern Recognition, 2000*, vol. 1 (IEEE, Washington, 2000), pp. 464–471
69. B. Moghaddam, T. Jebara, A. Pentland, Bayesian face recognition. Pattern Recogn. **33**(11), 1771–1782 (2000)

70. V. Nair, G.E. Hinton, Rectified linear units improve restricted Boltzmann machines, in *Proceedings of the 27th International Conference on Machine Learning (ICML-10)* (2010), pp. 807–814
71. H.-W. Ng, S. Winkler, A data-driven approach to cleaning large face datasets, in *2014 IEEE International Conference on Image Processing (ICIP)* (IEEE, Washington, 2014), pp. 343–347
72. H.V. Nguyen, L. Bai, Cosine similarity metric learning for face verification, in *Computer Vision—ACCV 2010* (Springer, Heidelberg, 2011), pp. 709–720
73. K. Nigam, A.K. McCallum, S. Thrun, T. Mitchell, Text classification from labeled and unlabeled documents using EM. Mach. Learn. **39**(2–3), 103–134 (2000)
74. E. Nowak, F. Jurie, Learning visual similarity measures for comparing never seen objects, in *IEEE Conference on Computer Vision and Pattern Recognition, 2007 (CVPR'07)* (IEEE, Washington, 2007), pp. 1–8
75. A. Ouamane, B. Messaoud, A. Guessoum, A. Hadid, M. Cheriet, Multi scale multi descriptor local binary features and exponential discriminant analysis for robust face authentication, in *2014 IEEE International Conference on Image Processing (ICIP)* (IEEE, Washington, 2014), pp. 313–317
76. S.J. Pan, Q. Yang, A survey on transfer learning. IEEE Trans. Knowl. Data Eng. **22**(10), 1345–1359 (2010)
77. P.J. Phillips, P.J. Flynn, T. Scruggs, K.W. Bowyer, J. Chang, K. Hoffman, J. Marques, J. Min, W. Worek, Overview of the face recognition grand challenge, in *IEEE Computer Society Conference on Computer Vision and Pattern Recognition, 2005 (CVPR 2005)*, vol. 1 (IEEE, Washington, 2005), pp. 947–954
78. N. Pinto, J.J. DiCarlo, D.D. Cox, How far can you get with a modern face recognition test set using only simple features? in *IEEE Conference on Computer Vision and Pattern Recognition, 2009 (CVPR 2009)* (IEEE, Washington, 2009), pp. 2591–2598
79. S. Rahimzadeh Arashloo, J. Kittler, Class-specific kernel fusion of multiple descriptors for face verification using multiscale binarised statistical image features. IEEE Trans. Inf. Forensics Secur. **9**(12), 2100–2109 (2014)
80. S.A. Rizvi, P.J. Phillips, H. Moon, The FERET verification testing protocol for face recognition algorithms, in *Proceedings of the Third IEEE International Conference on Automatic Face and Gesture Recognition, 1998* (IEEE, Washington, 1998), pp. 48–53
81. C. Sanderson, B.C. Lovell, Multi-region probabilistic histograms for robust and scalable identity inference, in *Advances in Biometrics* (Springer, Heidelberg, 2009), pp. 199–208
82. F. Schroff, D. Kalenichenko, J. Philbin, FaceNet: a unified embedding for face recognition and clustering, in *2015 IEEE Conference on Computer Vision and Pattern Recognition (CVPR)* (IEEE, Washington, 2015), pp. 815–823
83. H.J. Seo, P. Milanfar, Training-free, generic object detection using locally adaptive regression kernels. IEEE Trans. Pattern Anal. Mach. Intell. **32**, 1688–1704 (2010)
84. H.J. Seo, P. Milanfar, Face verification using the LARK representation. IEEE Trans. Inf. Forensics Secur. **6**(4), 1275–1286 (2011)
85. K. Simonyan, O.M. Parkhi, A. Vedaldi, A. Zisserman, Fisher vector faces in the wild, in *The British Machine Vision Conference*, vol. 60 (2013)
86. L. Sirovich, M. Kirby, Low-dimensional procedure for the characterization of human faces. J. Opt. Soc. Am. A **4**(3), 519–524 (1987)
87. C.J. Stone, Consistent nonparametric regression. Ann. Stat. 595–620 (1977)
88. Y. Sun, Y. Chen, X. Wang, X. Tang, Deep learning face representation by joint identification-verification, in *Advances in Neural Information Processing Systems* (2014), pp. 1988–1996
89. Y. Sun, D. Liang, X. Wang, X. Tang, DeepID3: face recognition with very deep neural networks (2015). ArXiv preprint arXiv:1502.00873
90. Y. Sun, X. Wang, X. Tang, Deep convolutional network cascade for facial point detection, in *2013 IEEE Conference on Computer Vision and Pattern Recognition (CVPR)* (IEEE, Washington, 2013), pp. 3476–3483

91. Y. Sun, X. Wang, X. Tang, Hybrid deep learning for face verification, in *2013 IEEE International Conference on Computer Vision (ICCV)* (IEEE, Washington, 2013), pp. 1489–1496
92. Y. Sun, X. Wang, X. Tang, Deep learning face representation from predicting 10,000 classes, in *2014 IEEE Conference on Computer Vision and Pattern Recognition (CVPR)* (IEEE, Washington, 2014), pp. 1891–1898
93. Y. Sun, X. Wang, X. Tang, Deeply learned face representations are sparse, selective, and robust (2014). ArXiv preprint arXiv:1412.1265
94. C. Szegedy, W. Liu, Y. Jia, P. Sermanet, S. Reed, D. Anguelov, D. Erhan, V. Vanhoucke, A. Rabinovich, Going deeper with convolutions (2014). ArXiv preprint arXiv:1409.4842
95. Y. Taigman, L. Wolf, Leveraging billions of faces to overcome performance barriers in unconstrained face recognition (2011). ArXiv preprint arXiv:1108.1122
96. Y. Taigman, L. Wolf, T. Hassner, Multiple one-shots for utilizing class label information, in *The British Machine Vision Conference* (2009), pp. 1–12
97. Y. Taigman, M. Yang, M. Ranzato, L. Wolf, Deepface: closing the gap to human-level performance in face verification. in *2014 IEEE Conference on Computer Vision and Pattern Recognition (CVPR)* (IEEE, Washington, 2014), pp. 1701–1708
98. B. Thomee, D.A. Shamma, G. Friedland, B. Elizalde, K. Ni, D. Poland, D. Borth, L.-J. Li, The new data and new challenges in multimedia research (2015). ArXiv preprint arXiv:1503.01817
99. M.A. Turk, A.P. Pentland, Face recognition using eigenfaces, in *Proceedings CVPR'91., IEEE Computer Society Conference on Computer Vision and Pattern Recognition, 1991* (IEEE, Washington, 1991), pp. 586–591
100. R. Urtasun, T. Darrell, Discriminative Gaussian process latent variable model for classification, in *Proceedings of the 24th International Conference on Machine Learning* (ACM, New York, 2007), pp. 927–934
101. P. Viola, M. Jones, Rapid object detection using a boosted cascade of simple features, in *Proceedings of the 2001 IEEE Computer Society Conference on Computer Vision and Pattern Recognition, 2001 (CVPR 2001)*, vol. 1 (IEEE, Washington, 2001), p. I-511
102. K.Q. Weinberger, J. Blitzer, L.K. Saul, Distance metric learning for large margin nearest neighbor classification, in *Neural Information Processing Systems* (2006)
103. Wikipedia, The Hours (Film)—Wikipedia, The Free Encyclopedia (2015). Accessed 30 June 2015
104. L. Wolf, T. Hassner, Y. Taigman, Descriptor based methods in the wild, in *Workshop on Faces in 'Real-Life' Images: Detection, Alignment, and Recognition*, 2008
105. L. Wolf, T. Hassner, Y. Taigman, Similarity scores based on background samples, in *Computer Vision—ACCV 2009* (Springer, Heidelberg, 2010), pp. 88–97
106. L. Wolf, T. Hassner, Y. Taigman, Effective unconstrained face recognition by combining multiple descriptors and learned background statistics. IEEE Trans. Pattern Anal. Mach. Intell. *33*(10), 1978–1990 (2011)
107. X. Xiong, F. De la Torre, Supervised descent method and its applications to face alignment, in *2013 IEEE Conference on Computer Vision and Pattern Recognition (CVPR)* (IEEE, Washington, 2013), pp. 532–539
108. D. Yi, Z. Lei, S.Z. Li, Towards pose robust face recognition, in *2013 IEEE Conference on Computer Vision and Pattern Recognition (CVPR)* (IEEE, Washington, 2013), pp. 3539–3545
109. D. Yi, Z. Lei, S. Liao, S.Z. Li, Learning face representation from scratch (2014). ArXiv preprint arXiv:1411.7923
110. Q. Yin, X. Tang, J. Sun, An associate-predict model for face recognition, in *2011 IEEE Conference on Computer Vision and Pattern Recognition (CVPR)* (IEEE, Washington, 2011), pp. 497–504
111. Y. Ying, P. Li, Distance metric learning with eigenvalue optimization. J. Mach. Learn. Res. *13*(1), 1–26 (2012)
112. M.D. Zeiler, R. Fergus, Visualizing and understanding convolutional networks, in *Computer Vision—ECCV 2014* (Springer, Heidelberg, 2014), pp. 818–833

113. E. Zhou, Z. Cao, Q. Yin, Naive-deep face recognition: touching the limit of LFW benchmark or not? (2015). ArXiv preprint arXiv:1501.04690
114. Z. Zhu, P. Luo, X. Wang, X. Tang, Recover canonical-view faces in the wild with deep neural networks (2014). ArXiv preprint arXiv:1404.3543
115. X. Zhu, Z. Lei, J. Yan, D. Yi, S.Z. Li, High-fidelity pose and expression normalization for face recognition in the wild, in *Proceedings of the IEEE Conference on Computer Vision and Pattern Recognition* (2015), pp. 787–796

Reference-Based Pose-Robust Face Recognition

Mehran Kafai, Kave Eshghi, Le An, and Bir Bhanu

Abstract Despite recent advancement in face recognition technology, practical pose-robust face recognition remains a challenge. To meet this challenge, this chapter introduces reference-based similarity where the similarity between a face image and a set of reference individuals (the "reference set") defines the reference-based descriptor for a face image. Recognition is performed using the reference-based descriptors of probe and gallery images. The dimensionality of the face descriptor generated by the accompanying face recognition algorithm is reduced to the number of individuals in the reference set. The proposed framework is a generalization of previous recognition methods that use indirect similarity and reference-based descriptors. The effectiveness of the proposed algorithm is shown by transforming multiple variations of the standard, yet powerful, local binary patterns descriptor into pose-robust face descriptors. Results are shown on several publicly available face databases. The proposed approach achieves good accuracy as compared to popular state-of-the-art algorithms, and it is computationally efficient due to its compatibility with orthogonal transform based indexing algorithms.

1 Introduction

Face recognition is an important computer vision and pattern recognition technology that is used in many different applications, from organizing photo albums to surveillance. In its most common form named *identification*, a database of images of individuals, one or more images per individual, called the gallery is available. There is also a set of images, called probes, taken of individuals some of whom

M. Kafai (✉) • K. Eshghi
Hewlett Packard Labs, Palo Alto, CA, USA
e-mail: mehran.kafai@hpe.com; kave.eshghi@hpe.com

L. An
BRIC, University of North Carolina at Chapel Hill, Chapel Hill, NC, USA
e-mail: lan004@unc.edu

B. Bhanu
CRIS, University of California, Riverside, CA, USA
e-mail: bhanu@cris.ucr.edu

© Springer International Publishing Switzerland 2016
M. Kawulok et al. (eds.), *Advances in Face Detection and Facial Image Analysis*,
DOI 10.1007/978-3-319-25958-1_9

249

may be the individuals in the gallery. The task is to identify which probe and gallery images correspond to the same individual. If a probe individual is not in the gallery (open-set identification) then the system should be able to infer it. In this chapter, we assume that the gallery contains a matching image for every probe (i.e., closed-set identification); however the proposed framework can also be used for open-set identification.

When the probe and gallery images are taken under the same pose (for example frontal or side-view), the task essentially boils down to a pattern recognition problem: the gallery and probe images are first processed to generate face descriptors (e.g., Local Binary Patterns (LBP) [1]). The probe descriptors are compared with the gallery descriptors using some distance measure, such as the Euclidean distance. The gallery images which are closest in the distance measure to the probe image are deemed to be of the same individual.

Face recognition across pose is the problem of recognizing a face from a new viewpoint which has not previously been seen (e.g., the probe image is a profile/side-view image, whereas the gallery image is frontal). The problem is that, using descriptors such as LBP and an associated distance measure, two images of the same person in different poses are typically more distant from each other than the images of two different individuals in the same pose. As a result, poor recognition accuracy may be observed. Most of the solutions to this problem assume that for each individual, multiple images in different poses are available (either in the gallery or among the probes). This assumption is not always valid, e.g., in surveillance [3], where only one image may be available in the gallery, and one image, in a possibly different pose, as the probe. There are algorithms that can match a single gallery image with a given probe image in a different pose after generating a 3-D model for every gallery image (e.g., [37]), but these algorithms are computationally expensive, particularly when the size of gallery is large.

Here we describe a new framework for face recognition across pose. The proposed approach is a computationally inexpensive solution to the two major challenges in face recognition: face recognition across pose, and face recognition with a single image per person gallery. This approach relies on generating a *reference-based (RB) descriptor* for each gallery and probe face image based on their similarity to a reference set. The reference set is utilized as a basis to generate RB descriptors for the probe and gallery images. Once the RB descriptors are computed, the similarity between the probe and gallery images is rated on the RB descriptors. The cosine measure is utilized for this purpose.

To make sure that the proposed approach is not database-specific and is not tuned to a certain database, we use a combination of seven face databases to obtain the face images. The seven databases include: LFW [20], FEI [47], RaFD [27], FERET [36], FacePix [30], CMU-PIE [43], and Multi-PIE [15]. We have chosen the gallery to contain only frontal face images; however, this is not a requirement and the gallery may contain any arbitrary pose image.

In the rest of this chapter, Sect. 2 describes the related work and contributions of this work. Section 3 introduces the reference set, presents the overall system framework, and discusses the technical approach. Section 4 puts forward a discussion

Table 1 Important term definitions

Term	Definition
Profile face image	The side view image of a face
Reference set	A set of images of various individuals. For each individual multiple images in different poses are stored
Reference face	An individual from the reference set. The reference set contains N reference faces
Gallery	A set of face images of multiple individuals. Each individual is represented with a **single** image
Probe	A query face image which matches with **one** individual from the gallery
First level descriptor (FLD)	A vector representing any gallery, probe, or reference set image. Here we use LBP as the FLD
First level similarity (FLS)	Similarity between the probe or gallery images and the reference set images using an existing method
Reference-based descriptor	A vector of FLS scores representing the similarity between a given face image and reference faces
Second level similarity (SLS)	Similarity between probe and gallery reference-based descriptors

on how the proposed framework is related to previous work on reference-based descriptors. Section 5 introduces the face databases and presents the experimental results. Concluding remarks are provided in Sect. 6. For better understanding of the terms and symbols, we summarize the definitions of the important terms in Table 1, and define the symbols in Table 2.

2 Related Work and Our Contributions

2.1 Related Work

Researchers approach face recognition across pose via three main methods [53]: 2-D algorithms, 3-D approaches, and general use methods.

2-D recognition algorithms cover three main subcategories: multi-view galleries [8, 25, 44], using pose tolerant control points [19], and probabilistic graphical models [4]. In multi-view gallery approaches, the gallery holds multiple views of every individual with different poses. The poses can be real images [44] or synthetically generated images [25]. In multi-view galleries that contain real images, multiple poses are available for each individual. This differs from the work being

Table 2 Definition of symbols

Symbol	Definition
A, B	Any individual from the gallery or probe set
I_j^A	Face image of individual A under pose j
N	Number of individuals in the reference set
R_u	The uth individual in the reference set. Individuals are labeled 1 to N
$FLS(I_j^A, I_i^{R_u})$	FLS between image of individual A under pose j and image of individual u from the reference set under pose i
F^A	An N-dimensional reference-based descriptor for individual A
f_u^A	Maximum FLS between individual A and all poses of individual u from the reference set
$SLS(F^A, F^B)$	Second level similarity between F^A and F^B
γ	Lower incomplete gamma function
k	Degrees of freedom of the Chi-sq distribution
Γ	Gamma function with closed form values
P, O	LBP histograms of an image
$\chi^2(P, O)$	Chi-square distance between P and O

addressed here for single-view galleries. Algorithms with synthetic-view galleries tend to have high computational complexity due to the process of generating virtual synthetic views for different poses for all individuals within the gallery. Approaches based on Markov random fields have also been developed for pose-invariant face recognition [4]; however, they are computationally expensive due to the iterative optimization stage (e.g., gradient descent, simulated annealing, or iterated conditional modes).

An et al. [2] use warped average face (WAF) for video based face recognition in which the small pose variations are rectified and the frontal view of a subject is generated by iterative averaging and template warping. Zhang et al. [56] propose a high-level feature learning scheme to extract pose-invariant identity features for face recognition. In [3] pose and person-specific dynamic information is encoded using a dynamic Bayesian network (DBN) and faces captured by multiple cameras are jointly utilized for improved recognition accuracy.

Prince et al. [38] describe face pose variations using a tied factor analysis model. The idea is that identity-independent linear transformations can be utilized to generate all poses of a face from a single vector in identity space. The identity vectors and the corresponding linear transformation parameters are determined from a dataset of face images with various poses. This approach shows promising results; however, image variation due to pose changes cannot be completely depicted by linear transformations [53]. Li et al. [28] learn a domain-specific cross-view classifier by using a regressor with a coupled bias-variance tradeoff, and stabilize the regressor against the pose variation for face recognition across pose.

3-D face recognition algorithms typically use a 3-D model to address across-pose recognition [5, 7, 37]. The model is either obtained by a 3-D scan or generated using facial features/textures from 2-D/3-D images.

Asthana et al. [5] propose an automatic 3-D face recognition system via 3-D pose normalization. The proposed system generates a synthetic frontal view face image for any given gallery or probe image, then performs recognition based on the frontal views. The 3-D pose normalization approach has major drawbacks; it only handles pose variation up to ±45° in yaw and ±30° in pitch angles. Further, it requires accurate detection of landmark points (eyes, nose, and mouth), and it is computationally expensive. Overall, due to their complexity, 3-D face recognition algorithms are not used typically for real-time face recognition across pose.

General use methods do not specifically aim at solving face recognition across pose but are able to perform under limited pose variations. Some commonly used examples of this type of face recognition approaches include local binary patterns [1], and Directional Corner Points (DCP) [14]. These approaches are rarely used for face recognition applications where recognition across pose is specifically desired. A summary of the related work for face recognition across pose is shown in Table 3.

Table 3 Related work summary for face recognition across pose

Principle and methodology	Advantages and limitations
2-D algorithm: pose tolerant control points [19]	Pros: simple, fast, works with single-view galleries Cons: small pose tolerance
2-D algorithm: multi-view galleries [25, 44]	Pros: good performance for pose variation Cons: high complexity when views are synthetically generated
2-D algorithm: use extra bridge dataset [52]	Pros: good performance for pose, expression, and illumination variation, works with single-view galleries Cons: first model assumes that identity data set has similar images to probe and gallery, second model requires training for person-specific classifier
3-D algorithm: 3-D scans [6]	Pros: good performance, no training required, works with single-view galleries Cons: inaccurate 3-D model from single view galley, computationally expensive
3-D algorithm: 3-D generated models [7, 37, 55]	Pros: high accuracy, works with single-view galleries Cons: extensive training is required, high computational complexity
General use algorithms: local features [21]	Pros: simple Cons: difficult to match local features between different poses
General use algorithms: holistic methods [14, 48]	Pros: fast, works with single-view galleries Cons: fail with large pose variation, usually require training
Reference-based: reference-based descriptors	Pros: simple, fast, no training required, pose-invariant recognition, can be used with any state-of-the-art face recognition algorithm and face descriptor, works with single-view galleries

Collecting multiple images for each individual is a tedious task for large size galleries and often, for most face recognition applications it is not possible. Therefore, single sample per person face recognition algorithms are of great importance. Single view per individual galleries are commonly used in real-world applications [46] due to the fact that image retrieval is more efficient and less complex. The most advanced algorithms in this area use local facial features [1] in conjunction with global (e.g., entire face) features [32] to perform recognition.

Duin et al. [13] discuss the basic idea of a representation set in which a dissimilarity space is a data-dependent mapping using a representation set. The representation set can be chosen externally or from the available training data. The work in [17, 40], and [33] are all direct applications of the representation set in [13]. This work is significantly different both in theory and applications in [13, 17, 33, 40]. Directly using the aforementioned methods would not work in our case since for practical face recognition, the variations in pose and illumination complicate the recognition process. In our reference framework, for each reference subject, multiple images representing different poses were enrolled and the similarity between a query subject and a reference subject is determined by the maximum similarity among different samples of this reference subject. In this way, the pose variation can be handled and it is proven to be effective for pose invariant face recognition in the experiments. In addition, we evaluate the reference set from a different perspective: we formulate it as an optimization problem to determine the diversity of the reference set.

In the following, we briefly discuss the existing work on face recognition via indirect similarity and reference-based descriptors. A detailed discussion on how the proposed reference-based framework relates with the previous work is presented in Sect. 4.

2.1.1 Recognition via Indirect Similarity

Recognition and retrieval via reference-based descriptors and indirect similarity has been well explored in the field of computer vision. Guo et al. [16] use exemplar-based embedding for vehicle matching. Rasiwasia et al. [39] label images with a set of pre-defined visual concepts, and then use a probabilistic method based on the visual features for image retrieval. Liu et al. [31] represent human actions by a set of attributes, and perform activity recognition via visual characteristics symbolizing the spatial-temporal evolution of actions in a video.

Kumar et al. [26] use attribute and simile classifiers for verification, where face-pairs are compared via their similes and attributes rather than a direct comparison. Experiments are performed on PubFig and Labeled Faces in the Wild (LFW) databases [20] only; however, these databases do not contain images with large pose variation. Also, attribute classifiers require extensive training for recognition across pose.

In the field of face recognition, methods based on rank-lists [41] have been investigated. Schroff et al. [41] propose describing a face image by an ordered list of similar faces from a face library, i.e., a rank list representation (Doppelgänger list) is generated for each image. Proximity between gallery and probes images is determined via the similarity between ordered rank lists of the corresponding images. The main drawback of this approach is the complexity of comparing the Doppelgänger lists. Also, no suitable indexing structure is available for efficient ranked-list indexing.

Likelihood-predict and associate-predict models [52] have been introduced for face recognition under different pose, expression, and illumination. The idea is to use a third data set called the identity data set as a bridge between any two images that ought to be matched. Recently, Cui et al. [11] introduced a quadratic programming approach based on reference image sets for video-based face recognition. Images from the gallery and probe video sequences are bridged with a reference set pre-defined and pre-structured to a set of local models for exact alignment. Once the image sets are aligned, the similarity is measured by comparing the local models.

Chen et al. [9] represent a face x by the sum of two independent Gaussian random variables: $x = \mu + \epsilon$, where μ represents the identity of the face and ϵ represents face variations within the same identity. The covariance matrices of μ and ϵ, S_μ and S_ϵ, are learned from the training data using an expectation maximization algorithm. Having learned these, two matrices A and G are derived, and the log-likelihood ratio for two input images x_1 and x_2 being of the same individual is shown to be

$$r(x_1, x_2) = x_1^T A x_1 + x_2^T A x_2 - 2 x_1^T G x_2. \tag{1}$$

By decomposing G and performing some offline computations, the online cost of computing the log-likelihood ratio can be further reduced.

Liao et al. [29] proposed a similar method for generating invariant representations. Compared to [29], the reference set in our method is much smaller, while achieving better results (e.g. on LFW). While [29] focuses more on theoretical study of invariant representations, we have provided thorough empirical study on different reference-based methods (i.e., different FLS methods), performed experiments on many more datasets, and compared with more recent methods. Our method is a simple and effective framework which intrinsically takes care of pose and other variations thanks to the reference set. It can be used in conjunction with existing features and recognition methods. Using RB achieves feature dimensionality normalization regardless of the original feature dimension. We have observed that RB can be used with indexing (e.g., DCT hashing) for accurate and efficient retrieval. Note that our method does not require training yet it achieves competitive results on multiple datasets as compared to many recent methods.

A summary of the most recent work for face recognition via indirect similarity is shown in Table 4. After presenting the details of our proposed reference-based framework in Sect. 3, we discuss, in Sect. 4, more technical details of the related work in Table 4 and how our framework is a generalization of previous work in [26, 42, 52], and [9].

Table 4 Recent work summary for face recognition via indirect similarity and reference-based descriptors

Methodology and authors	Comments
AP and LP models [52]	Training required for probe images during online processing, person specific classifiers need to be trained
Attribute and simile classifiers [26]	Requires trained classifier for each face part of each reference individual, facial key points need to be extracted, incapable of handling large pose variations
Doppelgänger lists [41]	Effective for across pose recognition, rank-lists comparison is expensive, slow retrieval
Message passing model [42]	Ineffective for recognition across pose, message passing algorithm is used to choose the best reference images
Joint Bayesian formulation [9]	Probabilistic reference-based method approach, incapable of handling large pose variations
Invariant representations [29]	Training and large reference set is required
Reference-based framework	Fast retrieval, compatible with efficient indexing structures

2.2   Contributions

Unlike the previous work, as shown in Table 4, the contributions of this work are:

1. *Unified reference-based face recognition*: We propose a unified reference-based face recognition framework. We introduce the *reference set* as the basis set for similarity matching. The reference set contains multiple poses of different individuals. A reference-based (RB) descriptor is computed for every gallery and probe image based on their similarity to the reference set images. We call this similarity, the first level similarity (FLS).
2. *Compatible with other face recognition algorithms and face descriptors*: Our reference-based face recognition framework is compatible with any existing face recognition algorithm. In other words, it can use any current recognition algorithm to compute the first level similarity between a given gallery/probe image and the reference set images. In addition to the descriptors used in our experiments, other descriptors (e.g., LE [52], histogram of gradients [12]) can also be used to improve the first level similarity.
3. *Dimensionality normalization*: The dimensionality of the reference-based descriptor is equal to the number of individuals in the reference set. It is independent of the face recognition algorithm being used, the image size, face descriptor type, etc.
4. *Efficient retrieval*: The reference-based descriptors generated from the proposed reference-based framework can be used in conjunction with state-of-the-art indexing algorithms (e.g., DCT locality sensitive hashing) for efficient retrieval.

5. *Practicality*: The proposed reference-based recognition framework is simple, efficient, and can be used for real-world face recognition applications. Also, it can be extended for applications other than face recognition across pose (e.g., facial expression recognition, non-facial image recognition, etc.).
6. *Generalizability*: Our method can handle open-set identification as a query's identity can be determined unknown when its similarity to the gallery is below certain threshold. The threshold can be learned from the reference set itself.

3   Technical Approach

Figure 1 presents the overall diagram for reference-based face recognition across pose. We start with two image sets: the gallery, and the reference set. The gallery consists of frontal face images. The goal is to rank the gallery images based on their similarity to a given probe image.

The reference set contains multiple individuals with several images under different poses for each individual. The reference set individuals are different from the individuals in the gallery and all probes; thus, no person from the gallery or probe set is present in the reference set. The reference set is built only once and it is used in conjunction with any gallery or probe set. The reference set images are collected and prepared offline before the testing (online) stage. We use images from FEI [47] and Multi-PIE [15] to build the reference set for all experiments except the experiment on the LFW database [20], for which we use a reference set with images from the WDRef dataset [9].

We refer to offline processing as all that is done before a probe enters the system and it is ready to be recognized. During offline processing, the gallery images are compared, using first level similarity (FLS) with the reference set images and a reference-based descriptor is generated for each gallery image. The dimensionality

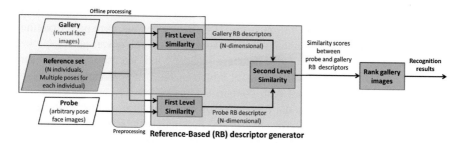

Fig. 1 The overall system diagram for reference-based face recognition across pose. After preprocessing, the reference-based descriptors for the gallery images are computed by measuring first level similarity between the gallery and reference set feature vectors. The reference-based descriptor for the probe image is generated by computing the first level similarity between the probe and reference set images. The gallery images are ranked based on the second level similarity scores between reference-based descriptors of the probe and gallery images

of each RB descriptor is equal to the number of individuals in the reference set. The offline processing stage is complete after the reference-based descriptors are generated for all gallery images.

During online processing a probe image enters the system and an RB descriptor is generated for it. At this stage, RB descriptors are available for each gallery image and the probe image. Recognition is performed using a similarity measure between the gallery and probe reference-based descriptors. We refer to this similarity as the *Second Level Similarity* (SLS).

More details of reference-based face recognition and how the reference-based descriptors are generated are discussed in Sect. 3.2.1.

For first level descriptors (FLD), we chose to use LBP due to the fact that LBP features are widely used in computer vision applications such as image retrieval [51], and especially face recognition [1].

The reference-based algorithm is independent of any specific preprocessing steps. For example, using a different facial feature extraction algorithm or replacing LBP with another face descriptor does not change how the reference-based algorithm operates.

3.1 Reference Set and Motivation

3.1.1 Reference Set

The reference set is a set of faces belonging to N individuals. We call these individuals reference individuals and represent them with $R_1, R_2, \ldots, R_N$. We use two reference sets. The first reference set includes 200 individuals from the FEI database [47] and 300 individuals from the Multi-PIE database [15]. For each individual from the FEI database, we select 13 images under various poses from full profile left to full profile right. Thus, each pose from the FEI database has one image. For each individual from the Multi-PIE database we select 13 poses from profile left to profile right with 15 degree increments, and select three images for each pose with varying illumination (images #01, #07, and #13). In other words, each pose from the Multi-PIE database is represented with three images in the reference set.

For the experiment on the LFW database in Sect. 5.4.5, we use a reference set with images from the WDRef [9] dataset which includes 99,773 images of 2995 individuals. LFW and WDRef both contain unconstrained face images collected from the internet; thus, we test our framework on the LFW database with a reference set from the WDRef dataset. We choose 500 individuals (which have 40 images each) from the WDRef dataset to build our reference set. There are no common individuals between the LFW test individuals and the reference set.

Our focus is on recognition across pose; however, the addition of images under different lighting conditions extends our approach's capability to illumination invariant recognition as well. The image of reference individual R_u in pose j is

Fig. 2 Sample images from the reference set

denoted as $I_j^{R_u}$. For example, $I_{10}^{R_2}$ is the frontal image of the reference individual R_2. Section 5.2 discusses the evaluation of the reference set selection. Sample images from the reference set are shown in Fig. 2.

3.1.2 Motivation for Reference-Based Descriptors

Our technique is based on the following two intuitions:

- A face image can be described by its degree of similarity to the images of a set of reference individuals. In this context, the reference individuals form a coordinate system in an N-dimensional space and each input face corresponds to a point in this space, the coordinates of which are determined by the similarity of the face image to the reference individuals.
- Two faces which are visually similar in one pose, for example profile, are also, to some extent, similar in other poses, for example frontal. In other words, we assume that visual similarity follows from underlying physical similarity in the real world. We take advantage of this phenomenon in the following way: compare B, a gallery/probe face image, with the images of all available poses of a reference individual from the reference set. The degree of similarity with the best matching image of the reference individual is a degree of the similarity of the two faces, B and the reference individual. By repeating this procedure for each one of the reference individuals, we create a descriptor for the face image B that reflects the degree of similarity between B and the reference set faces.

3.2   Reference-Based Face Recognition

Let us first define some terminology, and then we will describe the reference-based approach to face recognition across pose. In what follows, we use the following terminology: We use the term *input face* to denote the face for which we want to create a descriptor which may belong to the gallery or probe set. *Input image* denotes the image of the input face that we have at our disposal. For a given face A, we use the notation I_j^A to refer to an image of A where j is used to differentiate this image from other images of A. For example, j could refer to the pose under which the image was taken.

Reference-based face recognition consists of two stages: computing reference based descriptors and performing recognition using these descriptors. Table 5 presents the pseudocode for reference-based face recognition.

3.2.1   Computing Reference-Based Descriptors

Figure 3 illustrates the setup of generating the reference-based descriptor.

Typically, we have an image I_j^A (a probe or gallery image) of individual A for which we wish to construct a reference-based descriptor, which is an N-dimensional real vector. Let

$$F^A = < f_1^A, f_2^A, \ldots, f_N^A > \qquad (2)$$

Table 5 Pseudocode for reference-based face recognition

1	Generate RB descriptors for all gallery images
	Generate RB descriptor for a given probe image
2	Compute SLS score between each gallery RB descriptor and the probe RB descriptor
	Rank gallery images based on SLS scores

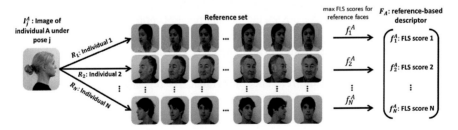

Fig. 3 Computing the reference-based descriptor. Any given gallery/probe image I_j^A is compared to all images of each reference face $R_1, R_2, \ldots, R_N$. The FLS score f_u^A is computed for I_j^A and each reference face R_u. f_u^A is the maximum similarity between I_j^A and all images of R_u. The RB descriptor for I_j^A, F^A, is defined as $< f_1^A, f_2^A, \ldots, f_N^A >$

be the RB descriptor for A. The idea is that the elements $f_1^A, f_2^A, \ldots, f_N^A$ represent the similarity of A to the reference individuals $R_1, R_2, \ldots, R_N$. We estimate the similarity of A to a given reference individual by selecting the maximum FLS score between A and all images of a given reference individual R_u. In other words, to compute f_u^A, we compare the image A with each one of the images of the reference individual R_u in all different poses by computing $\mathrm{FLS}(I_j^A, I_i^{R_u})$, which is a measure of the similarity of the two images. Thereafter, the largest FLS value obtained from this comparison is used as the value of f_u^A. The point is that the reference set would have some images that are close to the variation that the input image presents. Our framework does not require that the most similar reference set face image to a probe image has to be under the same pose. Thus,

$$f_u^A = \max(\mathrm{FLS}(I_j^A, I_1^{R_u}), \ldots, \mathrm{FLS}(I_j^A, I_K^{R_u})). \tag{3}$$

Equation (3) is computed for all reference individuals, i.e., given A, f_u^A is computed for $u = 1 \ldots N$. These N values represent the N-dimensional RB descriptor for A.

3.2.2 Face Recognition Using RB Descriptors

Images from probe and gallery are compared using the SLS measure between RB descriptors. The cosine between the RB descriptors is used as the SLS measure, i.e.,

$$SLS(F^A, F^B) = \cos(F^A, F^B) = \frac{F^A . F^B}{||F^A|| \times ||F^B||}. \tag{4}$$

In other words, given two images I_u^A and I_v^B, and their corresponding RB descriptors F^A and F^B, $\cos(F^A, F^B)$ represents the second level similarity between the two images. As mentioned previously, SLS is the measure we use for ranking the gallery images for their degree of similarity to the probe image.

3.3 The First Level Similarity Variants

We have investigated three variants for the first level similarity measure. The second level similarity computation is the same for all these variants. The three FLS variants are PCA-cosine method, Chi-square method, and pose-specific Chi-square method. All three variants have the following initial steps in common.

After preprocessing, LBP features are computed for each image. Our LBP setup is as follows. Before extracting the LBP features, the face images are cropped and resized to 200×200 and divided to 20×20 regions. For each region the uniform

LBP with a radius of 1 and 8 neighboring pixels are computed, and a histogram of 59 possible patterns is generated [1]. Histograms of all 100 regions are concatenated in a single histogram of length 5900 to describe the face image.

3.3.1 PCA-Cosine Method

PCA [49] is commonly used for decorrelation and dimensionality reduction. Utilizing PCA, the principal components are generated and then used to convert the high-dimensional feature subspace of the reference/probe/gallery face images to a smaller eigenspace subspace. We use it to speed up and improve the accuracy of the first level similarity calculation.

We use the LBP descriptors of the images in the reference set to derive the PCA coefficient matrix. Once the coefficient matrix is derived, it is applied to the LBP descriptors of all the images: the reference set, probes, and gallery. Thereafter, the reduced dimension vectors are used for computing first level similarity.

The cosine similarity measure is utilized to measure FLS in the PCA-cosine method. After projecting the LBP image histograms into eigenspace, they are no longer region-based histograms. Therefore, we choose to use cosine similarity instead of the commonly used Chi-square distance measure to measure FLS. The cosine measure works very well as the distance measure in this context, and is significantly cheaper to compute than Chi-square on the original LBP vectors.

3.3.2 Chi-Square Method

In the Chi-square method, unlike the PCA-cosine method, the original LBP feature vectors for the reference set, gallery, and probe face images are directly used to compute the FLS. Initially, to compute the FLS scores, the Chi-square distance

$$x = \chi^2(P, Q) = \sum_{b,w} \frac{(P_{b,w} - Q_{b,w})^2}{P_{b,w} + Q_{b,w}} \tag{5}$$

is measured between reference set LBP histograms and gallery/probe LBP histograms. In Eq. (5), P and Q are the LBP histograms, b denotes the histogram bin number, and w refers to the local region. We chose Chi-square distance because it is the most commonly used distance measure for LBP features [1] in the literature and has proven to be accurate.

The distance x is converted to similarity score using the Chi-square cumulative distribution function

$$\text{FLS} = 1 - \frac{\gamma(\frac{k}{2}, \frac{x}{2})}{\Gamma(\frac{k}{2})}, \tag{6}$$

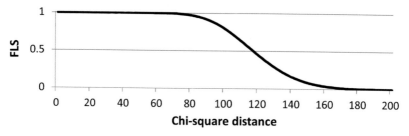

Fig. 4 FLS plot for a sample distance vector

where $\Gamma(\frac{k}{2})$ is the Gamma function with closed form values for half-integer arguments, γ is the lower incomplete gamma function, and k refers to the degrees of freedom (in this case 59). Figure 4 shows the plot for FLS computed for a sample distance vector with values from 1 to 200.

A limitation of the Chi-square method is its speed. The computation of the Chi-square distance is expensive, and yet it must be carried out to compare each probe/gallery image to all the images in the reference set. To overcome this problem, we investigate the pose specific Chi-square method. More detail on the processing times is discussed in Sect. 5.5.

3.3.3 Pose Specific Chi-Square Method

In the pose specific Chi-square method, each probe/gallery face image is only compared to a single image of each reference individual. The image chosen from each individual for generating the similarity feature descriptor should be of the same pose as the probe/gallery image for which the similarity face descriptor is to be generated. Doing such accelerates the processing time dramatically because it does not require comparing the probe/gallery images with every image of each reference individual. Another difference between this method and the Chi-square method is that when generating the reference-based descriptor, only one similarity score per reference individual will be available to choose from. This similarity score is used for the RB descriptor of the probe/gallery images. To use this technique, the pose of the probe and gallery images needs to be estimated.

4 Unified Reference-Based Framework

This chapter introduces a recognition algorithm based on indirect similarity and reference-based descriptors; it is the first work to introduce a unified generalization of reference-based methods. Other reference-based approaches (e.g., [9, 26, 42]) can be considered as specific instances of the proposed reference-based framework.

Section 2.1.1 briefly discussed the previous work on recognition via reference-based descriptors. In this section, we discuss how the proposed unified reference-based framework is a generalization of the previous work. It's important to note that all works discussed in this section perform face verification, whereas in this chapter both face verification and face identification are performed.

The main drawback of all the methods discussed in Sect. 2.1.1 is related to how the first level similarity scores are computed, because the similarity with each image in the reference set is mapped to a value in the reference-based descriptor. The assumption is that for a pair of test images A and B, if A is similar to a specific image of an individual in the reference set, then B is also similar to the same image. This approach is incapable of performing accurately in the case of large pose variation because two images of the same person in different poses (e.g., profile vs. frontal) will typically not have close similarity scores to any single image of a reference set individual. All aforementioned reference-based approaches [9, 26, 41, 42, 52] can adopt our general reference-based framework and overcome the pose variation.

Kumar et al. [26] propose face verification via attribute and simile classifiers. The attribute classifier is not relevant to our proposed framework, although using such a classifier as a filter would probably improve our results. For the simile classifier Kumar et al. train a classifier for parts of the faces of reference individuals. For example, there is a classifier for Harrison Ford's nose, another classifier for his mouth, etc. To create the descriptor for an input image, the image is first partitioned into the mouth, nose, eyes etc., and then the relevant classifiers are used for each of the face parts of each reference individual. The outputs of the classifiers are used to generate the reference-based trait vector. This differs from what our proposed reference-based framework performs in two ways: we use the entire face of the reference individual, and we do not use classifiers. Using the parts of the faces of reference individuals requires a fairly accurate detector for the fiduciary points on the face, which we do not require.

While [26] uses a SVM classifier for each face part of each reference individual, we use a training-free approach by directly comparing the input image with the images of the reference individual and choosing the best match. One way of looking at it is that we use a nearest neighbor classifier (we find the nearest neighbor to the input image for computing the similarity score) to determine the degree of membership in the class, whereas [26] uses a support vector machine classifier. Of course, we do not need to train any classifiers, and we can use advanced nearest neighbor search methods (such as the use of locality sensitive hash functions) to find the nearest neighbor. For face verification, a SVM classifier is trained on the trait vectors, and it is used to determine if the faces belong to the same individual. We use a simple cosine measure computation for the final verification step.

The approach discussed in [26] can adopt our reference-based framework to extend its capability for recognition across pose. First level similarity would be computed using the SVM classifier scores, and second level similarity would be measured between the trait vectors.

Shen et al. [42] propose a novel method for face verification. The reference set in [42] consists of a set of face images. The reference set in our proposed framework contains a set of individuals for each of whom we have many images at different poses. Both approaches create a descriptor that is a vector of first level similarity scores. In [42] the value of the ith coordinate in the descriptor is the similarity score between the input image and the reference image number i. In our work, the value of the ith coordinate in the descriptor is the highest similarity score between the input image and all the images of reference individual number i. This is a fundamental difference, since by comparing the input image to all the images of the reference individuals and choosing the best one we reduce the effects of pose variations.

We use PCA-cosine measure to compute the first level similarity between the input image and the images in the reference set. The method in [42] uses One Shot Similarity (OSS) to compute this value. Computation of the PCA-cosine measure is very efficient, and it is possible to further speed it up using locality sensitive hash functions. In our framework, we use a set of randomly chosen reference individuals, and evaluate our reference set via an optimization approach. In [42], a message passing algorithm is used to choose the best reference images. The work in [42] can be considered as a specific implementation of our reference-based framework where each reference set individual is represented with a single image in the reference set.

Similar to [26], the method in [9] is incapable of handling large pose variations. The technique in [9] can be considered a probabilistic reference-based method, with infinite references, with the assumption that the face features are distributed according to the Gaussian distribution. The experiments are performed on the LFW and WDRef databases; however neither database contains large pose variations. Applying our framework to [9] would result in the capability of recognition across pose in addition to more efficient retrieval via DCT hashing. To do so, the first level similarity would be defined as the conditional likelihood $P(x|\mu_i)$, and the second level similarity as the log likelihood ratio of the two Bayesian reference-based descriptors.

Likelihood-predict (LP) and associate-predict (AP) models were introduced in [52]. The way the identity set is used in [52] is different from the way we use the reference set. Let A and B be two images to be compared. In the LP model [52], a personalized classifier is trained for both A and B, using a set of similar and dissimilar images from the identity data set as the training set. These classifiers are used to determine whether A belongs to the same class as B and B belongs to the same class as A. Training a classifier for each individual is expensive, which is the main drawback of this approach. Notice that the classifier needs to be trained for the probe images as well as the gallery images; thus, classifier training is part of online processing. In the AP model [52], the person in the identity set most similar to A is determined, and the image of that person in the same pose as B, call it A', is selected. Then A' is compared with B using a standard appearance-based algorithm. The same is done with B, i.e., B' is found and compared with A. This assumes that the identity set includes individuals sufficiently similar in appearance to A and B to make the comparison meaningful, an assumption that is not easy to satisfy in reality. Our approach does not require training any classifiers, and does not assume that the reference set contains a person similar to the input image.

5 Experimental Results

5.1 Data

We use face images from seven face databases: LFW [20], FEI [47], RaFD [27], FacePix [30], CMU-PIE [43], Multi-PIE [15], and the Facial Recognition Technology (FERET) database [36].

The FEI [47] face database consists of 200 individuals. For each individual 14 images are available, 11 different poses from full profile left to full profile right, one image with a smiling expression, and two images with different illumination conditions. The RaFD face database [27] includes multiple images with various camera angles, expressions, and gaze directions for 67 individuals. For each individual we use five images with neutral expression and the following poses: profile left, half left, full frontal, half right, and profile right. The Multi-PIE [15] is a collection of 755,370 images from 337 individuals. Images are taken from 15 viewpoints under 19 illumination settings. The CMU-PIE database [43] includes 41,368 images of 68 individuals, with 13 poses, 43 illumination conditions, and 4 facial expressions. The FacePix [30] database holds images of 30 individuals, with images spanning the spectrum of $180°$ with increments of $1°$. The FERET face database [36] contains 14,126 images of 1199 individuals and is commonly used within the face recognition research community. The LFW database [20] includes 13,233 images gathered from the web of 5749 individuals.

The gallery consists of 1000 frontal face images chosen from all aforementioned databases, except for the LFW database which is discussed in Sect. 5.4.5. For the experiment in Sect. 5.2 the gallery size is increased to 10,000. We have chosen the gallery to contain only frontal face images but this is not a requirement for our framework. The gallery and probe sets may contain any arbitrary pose image. Note that, the reference set does not include any individuals from the gallery or probes.

Face alignment and illumination normalization are performed on all probe, gallery, and reference set images. Face alignment is executed in two steps: pose estimation and template alignment. Head pose estimation is performed via a Cluster-Classification Bayesian Network (CCBN) proposed in [23]. The face images are aligned with the template corresponding to their estimated pose using a set of extracted control points (eyes, mouth, and nose).

A reliable face recognition algorithm should be able to process face images under various lighting and shading conditions. For this reason, we employ the illumination normalization method proposed in [45] to provide robustness under lighting variations.

5.2 Reference Set Evaluation

In the proposed method, the reference set is utilized as a basis set, where gallery and probes images are projected into the reference space and RB descriptors are generated. To evaluate the reference set we reformulate our problem, following the method introduced in [35], as an image alignment problem by sparse and low-rank decomposition. For each pose in the reference set, we construct matrix D whose columns represent images $I_1, \ldots, I_N$ of all reference individuals. The goal is to determine the diversity of D and how effective it is as the basis function matrix. We formulate the problem as an optimization problem as follows [35]:

$$\min_{A,E,\tau} ||A||_\star + \lambda ||E||_1 \quad \text{s.t.} \quad D \circ \tau = A + E, \tag{7}$$

where A is the aligned version of D, $||A||_\star$ corresponds to the nuclear norm of A, E is the error matrix, λ is a positive weighting parameter, $||E||_1$ corresponds to the 1-norm of E, and τ is a set of transformations specified by,

$$D \circ \tau = [I_1 \circ \tau_1, \ldots, I_n \circ \tau_n]. \tag{8}$$

Images in D are better aligned if they are similar; thus, the error E is small. On the other hand, the more dissimilar images in D are, the error E is larger. In the reference set, dissimilar images are desired to define a more definitive basis set. Figure 5 presents the mean squared error of E over all poses. The results were averaged over randomly chosen individuals over multiple experimental runs. We use the reference set with 400 individuals reporting the highest $MSE(E)$ for the remaining experiments.

Figure 6 presents the average number of reference individuals required to achieve 85 % rank-5 recognition rate for galleries of different size. For galleries with more than 500 images the number of reference individuals needs not to be increased to achieve 85 % rank-5 recognition rate. In other words, Fig. 6 shows that when the reference set contains 400 or so individuals, there is not a perceptible change

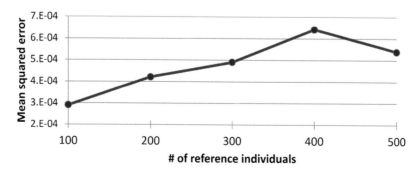

Fig. 5 Mean squared error for various reference set sizes

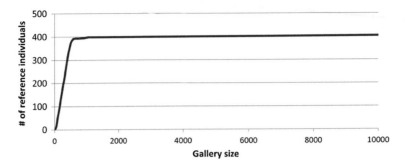

Fig. 6 Average number of reference individuals required to achieve 85 % rank-5 recognition rate

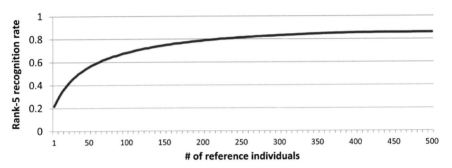

Fig. 7 Rank-5 recognition rate for increasing number of reference set individuals

in accuracy when the size of the gallery increases. Since the gallery/probes and reference set are from different datasets with different pose, illumination, and image settings, there is no reason to believe that if the number of gallery/probe images increases accuracy would decrease unless the reference and test sets are biased in some way (e.g., if they are both biased toward one type of face such as young male). If that is the case, choosing a reference set that is more representative of the type of individuals that can occur in the test set would solve the problem. The whole premise behind the reference-based approach is that while human faces vary, the amount of variation is limited and can be captured by a representative reference set. Figure 6 demonstrates the scalability of reference-based recognition, i.e., a 400 dimensional RB descriptor is able to distinctively represent large size galleries. Note that the number of reference individuals (N) determines the dimensionality of the RB descriptors. All results in this section are averaged over ten experiments with randomly selected reference set individuals from the Multi-PIE and FEI databases.

Figure 7 illustrates how the number of reference set individuals affects the rank-5 recognition rate on a gallery of size 1000. As the number of reference individuals increases over 400, the plot flattens around 85 %.

5.3 Recognition Results

The results in this section are reported on a gallery with 1000 frontal face images, and 200 randomly selected probe images. The results are averaged after repeating the experiments 10 times, each time randomly selecting 200 new probe images.

5.3.1 Reference-Based vs. Direct LBP

The Cumulative Match Characteristic (CMC) curve in Fig. 8 compares the PCA-cosine RB method with direct LBP. The plot clearly shows that our proposed method greatly improves the recognition rate compared to using LBP directly and extends LBP's capability to perform face recognition across pose.

5.3.2 Comparison of Reference-Based Methods

CMC curves comparing the three proposed methods are presented in Fig. 9. The results show that the PCA-cosine based method has better performance than the other two methods. The pose-specific Chi-square method reports lower recognition

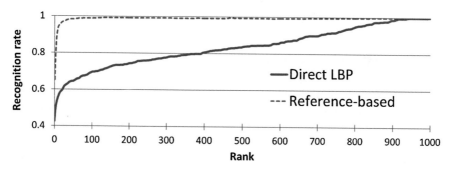

Fig. 8 CMC curve comparing the PCA-cosine RB vs. direct LBP

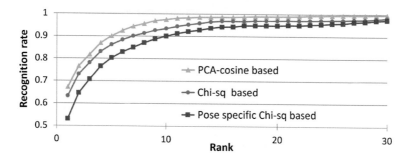

Fig. 9 Comparison of the three proposed RB methods

rates than the Chi-square method. This is because the pose estimation has great influence on the final recognition rate. The reference set images are labeled with the pose ID; however, the ID and the angle it corresponds to may be inaccurate. Therefore, when comparing the probe image with the reference set, comparison may be performed between images with different poses which affects the recognition rate.

5.4 Comparison with Other Methods

To evaluate our framework's performance, we compare our results with multiple state-of-the-art algorithms and three LDA-based approaches. Also, Sect. 5.4.5 demonstrates how the proposed framework performs on the LFW database [20].

As discussed in Sect. 4, the main reason that the proposed reference-based framework has superior performance compared to other reference-based approaches [26, 41, 42, 52] is related to how the first level similarity is computed. The reference-based descriptor generated in previous approaches holds a similarity score for each reference set image. On the contrary, in our proposed framework, each individual in the reference set is represented with a similarity score in the reference-based descriptor and not each image. Selecting the most similar image for each reference set individual to represent that specific individual in the reference-based descriptor increases our framework's capability in handling large pose variations.

5.4.1 Comparison with Doppelgänger Lists [41]

Table 6 presents results comparing our proposed method with the Doppelgänger list approach [41]. For a fair comparison, the experimental setup is similar to that performed in [41]. For this experiment, images are selected from the FacePix database [30], FPLBP features [50] are generated for all images, and 10 test sets each containing 500 positive and 500 negative pairs are used to perform face verification (pair-wise matching). The results show that reference-based recognition is superior to Doppelgänger list comparison for all experimented pose ranges. On the Multi-PIE dataset [15], Doppelgänger list comparison reports 74.5 % ± 2.6 whereas our reference-based recognition achieves 80.3 % ± 1.1.

Table 6 Classification accuracy over various pose ranges

Probe 1	Probe 2	Doppelgänger list [41]	Reference-based
−30° to 30°	−30° to 30°	74.5 % ± 2.6	80.3 % ± 0.9
−90° to 90°	−90° to 90°	68.3 % ± 1.0	75.2 % ± 0.8
−10° to 10°	Angle > |70°|	66.9 % ± 1.0	74.4 % ± 0.8

5.4.2   Comparison with Associate-Predict Model [52]

We compare reference-based recognition with the associate predict (AP) model [52] using images from the Multi-PIE database. The results in Fig. 10 demonstrate that reference-based (RB) recognition has better performance than both AP and likelihood-predict (LP) models.

5.4.3   Comparison with 3D Pose Normalization [5]

We also compare unified reference-based recognition with 3D pose normalization introduced in [5]. 3D pose normalization is limited to poses between $-45°$ to $45°$ in yaw, and is incapable of processing full profile face images. The results in Table 7 show that 3D pose normalization performs better than the proposed reference-based recognition for close to frontal poses (i.e., $-30°$ to $30°$); however, for larger pose angles (i.e., $-45°$ to $-31°$ and $31°$ to $45°$) reference-based recognition has better performance than 3D pose normalization. The reason for such results is that 3D pose normalization is unable to accurately reconstruct the frontal face when the pose angle is large and facial features differ significantly.

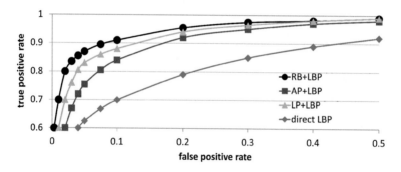

Fig. 10 Reference-based (RB) recognition vs. AP [52], LP [52], and direct LBP [1]

Table 7 Rank-1 recognition rates (%) for FacePix database

Pose →	−90 to −46	−45 to −31	−30 to −16	−15 to −1
3D [5]	–	71.6 %	90.0 %	97.3 %
RB+LGBP	70.6 %	78.9 %	84.7 %	90.1 %
RB+LGTF	75.1 %%	83.6 %	88.6 %	94.4 %

Pose →	1 to 15	16 to 30	31 to 45	46 to 90
3D [5]	95.8 %	92.7 %	74.8 %	–
RB+LGBP	89.9 %	87.2 %	81.3 %	73.2 %
RB+LGTF	94.2 %	91.0 %	85.8 %	77.6 %

In Table 7, for fair comparison, we use Local Gabor Binary Patterns (LGBP) [54] as the feature to compare our results with the 3D pose normalization method [5]. The performance of our proposed reference-based framework relies on the first level similarity; thus, we also show results using the LGTF (LBP+Gabor+TPLBP+FPLBP) feature introduced in [50]. Although the reference-based framework is proposed for recognition across pose, it also performs well for near frontal faces. Using multiple databases, especially the LFW database, shows that the proposed unified reference-based framework, works well for unconstrained face images and frontal face images as well as images with large pose variations.

5.4.4   Comparison with LDA-Based Methods

Table 8 compares reference-based recognition with three LDA-based approaches on the CMU PIE database [43]. For this experiment, a total of 2856 images (68 subjects, 48 images per subject) were randomly selected for testing.

The results in Table 8 show that reference-based recognition has better performance compared to a variety of LDA-based approaches. Holistic approaches (e.g., LDA-based) are more dependent on pixel-wise correspondence of gallery and probe images as compared to local feature-based approaches (e.g. local binary patterns). The outcome is usually higher sensitivity to pose variation for holistic approaches.

5.4.5   Comparison on LFW Database [20]

The face images in the LFW database [20] have a large degree of variability in pose and illumination settings. Compared to the other databases used in our experiments, LFW is the only database with images taken under unconstrained settings. We show our proposed method's performance on the LFW database by comparing it to the associate-predict (AP) and likelihood-predict (LP) models introduced in [52], reference-based verification with message passing model [42], and Cosine similarity metric learning [34]. For fair comparison, we have chosen those methods which use LBP features. We do not compare with other approaches using more advanced features (e.g., [10]).

Table 8 Comparison with LDA-based methods on CMU-PIE [43].

Method	Rank-1	Rank-2	Rank-3
Reference-based	**95.5 % ± 0.006**	**97.4 % ± 0.004**	**97.9 % ± 0.004**
KSLDA [19]	78.5 % ± 0.03	80.4 % ± 0.03	81.5 % ± 0.03
Direct-LDA [22]	75.0 % ± 0.03	77.5 % ± 0.03	78.5 % ± 0.03
SLDA [18]	76.0 % ± 0.03	77.6 % ± 0.02	78.9 % ± 0.03

The bold values highlight the highest value (accuracy) in each column

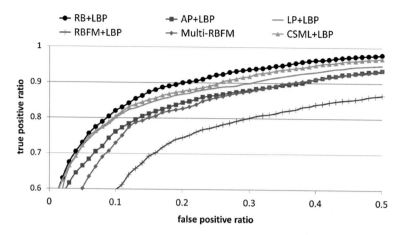

Fig. 11 Proposed reference-based (RB) recognition vs. other algorithms on LFW database [20]

For the experiment on the LFW database [20] we use a reference set with images taken from the WDRef [9] dataset which includes 99,773 images of 2995 individuals. LFW and WDRef both contain unconstrained face images collected from the Internet; thus, we test our framework on the LFW database with a reference set from the WDRef dataset. We choose 500 individuals (which have 40 images each) from the WDRef dataset to build our reference set. There are no common individuals between the LFW test individuals and the reference set. Figure 11 presents the results. Note that we are not strictly following the restricted or unrestricted setting due to the fact that we utilize an external reference set. However, the comparison is fair since the results of the competing methods were obtained using the same settings.

5.5 Processing Time Comparison

Figure 12 compares the processing time between the PCA-cosine RB and the other proposed RB recognition algorithms. The reported time in Fig. 12 refers to the processing time required for computing the first level similarity scores and generating the RB descriptors. The reason for not reporting the entire processing time is because the time to generate the LBP histograms and the time to measure the second level similarity scores is the same for all three methods. Clearly, using PCA not only improved recognition performance but also accelerated the recognition process dramatically. Our C++ and Matlab implementation was tested on an HP workstation laptop with Intel Core i7 CPU and 8 GB of RAM.

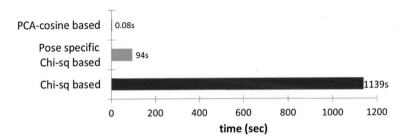

Fig. 12 Processing time comparison

Table 9 Pseudocode for RB retrieval using DCT hashing

Offline phase
– Compute DCT hash set for gallery RB descriptors
– Construct hash table I with U keys; insert DCT hashes into I

Online phase
– Compute DCT hashes for probe RB descriptors
– Generate list of gallery image IDs from accessed buckets
– Process frequency histogram and rank gallery images

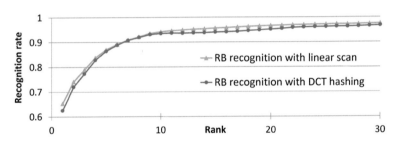

Fig. 13 RB recognition: linear scan vs. DCT hashing

5.6 Retrieval via DCT Hashes

An important advantage of the proposed RB recognition approach compared to other works in Table 4 is that reference-based descriptors are indexable for fast and efficient retrieval. We use Discrete Cosine Transform (DCT) hashing introduced in [24] for this purpose. Table 9 shows the pseudocode for RB retrieval using DCT hashing.

Figure 13 presents the CMC curve comparing linear scan vs. DCT hashing for RB recognition. The results show that RB recognition with DCT hashing, can achieve retrieval accuracy that is very close to the best result with linear scan. Linear scan requires calculating the distance of a probe image to each one of the gallery images which involves thousands of operations. The retrieval time per probe is 74 ms for RB recognition with linear scan, and 9 ms for RB recognition with DCT hashing which is significantly faster.

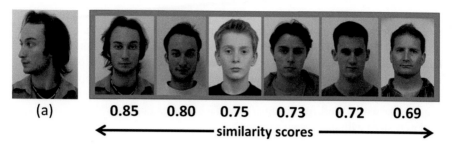

0.85   0.80   0.75   0.73   0.72   0.69

◀━━━━━━━━━ similarity scores ━━━━━━━━━▶

Fig. 14 Example of similar recognition results

5.7 Example Results

A recognition example is shown in Fig. 14. The probe is shown in Fig. 14a, and the top-6 similar images in the gallery returned by the RB method are displayed to the right of the probe. Under each gallery image, the corresponding cosine similarity score between that image and the probe is displayed. The left-most gallery image in Fig. 14 has the highest similarity score to the probe and the similarity scores decrease for the images from left to right.

6 Conclusions

In recognition across pose, the probe and gallery have very different face descriptors which make it a challenging problem. We overcome this challenge by introducing a unified reference-based face recognition framework. The key idea is to create a reference-based descriptor for the probe and gallery face images by comparing them to a reference set, instead of comparing the probe and gallery images directly.

The proposed algorithm normalizes the dimensionality of the face descriptor to the number of individuals in the reference set. That is, independent of the type of utilized face descriptor, size of the images, and any other parameters, the reference-based descriptor has a fixed dimensionality. An important property of the generated reference-based descriptors is that indexing techniques such as DCT locality sensitive hashing can be used for efficient retrieval.

The proposed algorithm was used in conjunction with various descriptors such as LBP, LGBP, and FPLBP, and it was compared with multiple state-of-the-art across-pose recognition algorithms and LDA-based approaches. In performing many comparisons and experiments on seven challenging databases (LFW, FERET, FEI, RaFD, FacePix, CMU-PIE, and Multi-PIE), we found that the proposed approach outperforms previous state-of-the art methods in terms of identification rate and verification accuracy.

While we have presented the variation in the images of human faces in terms of pose, our approach is more general: it can handle variations in expression, lighting, and tilt angle just as well, as long as the reference set includes pictures that cover these variations. Although we introduce reference-based recognition for across pose recognition, it also performs well for near frontal faces. Using multiple databases, especially the LFW database, shows that the proposed reference-based recognition framework, works well for unconstrained face images and frontal face images as well as images with large pose variations.

Reference-based recognition is computationally inexpensive and extremely fast; it took 0.08 s on average for each image to compute the FLS scores. The experiments were performed on an HP workstation laptop with Intel Core i7 CPU and 8 GB of RAM. Our selection of images for both reference sets ensures the repeatability of our experiments.

References

1. T. Ahonen, A. Hadid, M. Pietikainen, Face description with local binary patterns: application to face recognition. IEEE Trans. Pattern Anal. Mach. Intell. **28**(12), 2037–2041 (2006)
2. L. An, B. Bhanu, S. Yang, Boosting face recognition in real-world surveillance videos, in *AVSS* (2012)
3. L. An, M. Kafai, B. Bhanu, Dynamic Bayesian network for unconstrained face recognition in surveillance camera networks. IEEE J. Emerging Sel. Top. Circuits Syst. **3**(2), 155–164 (2013)
4. S. Arashloo, J. Kittler, Energy normalization for pose-invariant face recognition based on MRF model image matching. IEEE Trans. Pattern Anal. Mach. Intell. **33**(6), 1274–1280 (2011)
5. A. Asthana, T. Marks, M. Jones, K. Tieu, M. Rohith, Fully automatic pose-invariant face recognition via 3D pose normalization, in *ICCV* (2011), pp. 937–944
6. M. Bae, A. Razdan, G. Farin, Automated 3D face authentication & recognition, in *AVSS* (2007), pp. 45–50
7. C. Castillo, D. Jacobs, Using stereo matching for 2-D face recognition across pose, in *CVPR* (2007), pp. 1–8
8. X. Chai, S. Shan, X. Chen, W. Gao, Locally linear regression for pose-invariant face recognition. IEEE Trans. Image Process. **16**(7), 1716–1725 (2007)
9. D. Chen, X. Cao, L. Wang, F. Wen, J. Sun, Bayesian face revisited: a joint formulation, in *ECCV* (2012), pp. 566–579
10. D. Chen, X. Cao, F. Wen, J. Sun, Blessing of dimensionality: high-dimensional feature and its efficient compression for face verification, in *CVPR* (2013), pp. 3025–3032
11. Z. Cui, S. Shan, H. Zhang, S. Lao, X. Chen, Image sets alignment for video-based face recognition, in *CVPR* (2012), pp. 2626–2633
12. N. Dalal, B. Triggs, Histograms of oriented gradients for human detection, in *CVPR*, vol. 1 (2005), pp. 886–893
13. R.P. Duin, E. Pekalska, The dissimilarity space: bridging structural and statistical pattern recognition. Pattern Recogn. Lett. **33**(7), 826–832 (2012)
14. Y. Gao, Y. Qi, Robust visual similarity retrieval in single model face databases, Pattern Recogn. **38**, 1009–1020 (2005)
15. R. Gross, I. Matthews, J. Cohn, T. Kanade, S. Baker, Multi-PIE. Image Vis. Comput. **28**(5), 807–813 (2010)
16. Y. Guo, Y. Shan, H. Sawhney, R. Kumar, Peet: prototype embedding and embedding transition for matching vehicles over disparate viewpoints, in *CVPR* (2007), pp. 1–8

17. A. Gyaourova, A. Ross, Index codes for multibiometric pattern retrieval. IEEE Trans. Inf. Forensics Secur. **7**(2), 518–529 (2012)

18. J. Huang, P. Yuen, W.S. Chen, J. Lai, Component-based LDA method for face recognition with one training sample, in *IEEE International Workshop on Analysis and Modeling of Faces and Gestures* (2003)

19. J. Huang, P. Yuen, W.S. Chen, J.H. Lai, Choosing parameters of kernel subspace LDA for recognition of face images under pose and illumination variations. IEEE Trans. Syst. Man Cybern. B Cybern. **37**(4), 847–862 (2007)

20. G.B. Huang, M. Ramesh, T. Berg, E. Learned-Miller, Labeled faces in the wild: a database for studying face recognition in unconstrained environments. Technical Report 07–49, UMass Amherst (2007)

21. D. Huang, C. Shan, M. Ardabilian, Y. Wang, L. Chen, Local binary patterns and its application to facial image analysis: a survey, IEEE Trans. Syst. Man Cybern. Part C Appl. Rev. **41**(6), 765–781 (2011)

22. H.Y. Jie, H. Yu, J. Yang, A direct LDA algorithm for high-dimensional data – with application to face recognition. Pattern Recogn. **34**, 2067–2070 (2001)

23. M. Kafai, B. Bhanu, L. An, Cluster-classification Bayesian networks for head pose estimation, in *ICPR* (2012)

24. M. Kafai, K. Eshghi, B. Bhanu, Discrete cosine transform locality-sensitive hashes for face retrieval. IEEE Trans. Multimedia **16**(4), 1090–1103 (2014)

25. F. Kahraman, B. Kurt, M. Gokmen, Robust face alignment for illumination and pose invariant face recognition, in *CVPR* (2007)

26. N. Kumar, A. Berg, P. Belhumeur, S. Nayar, Attribute and simile classifiers for face verification, in *ICCV* (2009), pp. 365–372

27. O. Langner, R. Dotsch, G. Bijlstra, D. Wigboldus, S. Hawk, A. Van Knippenberg, Presentation and validation of the Radboud faces database. Cogn. Emot. **24**(8), 1377–388 (2010)

28. A. Li, S. Shan, W. Gao, Coupled bias-variance tradeoff for cross-pose face recognition. IEEE Trans Image Process. **21**(1), 305–315 (2012)

29. Q. Liao, J.Z. Leibo, T. Poggio, Learning invariant representations and applications to face verification, in *Advances in Neural Information Processing Systems* (NIPS), (2013), pp. 3057–3065

30. D. Little, S. Krishna, J. Black, S. Panchanathan, A methodology for evaluating robustness of face recognition algorithms with respect to variations in pose angle and illumination angle, in *ICASSP* (2005)

31. J. Liu, B. Kuipers, S. Savarese, Recognizing human actions by attributes, in *CVPR* (2011), pp. 3337–3344

32. A. Majumdar, R. Ward, Pseudo-Fisherface method for single image per person face recognition, in *IEEE International Conference on Acoustics, Speech and Signal Processing* (2008), pp. 989–992

33. Y. Mami, D. Charlet, Speaker recognition by location in the space of reference speakers. Speech Comm. **48**(2), 127–141 (2006)

34. H.V. Nguyen, L. Bai, Cosine similarity metric learning for face verification, in *ACCV* (2010), pp. 709–720

35. Y. Peng, A. Ganesh, J. Wright, W. Xu, Y. Ma, RASL: robust alignment by sparse and low-rank decomposition for linearly correlated images. IEEE Trans. Pattern Anal. Mach. Intell. **34**(11), 2233–2246 (2012)

36. P. Phillips, H. Moon, P. Rauss, S. Rizvi, The FERET evaluation methodology for face-recognition algorithms, in *CVPR* (1997)

37. U. Prabhu, J. Heo, M. Savvides, Unconstrained pose-invariant face recognition using 3D generic elastic models. IEEE Trans. Pattern Anal. Mach. Intell. **33**(10), 1952–1961 (2011)

38. S. Prince, J. Warrell, J. Elder, F. Felisberti, Tied factor analysis for face recognition across large pose differences. IEEE Trans. Pattern Anal. Mach. Intell. **30**(6), 970–984 (2008)

39. N. Rasiwasia, P. Moreno, N. Vasconcelos, Bridging the gap: query by semantic example. IEEE Trans. Multimedia **9**(5), 923–938 (2007)

40. K. Sakata, T. Maeda, M. Matsushita, K. Sasakawa, H. Tamaki, Fingerprint authentication based on matching scores with other data, in *International Conference on Advances in Biometrics* (2006), pp. 280–286
41. F. Schroff, T. Treibitz, D. Kriegman, S. Belongie, Pose, illumination and expression invariant pairwise face-similarity measure via Doppelgänger list comparison, in *ICCV* (2011), pp. 2494–2501
42. W. Shen, B. Wang, Y. Wang, X. Bai, L.J. Latecki, Face identification using reference-based features with message passing model. Neurocomputing **99**, 339–346 (2013)
43. T. Sim, S. Baker, M. Bsat, The CMU pose, illumination, and expression (PIE) database, in *FG* (2002), pp. 46–51
44. R. Singh, M. Vatsa, A. Ross, A. Noore, A mosaicing scheme for pose-invariant face recognition. IEEE Trans. Syst. Man Cybern. B Cybern. **37**(5), 1212–1225 (2007)
45. X. Tan, B. Triggs, Enhanced local texture feature sets for face recognition under difficult lighting conditions. IEEE Trans. Image Process. **19**(6), 1635–1650 (2010)
46. X. Tan, S. Chen, Z.H. Zhou, F. Zhang, Face recognition from a single image per person: a survey. Pattern Recogn. **39**(9), 1725–1745 (2006)
47. C.E. Thomaz, G.A. Giraldi, A new ranking method for principal components analysis and its application to face image analysis. Image Vis. Comput. **28**(6), 902–913 (2010)
48. M. Turk, A. Pentland, Face recognition using eigenfaces, in *CVPR* (1991)
49. S. Wold, K. Esbensen, P. Geladi, Principal component analysis. Chemom. Intell. Lab. Syst. **2**(1–3), 37–52 (1987)
50. L. Wolf, T. Hassner, Y. Taigman, Effective unconstrained face recognition by combining multiple descriptors and learned background statistics. IEEE Trans. Pattern Anal. Mach. Intell. **33**(10), 1978–1990 (2011)
51. X. Xianchuan, Z. Qi, Medical image retrieval using local binary patterns with image Euclidean distance, in *International Conference on Information Engineering and Computer Science* (2009), pp. 1–4
52. Q. Yin, X. Tang, J. Sun, An associate-predict model for face recognition, in *CVPR* (2011), pp. 497–504
53. X. Zhang, Y. Gao, Face recognition across pose: a review. Pattern Recogn. **42**(11), 2876–2896 (2009)
54. W. Zhang, S. Shan, W. Gao, X. Chen, H. Zhang, Local gabor binary pattern histogram sequence (LGBPHS): a novel non-statistical model for face representation and recognition, in *ICCV* (2005), pp. 786–791
55. X. Zhang, Y. Gao, M. Leung, Recognizing rotated faces from frontal and side views: an approach toward effective use of mugshot databases. IEEE Trans. Inf. Forensics Secur. **3**(4) 684–697 (2008)
56. Y. Zhang, M. Shao, E. Wong, Y. Fu, Random faces guided sparse many-to-one encoder for pose-invariant face recognition, in *ICCV* (2013), pp. 2416–2423

On Frame Selection for Video Face Recognition

Tejas I. Dhamecha, Gaurav Goswami, Richa Singh, and Mayank Vatsa

Abstract In surveillance applications, a video can be an advantageous alternative to traditional still image face recognition. In order to extract discriminative features from a video, frame based processing is preferred; however, not all frames are suitable for face recognition. While some frames are distorted due to noise, blur, and occlusion, others might be affected by the presence of covariates such as pose, illumination, and expression. Frames affected by imaging artifacts such as noise and blur may not contain reliable facial information and may affect the recognition performance. In an ideal scenario, such frames should not be considered for feature extraction and matching. Furthermore, video contains a large amount of frames and adjacent frames contain largely redundant information and processing all the frames from a video increases the computational complexity. Instead of utilizing all the frames from a video, frame selection can be performed to determine a subset of frames which is best suited for face recognition. Several frame selection algorithms have been proposed in the literature to address these concerns. In this chapter, we discuss the role and importance of frame selection in video face recognition, provide an overview of existing techniques, present an entropy based frame selection algorithm with the results and analysis on Point-and-Shoot-Challenge which is a recent benchmark database, and also propose a new paradigm for frame selection algorithms as a path forward.

1 Introduction

Face recognition is an extensively studied research area which continues to garner attention from the research community due to its challenging nature in the presence of covariates such as pose, illumination, expression, and low-resolution [13]. Traditional methods rely on matching still face images of different individuals using

Equal contributions by the student authors (Tejas I. Dhamecha and Gaurav Goswami).

T.I. Dhamecha (✉) • G. Goswami • R. Singh • M. Vatsa
IIIT Delhi, New Delhi, India
e-mail: tejasd@iiitd.ac.in; gauravgs@iiitd.ac.in; rsingh@iiitd.ac.in; mayank@iiitd.ac.in

© Springer International Publishing Switzerland 2016
M. Kawulok et al. (eds.), *Advances in Face Detection and Facial Image Analysis*,
DOI 10.1007/978-3-319-25958-1_10

279

a feature representation. However, processing still images poses challenges as they capture limited information and are not sufficient to uniquely represent a face when robustness to the aforementioned covariates is desired. Face recognition in videos is a promising approach which, recently, has received attention in the literature [2] but it presents its own set of challenges such as increased processing requirements and noisy data.

In order to apply existing methodologies to video face recognition application, generally individual frames are extracted, thereby converting it into a collection of still images. A video offers the advantage of capturing the face images under different conditions to facilitate the extraction of representative features. However, the abundance of data in a video also increases the chances of spurious matches since not all frames are informative for the purpose of recognition. Depending on the capture device and encoding scheme, the quality of the video and the extracted frames might suffer, leading to poor fidelity frames. Further, some frames might be unsuitable due to natural factors such as pose and expression that can cause a high degree of variation in the appearance of individual frames. If such frames are utilized to compute a face representation, its coherence and uniqueness can be easily compromised leading to poor performance. To further accentuate this aspect, consider the frames extracted from two videos in the Point and Shoot Challenge (PaSC) database [6] illustrated in Fig. 1. While one video contains overall good quality frames that capture multiple variations and provide discriminative information, the other video contains frames with motion and sensor blur, extreme pose variations and suffers from poor quality frames. If all the frames of both videos are considered to construct respective face representations, the second video would suffer from a poor representation. Surveillance videos also present a unique challenge for frame selection algorithms. A single video may contain multiple identities, and utilizing a video sequence with information of more than a single subject might lead to spurious identification results. Therefore, prior to processing such a video, different identities have to be sorted and only frames pertaining to a single identity should be used for matching with a gallery or another such sequence. This step has to be performed either manually or automatically, as recently proposed by Marsico et al. [8]. Finally, videos are also inherently costly in terms of storage and processing power due to the large volume of data they encompass.

As mentioned previously, utilizing all the frames is a popular strategy for existing video face recognition algorithms but it suffers from the aforementioned challenges. Frame selection is a process by which a subset of frames is selected instead of utilizing all the frames. In the example shown in Fig. 1, if frontal frame selection is utilized, the first video will still result in a good representation since all the frames are of sufficient fidelity [7]. However, the second video may suffer from a poor representation depending on which frames are chosen, since many frames have degrading factors such as blur. Most of the frontal frame selection algorithms would also not work optimally for either video since very few of the frames satisfy the frontal criteria. Thus, frontal frame selection may not necessarily translate in good recognition performance. On the other hand, intuitively there are non-frontal

Fig. 1 Sample frames from two videos in the Point and Shoot Challenge (PaSC) video face database [6]

face frames that may be representative of the person's identity. Therefore, it is evident that besides reducing the storage and processing requirements, intelligent frame selection can improve the quality of the extracted face representation, thereby improving the recognition performance. In this chapter, we discuss

1. a taxonomy of literature pertaining to frame selection algorithms
2. entropy based frame selection approach, and
3. analyze the impact of frame selection on the benchmark Point and Shoot Challenge (PaSC) database [6] using two face recognition algorithms. Experimental results suggest that entropy based frame selection reduces computational time while improving the recognition performance.

2 Literature Review

On basis of the primary methodology involved, existing frame selection algorithms can be classified into three major categories: (a) Clustering based, (b) optical flow based, and (c) quality based. We present an overview of existing algorithms pertaining to each of these categories:

2.1 *Clustering Based Frame Selection Algorithms*

Hadid and Pietikainen [11] proposed a clustering based solution to frame selection. They considered each face video as a collection of various face models, each model consisting of several frames that were related by certain covariates such as pose and expression. They converted the high-dimensional face images to a low-dimensional space using the Locally Linear Embedding (LLE) [27] algorithm and then applied the K-means clustering algorithm to obtain face clusters which were an approximation of the face models in the video. After that exemplar frames were selected from each of these models by selecting feature vectors close to the center of each cluster. The set of exemplar frames obtained were then utilized to represent each video. Berrani et al. [4] proposed a frame selection algorithm based on outlier removal. They utilized the Robust Principal Component Analysis (RobPCA) algorithm [12] for this purpose. The aim of the algorithm was to consider each video as a set of face images and remove any face image that corresponded to an outlier feature vector in the feature matrix for that video. They constructed a sub-feature space using the mean and covariance matrix computed by RobPCA and computed two types of distances for each of the faces relative to the center of the new sub-feature space. A threshold was applied to both of these distances to determine the outliers which were eliminated to perform frame selection.

Fan and Yeung [9] proposed another clustering based algorithm for frame selection. They computed a geodesic distance matrix for all the frames extracted from a video. In order to construct this matrix, they represent each $m \times n$ face in a mn-dimensional space. Euclidean distance was used to approximate the geodesic distance between neighboring faces and aggregated euclidean distance along short hops was used to approximate geodesic distance in other scenarios. A hierarchical agglomerative clustering (HAC) algorithm was then applied directly on the geodesic distance matrix to obtain the face clusters. The center feature vectors of each of these clusters were selected as the exemplars and used to construct the frame subset for each video. Liu et al. [16] proposed a frame selection algorithm which partitioned the video in the temporal domain and formed frame clusters based on temporal information. First, they reduced the dimensionality of faces by using PCA on the collection of training frames and performed K-means clustering on these frames with additional constraints on the distance metric to promote frames occurring close to each other in the temporal space to belong to the same cluster. Each cluster was modeled by a joint probability distribution and keyframes were selected from each cluster by maximizing the conditional probability of each frame belonging to the particular cluster.

Category	Authors and year	Algorithm
Clustering	Hadid and Pietikainen, 2004 [11]	Locally linear embedding (LLE) followed by K-means. Exemplar frames from each cluster are selected
	Berrani and Garcia, 2005 [4]	Robust PCA algorithm for outlier detection and removal. Remaining frames are selected
	Fan and Yeung, 2006 [9]	Geodesic distance based hierarchical agglomerative clustering (HAC). Central frames of each cluster are selected
	Liu et al., 2006 [16]	Temporal and spatial clustering using PCA and K-means. Frames with maximum conditional probability are selected from each cluster
Optical flow	Jillela and Ross, 2009 [14]	Lucas-Kanade algorithm to quantify inter-frame motion. Sequences of frames with large inter-frame motion are rejected
	Saeed and Dugelay, 2010 [25]	Lucas-Kanade algorithm to quantify speech induced motion by using mouth ROI. Frames with maximum motion are selected
Quality	Park and Jain, 2007 [20]	3D shape models for pose categorization and DCT for motion blur. Frames with different poses and low motion blur are selected
	Anantharajah et al., 2012 [1]	Face symmetry, sharpness, contrast, closeness of mouth, brightness, and openness of the eye quality measures combined using neural network
	Best-Rowden et al., 2013 [5]	Faceness and frontal-ness quality measures computed using COTS. Frames with highest faceness and frontal-ness are selected

2.2 Optical Flow Based Frame Selection Algorithms

Jillela and Ross [14] proposed an adaptive frame selection algorithm specifically targeted for low-resolution videos. Their primary objective was to eliminate frames with large pose variations in the selected frames so that a super-resolution technique could be applied for face recognition. To this end, they quantified inter-frame motion by computing optical flow matrices for two frames using the Lucas-Kanade algorithm [17], computing the L_2 norm of these matrices, and aggregating the largest k values in the resulting matrix. The value thus obtained was termed as the inter-frame motion parameter β; normalized to [0,1] using min-max normalization and computed between every pair of consecutive frames. A threshold T was then applied on β to eliminate frames with large inter-frame motion. In 2010, Saeed and Dugelay [25] proposed a frame selection algorithm to handle variations in faces caused due to speech. Each face frame was segmented into different regions and a Region of Interest (ROI) centered around the mouth region was obtained. The Lucas-Kanade algorithm [17] was used to measure optical flow in this ROI and frames with a local maxima of motion were selected as the keyframes.

2.3 *Quality Based Frame Selection Algorithms*

Park et al. [21] proposed a pose and motion blur based frame selection algorithm. A training set of images was used to learn multiple Active Appearance Models (AAMs), one for each facial pose. For each frame in a given video, the learned AAMs were used to determine facial keypoints. These keypoints were used to create a 3D shape model using the factorization algorithm [29] which was then compared with a set of generic 3D models corresponding to different poses. The pose of each frame was determined by determining the generic model with the least distance from the reconstructed 3D model. Motion blur was characterized using Discrete Cosine Transform (DCT) coefficients of the images to determine the high frequency components in each frame and frames with motion blur were eliminated.

Anantharajah et al. [1] proposed a quality based frame selection algorithm which obtained a face quality measure using the following cues: face symmetry, sharpness, contrast, closeness of mouth, brightness, and openness of the eye. The face symmetry feature [24] quantified face frontal-ness and uprightness. The sharpness feature eliminated the frames that were affected by motion blur and the contrast feature eliminated the dull frames. The brightness measure ensured that frames without sufficient illumination were not selected. Closeness of mouth was characterized using the height to width ratio of the detected mouth region and essentially provided a measure of expression neutrality. Otsu's thresholding method [19] was utilized to encode openness of the eyes to preserve the details around the eye region. Finally, all of these scores were normalized and fused using a [7-10-1] neural network architecture which was learnt to produce a high score for a high quality face image. Best-Rowden et al. [5] proposed another quality based frame selection algorithm. They considered two quality measures: *faceness* and frontal-ness. Faceness represents the confidence with which a face detector can affirm that the frame contains a face object and frontal-ness is a measure of how frontal the face pose is in a frame. These two measures were computed using a Commercial Off the Shelf (COTS) face recognition algorithm and combined such that frames with highest faceness and frontal-ness were selected for each video.

3   Entropy Based Frame Selection

Human face recognition has served as a benchmark for evaluating the performance of face recognition algorithms in the literature. Several cognitive studies have highlighted different aspects of the face recognition pipeline in humans and results have indicated the possibilities of the existence of brain functions designed specifically for faces and face-like objects [28]. Therefore, research efforts in the area have also been directed towards developing human-inspired algorithms and systems for face recognition by machines, especially in challenging scenarios [18]. The proposed frame selection algorithm is inspired by a similar characteristic, termed as *memorability*. Memorability, as the name suggests, is the property of an image stimulus by virtue of which it is easier or harder for a human observer to remember compared to other images of the same kind. For example, in the case of faces, humans might find certain faces to be easier to remember and recall compared to others. Some people tend to easily get lost in the crowd, whereas others stand out. Usually this property is associated with an individual and hence cannot be utilized for frame selection in this form. Recent research in the area reveals interesting results and provides another perspective for memorability of faces.

Khosla et al. [15] have shown that face memorability is also dependent on the face image itself. Compared to a real life scenario where a human gets a large number of face samples of a person, the memorability of a face image matters a lot more in situations where a far lesser number of samples are presented to the observer, such as in case of a recognition algorithm. With the help of a memory game undertaken by several human volunteers, Khosla et al. [15] have demonstrated that by using certain techniques they are able to modify human recognition rates for face images with 74 % accuracy. These results support the notion that memorability of a face in limited sample scenario is linked with the face image itself. In accordance with these observations, the proposed frame selection algorithm selects a subset of the most memorable frames from a face video. It is our hypothesis that the memorability of a face essentially captures the information encapsulated within the image for purpose of discriminative representation and future recall. Therefore, it is our assertion that memorable frames can enable representative feature extraction when utilized with a feature extraction and matching algorithm, thereby preserving the important details of a face video in lesser number of frames. Moreover, such an approach should also be able to reject spurious frames which do not contain useful information from a matching perspective and eliminate frames with motion and sensor blur, and other artifacts.

The proposed algorithm computes memorability by computing a measure of the *feature richness* of the image by a local analysis of the image content [10]. The image is represented as a sum of small parts which resembles the way visual data is processed by the human visual cortex [31]. An input frame I is preprocessed to a fixed pre-decided size (such as 100×100 pixels) and is converted to the HSV (Hue, Saturation, Value) color space. One of the common covariates concerning face recognition is illumination and utilizing only the Hue channel in the converted image allows for robustness to illumination variations. Thereafter, the local visual entropy [23] of the image is computed using 2×2 overlapping windows over the entire image. Visual entropy is a good encoder of image texture and can provide an estimate of the feature-richness of an image region since it captures a measure of the variation in pixel intensity values. Local neighborhoods which do not contain any variations are least likely to contain discriminative information and vice versa. Visual entropy, $H(\mathcal{X})$ of a window $\mathcal{X}$ is computed according to Eq. (1):

$$H(\mathcal{X}) = -\sum_{i=1}^{n} p(g_i) log_2 p(g_i) \tag{1}$$

where, $p(g_i)$ is the value of the probability mass function for g_i which signifies the probability that the grayscale value g_i exists in the neighborhood and n is the maximum possible number of such grayscale values. If the size of window $\mathcal{X}$ is $M_H \times N_H$ then

$$p(g_i) = \frac{n_{g_i}}{M_H \times N_H} \tag{2}$$

Here, n_{g_i} represents the number of pixels in the window with the grayscale value g_i, and $M_H \times N_H$ is the total number of pixels. The visual entropy value of each window is aggregated to obtain the memorability score of frame I according to Eq. (3):

$$\mathbf{H} = \sum_{i=1}^{n} (|H(\mathcal{X}_i)|) \tag{3}$$

where, $\mathbf{H}$ is the memorability score of the frame, n is the number of windows in I, and $H(\mathcal{X}_i)$ is the entropy of the ith window in frame I.

The memorability score of a frame is thus the cumulative value of local visual entropy. A high value denotes a frame with variation-rich regions which might be efficiently encoded by a feature extractor whereas a low value signifies that a frame is less likely to contain discriminative information. Therefore, to maximize the possibility of extracting a definitive representation in a lower number of frames, memorability of each frame is computed using the proposed algorithm and top $\mathcal{F}$ frames are selected according to high memorability scores. $\mathcal{F}$ is a parameter of the algorithm which dictates the number of frames selected per video. This parameter can be fixed to a particular value or determined dynamically by normalizing the memorability scores of each video sequence and applying a threshold. All frames with a memorability value above the threshold are selected and the others are rejected. Such a dynamic approach also has the added benefit of accounting for the fact that not all videos have equal facial content and some videos might require more frames to be captured compared to others where fewer number of frames might be optimal. Figure 2 shows samples of most memorable, least memorable, and randomly chosen frames.

Fig. 2 Selected frames for a few videos from the PaSC database. (**a**) Top ten most memorable frames, (**b**) ten random frames, (**c**) ten least memorable frames

4 Evaluation, Results, and Analysis

This section provides the details of the empirical evaluation and analysis of the entropy based frame selection approach. First, the details pertaining to the face recognition algorithms used, the dataset and protocol are provided followed by the results.

4.1   Face Recognition Algorithms

The proposed frame selection algorithm is evaluated with two different recognition methodologies, namely deep learning based MDLFace [10] and local region principal component analysis (LRPCA) [22].

MDLFace The MDLFace algorithm [10] first computes the memorable frames for each video, after which the most memorable frames from the training videos are utilized to train the Stacked Denoising AutoEncoder (SDAE) + Deep Boltzmann Machine (DBM) architecture illustrated in Fig. 3. Let $\mathbf{x} \in R^a$ be the input data; an autoencoder [3] maps the data into a feature representation $\mathbf{f}$ using a deterministic encoding function g_p

$$\mathbf{f} = g_p(x) = s(\mathbf{w} \cdot \mathbf{x} + \mathbf{b}) \tag{4}$$

where p is the parameter set; $p = \{\mathbf{w}, \mathbf{b}\}$, s represents the sigmoid activation function, $\mathbf{w}$ is the $a' \times a$ weight matrix, and $\mathbf{b}$ is the bias vector of size a'. The decoder function $g'_{p'}$ maps the feature f to $\hat{\mathbf{x}}$, a feature vector of dimensionality a such that:

$$\hat{\mathbf{x}} = g'_{p'}(\mathbf{f})g'_{p'}(\mathbf{f}) = s(\mathbf{w}' \cdot f + \mathbf{b}') \tag{5}$$

where $p' = \{\mathbf{w}', \mathbf{b}'\}$ is the appropriately sized parameter set. The parameters are optimized by utilizing the unsupervised training data and computing $\arg\min_{w,w'} ||\mathbf{x} - \hat{\mathbf{x}}||_2$. Denoising autoencoders [30], a variant of autoencoders, operate on the noisy input data $\mathbf{x_n}$ and attempt to reconstruct $\hat{\mathbf{x}}$. It is observed that these autoencoders are robust to noisy data and have good generalizability. If these autoencoders are stacked

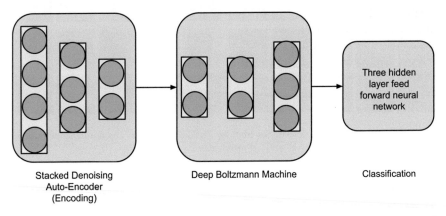

Stacked Denoising   Deep Boltzmann Machine   Classification
Auto-Encoder
(Encoding)

Fig. 3 An overview of the MDLFace deep learning architecture [10] for feature extraction and matching

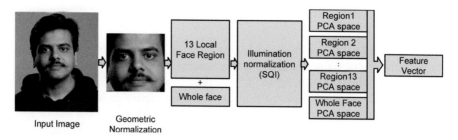

Fig. 4 An overview of the LRPCA algorithm [22] for face recognition

in a layered manner, they are called stacked denoising autoencoders and form a deep learning architecture. Deep Boltzmann Machine is an undirected graphical model, a deep network architecture, with symmetrically coupled binary units [26]. It is designed by layer-wise training of Restricted Boltzmann Machine (RBM) and stacking them together in an undirected manner. The purpose of the SDAE+DBM module is to extract meaningful and discriminative features from the frames which are used as face representation for matching. During training, the SDAE learns to accurately reconstruct the given face images (frames) in an unsupervised manner using a more concise representation. On the other hand, the DBM module further refines the extracted features for discrimination by utilizing a fine-tuning process. This process is performed with a small number of labelled samples and a large number of unlabeled samples. A part of the training data is also utilized to train the classification feed-forward neural network. For a probe video frame, the learned SDAE model extracts progressively more complex and concise features with every hidden layer. These features are further refined for discrimination using the DBM architecture and the final extracted feature is processed by the classifier to obtain the decision.

LRPCA LRPCA based algorithm [22] operates on face patches corresponding to 13 face regions and the full face. Figure 4 explains the procedure of feature extraction in LRPCA algorithm. On all the 14 regions (13 patches and full face), self-quotient normalization is applied to reduce the effect of illumination variation. Further, pixel intensities of each region are normalized such that mean and standard deviation of all the regions is zero and one respectively. Using the training dataset, one PCA subspace is learned individually for each region. Since the top principal components (PC) may carry covariate specific information, top two components are discarded, and the following 250 components are preserved. These principal components are further whitened, and weight of each dimension is adjusted according to Fisher's criterion. For every test image, at first the pre-processed face regions are obtained, which are then projected on to a 250 dimensional space. Thus, the dimensionality of each test face image is 3500 (250×14). The final matching is performed using Pearson's correlation.

4.2   Dataset and Protocol

The Point and Shoot Challenge video face database [6] is utilized for evaluating the performance of frame selection. It is a benchmark video face database which contains 2802 videos pertaining to 265 individuals with an average of 93 frames per video. The videos are captured using a handheld (referred as handheld videos) and a high quality video camera (referred to as control videos) in both indoor and outdoor locations. The subjects were asked to perform a given task and the videos were recoded. The database thus contains large variations in pose, expression, illumination, and resolution. Moreover, the overall quality of the videos captured using the handheld camera in unfavorable conditions such as low illumination and subject motion is quite low resulting in additional recognition challenges. The dataset has two predefined protocols, handheld and control, for benchmarking. As the names suggest, the protocols are designed to evaluate the recognition performance on videos captured from handheld and fixed devices separately. Face verification results are reported in terms of Receiver Operating Characteristic (ROC) curves.

4.3   Results and Analysis

Figures 5 and 6 present the results (ROC curves) on the handheld and control subsets of the PaSC database using random and entropy based frame selection with MDLFace deep learning algorithm and LRPCA, respectively. The key observations are as following:

- **Recognition performance:** Using both the face recognition algorithms, we have observed that the frame selection approach helps in improving the computational time required for matching. The verification performance is observed to be equal or better than no-frame selection and random frame selection.

  - In case of MDLFace algorithm, it is observed that, entropy based selection performs much better than random frames in both cases at all false accept rates. While experiments on both handheld and control videos have shown a large performance difference, it is more profound in the case of handheld videos. This observation supports our hypothesis that memorable frames are able to avoid most low quality and detrimental frames in handheld videos. Due to the overall better quality of the control subset videos, the difference in performance is smaller but still noticeable which indicates that even among higher quality frames, entropy based memorability approach provides a metric that can improve the recognition performance.
  - In case of LRPCA algorithm, although there is no improvement in verification accuracy; the performance obtained using only memorable frames is almost

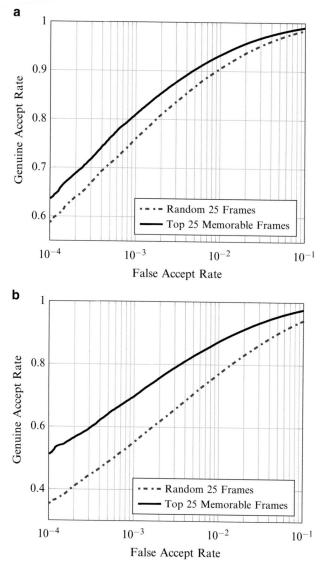

Fig. 5 Comparison of verification performance obtained using all frames and using only top 50 memorable frames. Frames are matched using the MDLFace algorithm [10]. (**a**) Control videos (**b**) Handheld videos

same as using all frames. This suggests that, an appropriate frame selection should not decrease the verification accuracy.

– Further analysis of LRPCA reveals that the correlation coefficient between similarity scores obtained with and without frame selection is 0.99 for both the protocols. Moreover, we have observed that the *p*-value is ≈0 which leads

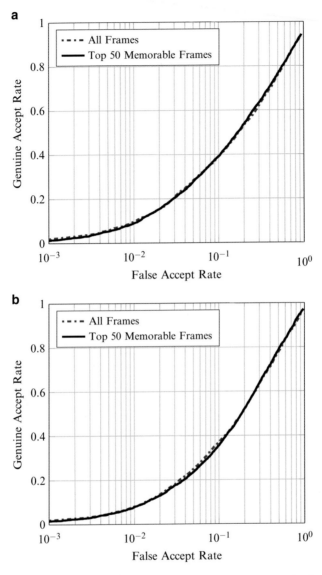

Fig. 6 Comparing the verification performance obtained using all frames and using only top 50 memorable frames. Frames are matched using LRPCA [22]. (**a**) Control videos (**b**) Handheld videos

to rejecting the hypothesis that *this correlation value can be obtained by a random chance*. Hypothesis testing indicates that correlation is significant; thus, the similar verification performance is originating from alike similarity scores. This demonstrates that the frames which are useful for LRPCA matching, are preserved with the entropy based frame selection algorithm.

Table 1 Summarizing the number of frame comparisons required for handheld and control protocols in the PaSC database

	Number of frame comparisons		
	All frames	Top 50 frames	Top 25 frames
Handheld	7,957,624,137	1,797,138,709	522,479,289
Control	8,392,015,867	1,853,240,838	512,279,873

- **Computational complexity:** In case of both the face recognition methodologies, with frame selection there is a guaranteed advantage of reduced computation complexity. Since each of the videos, at an average, contains 93 frames, the matcher that uses only $\mathcal{F}$ frames, would save $(93^2 - \mathcal{F}^2)$ frame comparisons on an average per video pair. As summarized in Table 1, by selecting 50 and 25 frames, the number of frame comparisons required is approximately 22 and 6 % of the number of comparisons required in absence of frame selection, respectively.
- **Number of selected frames:** It is empirically observed that selecting 25 memorable frames is more suitable for MDLFace recognition algorithm, whereas for LRPCA it is 50. It is our observation that minimum number of selected frames may be a function of the effectiveness of the recognition algorithm to obtain robust facial representation from the data.

5 Pair-Wise Frame Selection: A Path Forward

This section discusses the concept of pair-wise frame selection. Traditionally, frame selection has been explored as a *no reference* procedure, as the frames in one video do not affect the frame selection outcomes of other video. However, it is our assertion that frame(s) should be selected from a (probe) video taking into account the (gallery) video with which it is to be matched. For example, if frames in one video have a certain covariate present in them, it might be more appropriate to select the frames with similar covariates from another video. With this assertion, we believe that the future direction for frame selection should be *pair-wise frame selection*. Specifically, if we have a probe video $\mathcal{P}$ and gallery videos $\mathcal{G}_1$ and $\mathcal{G}_2$, when matching $\mathcal{P}$ with $\mathcal{G}_1$, a different set of frames may be selected from the probe video as compared to when $\mathcal{P}$ is matched with $\mathcal{G}_2$.

To accentuate our assertion and to understand the potential effectiveness of such pair-wise frame selection, we present an experimental analysis. For matching a video consisting of m_1 frames with a video consisting of m_2 frames, a block matrix of size $m_1 \times m_2$ is obtained in which each element (i, j) of matrix represents the match score between the ith frame of the first video and the jth frame of the second video. The PaSC challenge aggregates these all-frame-to-all-frame scores using the max-rule. In other worlds, maximum similarity value obtained from the block matrix value is used as the estimate of similarity between individuals in the two videos. The best matching frame pair is used as the representative of the video pair. This may be seen as one way of pair-wise frame selection. We analyze it further to understand the upper bound of the performance that can be obtained with pair-wise frame selection. In order to obtain the upper bound, we use the following approach:

- if the block matrix corresponds to a genuine video pair, use the frame-pair with maximum similarity, and
- if the block matrix corresponds to an impostor video pair, use the frame-pair with minimum similarity.

In this experiment, we are using *genuine* and *impostor* labels for frame-pair selection, this approach does not simulate a real world scenario; however, via this study, we are not reporting a state-of-the-art recognition performance, rather the *idea* is to find out the best achievable performance with an algorithm, if the ideal frame-pairs are selected. Therefore, this study emphasizes the importance of the frame-pair selection for any given feature extraction and matching algorithm. It is to be noted that even though this frame-pair selection approach utilizes the genuine/impostor labels, the best achievable performance is still within the matching capability of the verification algorithm. Using LRPCA based face recognition algorithm on the PaSC video dataset, we observe that the best achievable performance is more than 80 % genuine accept rate at 0.0001 % false accept rate. The corresponding ROC curves are shown in Fig. 7. Comparing the results of the max-rule (Fig. 6) with this study (Fig. 7), it can be inferred that a sophisticated frame-pair selection approach may yield significantly better results; which we believe is the path forward for future video face recognition algorithms.

6 Conclusion

In this chapter, we have discussed the problem of frame selection in video face recognition. Frame selection presents an intuitive approach towards filtering frames and extracting the most meaningful information for face recognition. We have presented an entropy based frame selection algorithm and evaluated its impact on a recent benchmark video database. Experimental results with two face recognition algorithms suggest that even with a much smaller subset of frames, recognition performance can be maintained and even improved. We further explore a novel pair-wise frame selection paradigm and conduct preliminary experiments. It is our opinion that future video face recognition algorithms can be benefitted significantly if *intelligent* pair-wise frame selection algorithms are utilized.

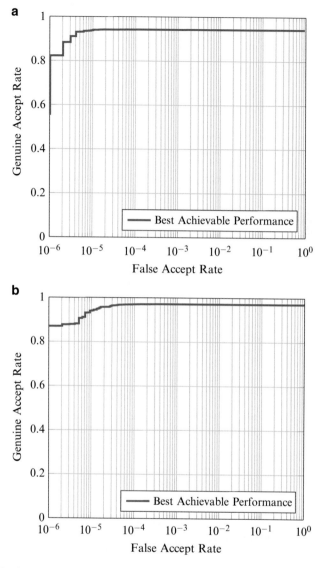

Fig. 7 Verification performance of LRPCA on the PaSC dataset using pair-wise frame selection. (**a**) Control (**b**) Handheld

References

1. K. Anantharajah, S. Denman, S. Sridharan, C. Fookes, D. Tjondronegoro, Quality based frame selection for video face recognition, in *International Conference on Signal Processing and Communication Systems* (2012), pp. 1–5
2. J.R. Barr, K.W. Bowyer, P.J. Flynn, S. Biswas, Face recognition from video: a review, in *International Journal of Pattern Recognition and Artificial Intelligence*, **26**(5), 1266002 (2012), doi 10.1142/S0218001412660024. http://www.worldscientific.com/doi/abs/10.1142/S0218001412660024
3. Y. Bengio, P. Lamblin, D. Popovici, H. Larochelle, Greedy layer-wise training of deep networks, in *Advances in Neural Information Processing Systems*, vol. 19 (MIT Press, Cambridge, 2007), pp. 153–160
4. S.A. Berrani, C. Garcia, Enhancing face recognition from video sequences using robust statistics, in *IEEE Conference on Advanced Video and Signal Based Surveillance* (2005), pp. 324–329
5. L. Best-Rowden, B. Klare, J. Klontz, A.K. Jain, Video-to-video face matching: establishing a baseline for unconstrained face recognition, in *International Conference on Biometrics: Theory, Applications and Systems* (2013), pp. 1–8
6. J. Beveridge, P. Phillips, D. Bolme, B. Draper, G. Given, Y.M. Lui, M. Teli, H. Zhang, W. Scruggs, K. Bowyer, P. Flynn, S. Cheng, The challenge of face recognition from digital point-and-shoot cameras, in *IEEE Conference on Biometrics: Theory, Applications and Systems* (2013), pp. 1–8
7. S. Bharadwaj, M. Vatsa, R. Singh, Biometric quality: a review of fingerprint, iris, and face. EURASIP J. Image Video Process. **2014**(1) (2014), doi 10.1186/1687-5281-2014-34. http://link.springer.com/article/10.1186%2F1687-5281-2014-34
8. M. De Marsico, M. Nappi, D. Riccio, ES-RU: an entropy based rule to select representative templates in face surveillance. Multimedia Tools Appl. **73**(1), 109–128 (2014)
9. W. Fan, D.Y. Yeung, Face recognition with image sets using hierarchically extracted exemplars from appearance manifolds, in *International Conference on Automatic Face and Gesture Recognition* (2006), pp. 177–182
10. G. Goswami, R. Bhardwaj, R. Singh, M. Vatsa, MDLFace: memorability augmented deep learning for video face recognition, in *International Joint Conference on Biometrics* (2014), pp. 1–7
11. A. Hadid, M. Pietikainen, Selecting models from videos for appearance-based face recognition, in *International Conference on Pattern Recognition* (2004), pp. 304–308
12. M. Hubert, P.J. Rousseeuw, K. Vanden Branden, ROBPCA: a new approach to robust principal component analysis. Technometrics **47**(1), 64–79 (2005)
13. A.K. Jain, S.Z. Li, *Handbook of Face Recognition* (Springer, New York, 2005)
14. R.R. Jillela, A. Ross, Adaptive frame selection for improved face recognition in low-resolution videos, in *International Joint Conference on Neural Networks* (2009), pp. 1439–1445
15. A. Khosla, W.A. Bainbridge, A. Torralba, A. Oliva, Modifying the memorability of face photographs, in *IEEE International Conference on Computer Vision* (2013), pp. 3200–3207
16. W. Liu, Z. Li, X. Tang, Spatio-temporal embedding for statistical face recognition from video, in *European Conference on Computer Vision* (2006), pp. 374–388
17. B.D. Lucas, T. Kanade, et al., An iterative image registration technique with an application to stereo vision, in *International Joint Conference on Artificial Intelligence*, vol. 81 (1981), pp. 674–679
18. E. Meyers, L. Wolf, Using biologically inspired features for face processing. Int. J. Comput. Vis. **76**(1), 93–104 (2008)
19. N. Otsu, A threshold selection method from gray-level histograms. Automatica **11**(285–296), 23–27 (1975)
20. U. Park, A.K. Jain, 3D model-based face recognition in video, in *Advances in Biometrics* (Springer, Berlin, 2007), pp. 1085–1094

21. U. Park, A.K. Jain, A. Ross, Face recognition in video: adaptive fusion of multiple matchers, in *IEEE Conference on Computer Vision and Pattern Recognition* (2007), pp. 1–8
22. P.J. Phillips, J.R. Beveridge, B.A. Draper, G. Givens, A.J. O'Toole, D.S. Bolme, J. Dunlop, Y.M. Lui, H. Sahibzada, S. Weimer, An introduction to the good, the bad, & the ugly face recognition challenge problem, in *IEEE International Conference on Automatic Face & Gesture Recognition Workshops* (2011), pp. 346–353
23. A. Renyi, On measures of entropy and information, in *Berkeley Symposium on Mathematical Statistics and Probability* (1961), pp. 547–561
24. E.A. Rúa, J.L.A. Castro, C.G. Mateo, Quality-based score normalization and frame selection for video-based person authentication, in *Biometrics and Identity Management* (Springer, Berlin, 2008), pp. 1–9
25. U. Saeed, J.L. Dugelay, Temporally consistent key frame selection from video for face recognition, in *European Signal Processing Conference* (2010), pp. 1311–1315
26. R. Salakhutdinov, G.E. Hinton, Deep Boltzmann machines, in *International Conference on Artificial Intelligence and Statistics* (2009), pp. 448–455
27. L.K. Saul, S.T. Roweis, Think globally, fit locally: unsupervised learning of low dimensional manifolds. J. Mach. Learn. Res. **4**, 119–155 (2003)
28. P. Sinha, B. Balas, Y. Ostrovsky, R. Russell, Face recognition by humans: nineteen results all computer vision researchers should know about. Proc. IEEE **94**(11), 1948–1962 (2006)
29. C. Tomasi, T. Kanade, Shape and motion from image streams under orthography: a factorization method. Int. J. Comput. Vis. **9**(2), 137–154 (1992)
30. P. Vincent, H. Larochelle, I. Lajoie, Y. Bengio, P. Manzagol, Stacked denoising autoencoders: learning useful representations in a deep network with a local denoising criterion. J. Mach. Learn. Res. **11**(12), 3371–3408 (2010)
31. D. Yoshor, W.H. Bosking, G.M. Ghose, J.H.R. Maunsell, Receptive fields in human visual cortex mapped with surface electrodes. Cereb. Cortex **17**(10), 2293–2302 (2007)

Modeling of Facial Wrinkles for Applications in Computer Vision

Nazre Batool and Rama Chellappa

Abstract Analysis and modeling of aging human faces have been extensively studied in the past decade for applications in computer vision such as age estimation, age progression and face recognition across aging. Most of this research work is based on facial appearance and facial features such as face shape, geometry, location of landmarks and patch-based texture features. Despite the recent availability of higher resolution, high quality facial images, we do not find much work on the image analysis of local facial features such as wrinkles specifically. For the most part, modeling of facial skin texture, fine lines and wrinkles has been a focus in computer graphics research for photo-realistic rendering applications. In computer vision, very few aging related applications focus on such facial features. Where several survey papers can be found on facial aging analysis in computer vision, this chapter focuses specifically on the analysis of facial wrinkles in the context of several applications. Facial wrinkles can be categorized as subtle discontinuities or cracks in surrounding inhomogeneous skin texture and pose challenges to being detected/localized in images. First, we review commonly used image features to capture the intensity gradients caused by facial wrinkles and then present research in modeling and analysis of facial wrinkles as aging texture or curvilinear objects for different applications. The reviewed applications include localization or detection of wrinkles in facial images, incorporation of wrinkles for more realistic age progression, analysis for age estimation and inpainting/removal of wrinkles for facial retouching.

N. Batool (✉)
Center for Medical Image Science and Visualization (CMIV),
Linköpings Universitet/US, 58185 Linköping, Sweden
e-mail: nazre.batool@liu.se

R. Chellappa
Department of Electrical and Computer Engineering and the Center for Automation Research,
UMIACS, University of Maryland, College Park, MD 20742, USA
e-mail: rama@umiacs.umd.edu

© Springer International Publishing Switzerland 2016
M. Kawulok et al. (eds.), *Advances in Face Detection and Facial Image Analysis*,
DOI 10.1007/978-3-319-25958-1_11

1 Introduction

Facial skin wrinkles are not only important features in terms of facial aging but can also provide cues to a person's lifestyle. For example, facial wrinkles can indicate the history of a person's expressions (smiling, frowning, etc.) [15], or whether the person has been a smoker [29], or has had sun-exposure [35]. Some of the factors influencing facial winkles are a person's lifestyle, overall health, skin care routines, genetic inheritance, ethnicity and gender. Hence, computer-based analysis of facial wrinkles has great potential to exploit this underlying information for relevant applications.

Face analysis is one of the main research problems in computer vision and facial features such as shape, geometry, eyes, nose, mouth, are analyzed in one way or another for different applications. However, research has been lacking in image-based analysis of facial wrinkles specifically. For example, a review of two good survey papers on facial aging analysis [14, 34] points to the absence of wrinkle analysis in facial aging research. As our review suggests, this can most probably be attributed to the following reasons:

Image quality: Lack of publicly available benchmark aging datasets with high resolution/high quality images clearly depicting facial wrinkles.

Age period: Lack of proper age period covered in aging datasets; most of these datasets do not have sufficient number of sample images of subjects with age 40 and more.

Challenges in wrinkle localization: Even in case of availability of high quality images of aged skin, facial wrinkles are difficult facial features to localize and hence are not commonly incorporated as curvilinear objects in image analysis algorithms.

Physically, skin wrinkles are 3D features on skin surface along with other features such as pores, moles, scars, dark spots and freckles. Most of these features are visible in 2D images due to their color or the particular image intensities they create. Image processing techniques interpret such image components as edges, contours, boundaries, texture, color space, etc. to infer information. The challenge arises when skin wrinkles cannot be categorized strictly as one of these categories. For example, despite causing image intensity gradients, wrinkles are not continuous as typical edges or contours. Wrinkles cannot be categorized as texture because they do not depict repetitive image patterns which is the defining characteristics of image textures. Wrinkles cannot be categorized as boundaries between two different textures as well as they appear in skin. The closest description of how wrinkles appear in a skin image can be as irregularities, discontinuities, cracks or sudden changes in the surrounding/background skin texture. A parallel can be drawn between the skin texture discontinuities caused by wrinkles in images and the cracks present in industrial objects like roads, steel slabs, rail tracks, etc. However, only in this case, more often than not, the background skin texture is not as smooth or homogeneous as that of a steel slab or road surface. The granular/rough/irregular

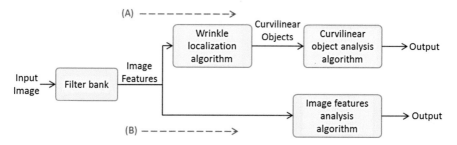

Fig. 1 A block diagram of the two approaches commonly employed to analyze facial aging

3D surface of skin appears as nonuniform or inhomogeneous image texture making it more difficult to localize wrinkles in surrounding skin texture. Although, a framework based on 3D analysis of skin surface would be better suited to draw conclusions based on facial wrinkles, such setups are not readily available to be used frequently.

In this chapter, we focus on research conducted on the analysis of facial wrinkles for applications in computer vision and leave out those in computer graphics. This research can be loosely categorized as following one of the two approaches. In the first and relatively more popular approach, wrinkles are considered as so-called 'aging skin texture' and analyzed as image texture or intensity features. In the second approach, wrinkles are analyzed as curvilinear objects, localized automatically or hand-drawn. Figure 1 depicts a block diagram of the two approaches. Each approach starts with an analysis of input image to obtain image features which can be simple image intensity values or image features obtained after some sort of filtering. Then, in texture-based approaches, image features are analyzed directly as illustrated by path 'B' in the diagram. In approaches based on wrinkles as curvilinear objects, an intermediate step is included in path 'A' for the extraction of curvilinear objects or localization of wrinkles before any other analysis. In Sect. 3 we will review work following the first approach, incorporating wrinkles as image texture, and in Sect. 4 we will review work following the second approach, incorporating wrinkles as curvilinear objects. However, first of all, we will mention early work on image-based analysis of facial wrinkles. Then, in Sect. 2, we will review briefly image filtering techniques applied to highlight intensity gradients caused by wrinkles. Table 1 presents a summary of the work reviewed in this chapter, the corresponding analysis approaches and applications.

Earlier Work As mentioned earlier, modeling of facial wrinkles and finer skin texture has been done commonly in computer graphics to obtain more realistic appearances of skin features. Specifically significant efforts have been reported on photo-realistic and real-time rendering of skin texture and wrinkles on 3D animated objects. This work typically follows the main approach of generating a pattern of skin texture/wrinkles based on some learned model and then render the resulting texture on 3D objects. Hence, most of the earlier work focused on

Table 1 Summary of research work reviewed in this chapter

Representative work	Approach	Image features	Application
Chen et al. [7], Luu et al. [24], Patterson et al. [31] and Sethuram et al. [36]	Image features	AAM, LBP, Gabor	Age estimation/synthesis
Suo et al. [38, 40]	Image features	AAM	Age synthesis
Suo et al. [37, 39]	Curvilinear objects	Curves	Age synthesis
Cula et al. [9, 10]	Curvilinear objects	Gabor filters	Assessment of wrinkle severity
Kwon and da Vitoria Lobo [21, 22]	Curvilinear objects	Deformable snakelets	Age group determination
Batool and Chellappa [1, 2]	Curvilinear objects	LoG	Wrinkle localization
Batool and Chellappa [4]	Curvilinear objects	Gabor filters	Wrinkle localization
Batool et al. [5]	Curvilinear objects	LoG	Soft biometrics
Batool and Chellappa [3]	Image features	Gabor filters	Wrinkle inpainting
Gyun et al. [17, 18]	Curvilinear objects	Steerable filters	Wrinkle localization
Ng et al. [27]	Curvilinear objects	Hessian filters	Wrinkle localization
Jiang et al. [19]	Curvilinear objects	Image intensity	Assessment of wrinkle severity
Fu and Zheng [13]	Image features	Ratio image	Age/expression synthesis
Mukaida and Ando [26]	Curvilinear/blob objects	Image luminance	Facial retouching
Liu et al. [23]	Image features	Ratio image	Facial expression synthesis
Tian et al. [42]	Curvilinear objects	Canny edge detector	Facial expression analysis
Ramanathan and Chellappa [33]	Image features	Image gradients	Age synthesis
Yin et al. [44]	Image features	Image intensity	Facial expression analysis
Zang and Ji [45]	Curvilinear objects	Edge detection	Facial expression analysis

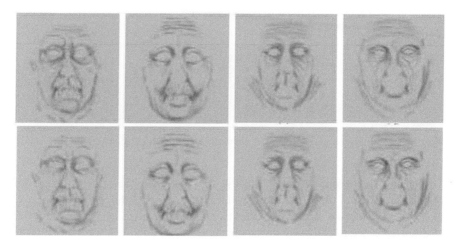

Fig. 2 Eight basic wrinkle masks corresponding to different gender, shape of the face and smiling history (reproduced from [6])

developing generic skin models for 3D rendering. The research work focusing on other applications include work by Kwon and da Vitoria Lobo [21, 22] on localization of wrinkles for age determination (described in detail in Sect. 4.2). Magnenat-Thalmann et al. [25] and Wu et al. [43] presented a computational model for studying the mechanical properties of skin with aging manifested as wrinkles. The model was intended to analyze different characteristics of wrinkles such as location, number, density, cross-sectional shape, and amplitude, as a consequence of skin deformation caused by muscle actions. Boissieux et al. [6] presented 8 basic wrinkle masks (Fig. 2) for aging faces corresponding to different gender, shape of the face and smiling history after analyzing skincare industry data. Figure 2 illustrates the eight patterns included in their work.

Cula et al. [11] presented a novel skin imaging method called bidirectional imaging based on quantitatively controlled image acquisition settings. The proposed imaging setup was shown to capture significantly more properties of skin appearance than standard imaging. The observed structure of skin surface and its appearance were modeled as a bidirectional function of the angles of incident light, illumination and observation. The enhanced observations about skin structure were shown to improve results for dermatological applications. Figure 3 depicts the variations in the appearance of a skin patch due to different illumination angles.

2   Image Features for Aging Skin Texture

In this section, we review image filtering techniques commonly applied to highlight intensity gradients caused by wrinkles as well as image features based on aging appearance and texture. Most of the applications reviewed in the later sections make use of one or more of these features.

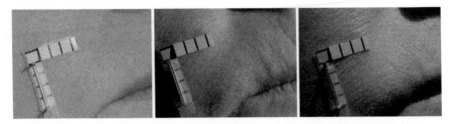

Fig. 3 A skin patch imaged using different illumination angles (reproduced from [11])

Laplacian of Gaussian The Laplacian is a 2-D isotropic measure of the second spatial derivative of an image. The Laplacian of an image highlights regions of rapid intensity change and is therefore often used for edge detection (e.g. zero crossing edge detectors). Since image operators approximating a second derivative measurement are very sensitive to noise, the Laplacian is often applied to an image that has first been smoothed with something approximating a Gaussian smoothing filter in order to reduce its sensitivity to noise, and hence, when combined, the two variants can be described together as Laplacian of Gaussian operator. The operator normally takes a single gray level image as input and produces another gray level image as output. Because of the associativity of the convolution operation, the Gaussian smoothing filter can be convolved with the Laplacian filter first, and then convolved with the image to achieve the required result. The 2D LoG function centered on zero and with Gaussian standard deviation σ has the form:

$$LoG(x, y; \sigma) = -\frac{1}{\pi \sigma^4} \left[1 - \frac{x^2 + y^2}{2\sigma^2} \right] \exp(-\frac{x^2 + y^2}{2\sigma^2}). \tag{1}$$

Hessian Filter The Hessian filter is a square matrix of second-order derivative and is capable of capturing local structure in images. The eigenvalues of the Hessian matrix evaluated at each image point quantify the rate of change of the gradient field in various directions. A small eigenvalue indicates a low rate of change in the field in the corresponding eigen-direction, and vice versa. The Hessian matrix $\mathbf{H}$ of the input image I, consisting of 2nd order partial derivatives at scale σ, is given as:

$$\mathbf{H} = \begin{bmatrix} \frac{\partial^2 I}{\partial x^2} & \frac{\partial^2 I}{\partial x \partial y} \\ \frac{\partial^2 I}{\partial y \partial x} & \frac{\partial^2 I}{\partial y^2} \end{bmatrix} = \begin{bmatrix} \mathbf{H}_a & \mathbf{H}_b \\ \mathbf{H}_b & \mathbf{H}_c \end{bmatrix}. \tag{2}$$

In order to extract the eigen-direction in which a local structure of the image is decomposed, eigenvalues λ_1, λ_2 of the Hessian matrix are defined as:

$$\lambda_1(x, y : \sigma) = \frac{1}{2} \left[H_a + H_c + \sqrt{(H_a - H_c)^2 + H_b^2} \right], \tag{3}$$

$$\lambda_2(x, y : \sigma) = \frac{1}{2}\left[H_a + H_c - \sqrt{(H_a - H_c)^2 + H_b^2}\right].$$

Different Hessian filters vary in ways the eigenvalues are analyzed to test a hypothesis about image structure. For example, to determine if a pixel corresponded to a facial wrinkle or not Ng et al. [27] (described in Sect. 4) defined the following similarity measures R and S to test the hypotheses:

$$R(x, y : \sigma) = (\frac{\lambda_1}{\lambda_2})^2, \tag{4}$$

$$S(x, y : \sigma) = \lambda_1^2 + \lambda_2^2.$$

Steerable Filter Bank Freeman and Adelson proposed a steerable filter [12, 17] to detect local orientation of edges. For any arbitrary orientation, a steerable filter can be generated from a linear combination of basis filters where the basis filter set for a pixel p is given by:

$$\mathbf{G}(p) = \left[\frac{\partial^2 g(p)}{\partial x^2} + \frac{\partial^2 g(p)}{\partial x \partial y} + \frac{\partial^2 g(p)}{\partial y^2}\right], \tag{5}$$

where $g(p)$ denotes Gaussian function of $\mathbf{R}^2$, the most used example of steerable filters. Let the interpolating function of orientation θ be given as:

$$\mathbf{k}(\theta) = \left[\cos^2(\theta) - \sin 2\theta \sin^2 \theta\right]^T. \tag{6}$$

Then the steerable filter associated with the orientation θ can be obtained as $g_\theta(p) = \mathbf{k}^T \mathbf{G}(p)$ and can be used to extract image structure in that orientation using convolution.

Gabor Filter Bank Gabor operator is a popular local feature-based descriptor due to its robustness against variation in pose or illumination. The real Gabor filter kernel oriented in a 2D image plane at angle α is given by:

$$Gab(x, y) = \frac{1}{2\pi\sigma_x\sigma_y} \exp\left[\frac{-1}{2}\left(\frac{x'}{\sigma_x^2} + \frac{y'}{\sigma_y^2}\right)\right] \cos(2\pi f x'), \tag{7}$$

where

$$\begin{bmatrix} x' \\ y' \end{bmatrix} = \begin{bmatrix} \cos\alpha & \sin\alpha \\ -\sin\alpha & \cos\alpha \end{bmatrix} \begin{bmatrix} x \\ y \end{bmatrix}. \tag{8}$$

Let $\{Gab_k(x, y), k = 0, \cdots, K - 1\}$ denote the set of real Gabor filters oriented at angles $\alpha_k = -\frac{\pi}{2} + \frac{\pi k}{K}$ where K is the total number of equally spaced filters over the angular range $\left[\frac{-\pi}{2}, \frac{\pi}{2}\right]$. Then Gabor features can be obtained by convolving this Gabor filter bank with the given image.

Local Binary Pattern (LBP) Ojala et al. [28] introduced the Local Binary Patterns (LBPs) to represent local gray-level structures. LBPs have been used widely as powerful texture descriptors. The LBP operator takes a local neighborhood around each pixel, thresholds the pixels of the neighborhood at the value of the central pixel and uses the resulting binary-valued integer number as a local image descriptor. It was originally defined for 3-pixel neighborhoods, giving 8-bit integer LBP codes based on the eight pixels around the central one. Considering a circular neighborhood denoted by (P, R) where P represents the number of sampling points and R is the radius of the neighborhood, the LBP operator takes the following form:

$$f_{(P,R)}(p_c) = \sum_{i=0}^{P-1} s(p_i - p_c)2^i, \tag{9}$$

$$\text{where } s(x) = \begin{bmatrix} 1 \text{ if } x \geq 0 \\ 0 \text{ otherwise} \end{bmatrix}, \tag{10}$$

and p_i is one of the neighboring pixels around the center pixel p_c on a circle or square of radius R. Several extensions of the original operator have been proposed. For example including LBPs for the neighborhoods of different sizes makes it feasible to deal with textures at different scales. Another extension called 'uniform patterns' has been proposed to obtain rotationally invariant features from the original LBP binary codes (see [28] for details). The uniformity of an LBP pattern is determined from the total number of bitwise transitions from 0 to 1 or vice versa in the LBP bit pattern when the bit pattern is considered circular. A local binary pattern is called uniform if it has at most 2 bitwise transitions. The uniform LBP patterns are used to characterize patches that contain primitive structural information such as edges and corners. Each uniform pattern, which is also a binary pattern, has a corresponding integer value. The uniform patterns and the corresponding integer values are used to compute LBP histograms where each uniform pattern is represented by a unique bin in the histogram and all the non-uniform patterns are represented by a single bin only. For example, the 58 possible uniform patterns in a neighborhood of 8 sampling points make a histogram of 59 bins where 59th bin represents the non-uniform patterns. It is common practice to divide an image in sub-images and then use the normalized LBP histograms gathered from each sub-image as image features.

An extension of LBPs, called Local Ternary Patterns (LTP) [41], has also been used in analyzing aging skin textures. LBPs tend to be sensitive to noise, because of the selection of the threshold value to be the same as that of the central pixel, especially in near uniform image regions. LTPs were proposed to introduce robustness to noise in LBPs by introducing a threshold value r other than that of the central pixel. Since many facial regions are relatively uniform, LTPs were shown to produce better results as compared to LBP. An LTP operator is defined as follows:

$$f_{(P,R)}^{LTP}(p_c) = \sum_{i=0}^{P-1} s(p_i, p_c)2^i, \tag{11}$$

$$\text{where } s_{LTP}(x, p_c) = \begin{bmatrix} 1 \text{ if } x \geq p_c + r \\ 0 \text{ if } |x - p_c| < r \\ -1 \text{ if } x \leq p_c - r \end{bmatrix}.$$

Each ternary pattern is split into positive and negative parts. These two parts are then processed as two separate channels of LBP codes. Each channel is used to calculate LBP histograms from LBP codes and the resulting LBP histograms from two channels are used as image features.

Active Appearance Model (AAM) The Active Appearance Model (AAM) was proposed in [8] to describe a statistical generative model of face shape and texture/intensity. It is a popular facial descriptor which makes use of Principle Component Analysis (PCA) in a multi-factored way for dimension reduction while maintaining important structure and texture elements of face images. To build an AAM model, a training set of annotated images is required where facial landmark points have been marked on each image. AAMs model shape and appearance separately. The shape model is learnt from the coordinates of the landmark points in annotated training images. Let N_T and N_L denote the total number of training images and the number of landmark points in each training facial image. Let $\mathbf{p} = [x_1, y_1, x_2, y_2, \ldots, x_{N_L}, y_{N_L}]^T$ be a vector of length $2N_L \times 1$ denoting the planar coordinates of all landmarks. The shape model is constructed by first aligning the set of N_T training shapes using Generalized Procrustes Analysis and then applying PCA on the aligned shapes to find an orthonormal basis of N_T eigenvectors, $\mathbf{E}_s \in \mathcal{R}^{2N_L \times N_T}$ and the mean shape $\bar{\mathbf{p}}$. Then the training images are warped onto the mean shape in order to obtain the appearance model. Let N_A denote the number of pixels that reside inside the mean shape $\bar{\mathbf{p}}$. For the appearance model, let $\mathbf{l}(\mathbf{x}), \mathbf{x} \in \mathbf{p}$ be a vector of length $N_A \times 1$ denoting the intensity/appearance values of the N_A pixels inside the shape model. The appearance model is trained in a similar way to the shape model to obtain N_T eigenvectors, $\mathbf{E}_a \in \mathcal{R}^{N_A \times N_T}$ and the mean appearance $\bar{\mathbf{l}}$.

Once the shape and appearance AAM models have been learnt from the training images, any new instance $(\mathbf{p}*, \mathbf{l}*)$ can be synthesized or represented as a linear combination of the eigenvectors weighted by the model parameters as follows:

$$\mathbf{p}* = \bar{\mathbf{p}} + \mathbf{E}_s \mathbf{a}, \tag{12}$$

$$\mathbf{l}* = \bar{\mathbf{l}} + \mathbf{E}_a \mathbf{b},$$

where $\mathbf{a}$ and $\mathbf{b}$ are the shape and appearance parameters respectively.

3 Applications Incorporating Wrinkles as Texture

Most computer vision applications involving facial aging incorporate wrinkles as aging texture where the specific appearance of the texture is incorporated as image texture features of choice. In this section, we present a review of the research work incorporating aging skin texture as image texture features.

3.1 Synthesis of Facial Aging and Expressions

Synthesis of aged facial images from younger facial images of an individual has several real world applications e.g. looking for lost children or wanted fugitives, developing face recognition systems robust to age related variations, facial retouching in entertainment and recently in healthcare to assess the long term effects of an individual's lifestyle. Facial aging causes changes in both the geometry of facial muscles and skin texture. The synthesis of facial aging is a challenging problem because it is difficult to synthesize facial changes in geometry and texture which are specific to an individual. Furthermore, the availability of only a limited number of prior images at different ages, mostly low-resolution, for an individual poses additional challenge.

In the absence of long term (i.e. across 3–4 decades) face aging sequences, Suo et al. [38, 40] made two assumptions. First, similarities exist among short term aging patterns in the same time span, especially for individuals of the same ethnic group and gender. Second, the long term aging pattern is a smooth Markov process composed of a series of short term aging patterns. In their proposed method, AAM features were used to capture and generate facial aging. Guided by face muscle clustering, a face image was divided into 13 sub-regions. An extended version of AAM model was then used to include a global active shape model and a shape-free texture model for each sub-region. Thus the shape-free texture component of the AAM model described changes in skin texture due to wrinkling (Fig. 4). The principle components of the extended AAM model were also analyzed to extract age-related components from non-age-related components.

With a large number of short term face aging sequences from publicly available face aging databases, such as FG-NET and MORPH, Suo et al. used their defined AAM model features to learn short term aging patterns from real aging sequences. A sequence of overlapping short term aging patterns in latter age span was inferred

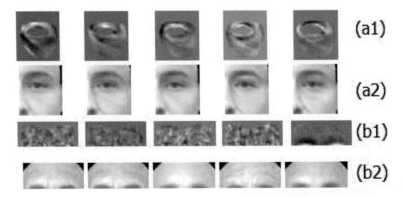

(a1)

(a2)

(b1)

(b2)

Fig. 4 Representation of aging texture in [40]; (**a1, a2**) depict the shape-free texture in the region around eye and the corresponding synthesized images and (**b1, b2**) depict the same for the forehead region (reproduced from [40]).

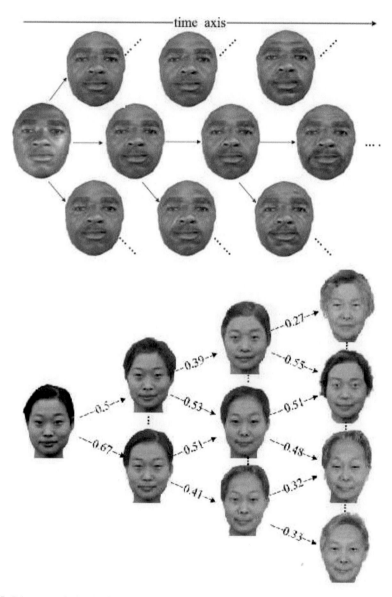

Fig. 5 Inherent variation in different instances of synthesized aged images for the same age (*top* reproduced from [40], *bottom* reproduced from [39])

from the overlapping short term aging patterns in current age span. The short term aging patterns for the later age were then concatenated into a smooth long term aging pattern. The diversity of aging among individuals was simulated by sampling different subsequent short term patterns probabilistically. For example, Fig. 5 shows inherent variations in terms of the aging of a given face using their method based

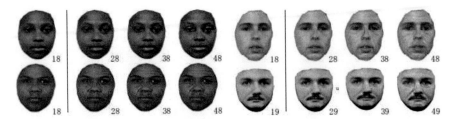

Fig. 6 Simulation of age synthesis in [38]; the *left most column* shows the input images, and the following *three columns* are synthesized images at latter ages (reproduced from [38])

on AAM and on And-Or graphs (described later). It can be observed that the appearance of a synthesized aged face varies with increase in age. Figure 6 shows examples of age synthesis for four subjects using their AAM features.

In a different approach to aging synthesis Suo et al. [37, 39] presented a hierarchical And-Or graph based generative model to synthesize aging. Each age group was represented by a specific And-Or graph and a face image in this age group was considered to be a transverse of that And-Or graph, called parse graph. The And-Or graph for each age group consisted of And-nodes, Or-nodes and Leaf nodes. The And nodes represented different parts of face in three levels—coarse to fine—where wrinkles and skin marks were incorporated at the third, finer level. Or nodes represented the alternatives learned from a training dataset to represent the diversity of face appearance at each age group. By selecting alternatives at the Or-nodes, a hierarchical parse graph was obtained for a face instance whose face image could then be synthesized from this parse graph in a generative manner. Based on the And-Or graph representation, the dynamics of face aging process were modeled as a first-order Markov chain on parse graphs which was used to learn aging patterns from annotated faces of adjacent age groups.

To incorporate wrinkles in synthesized images, parameters of curves were learned in 6 wrinkle zones from the training dataset. Wrinkle curves were then stochastically generated in two steps to be rendered on synthesized face images: generation of curve shapes from a probabilistic model and calculation of curve intensity profiles from the learned dictionary. After warping the intensity profiles to the shape of wrinkle curves, Poisson image editing was used to synthesize realistic wrinkles on a face image. Figure 7 shows a series of generated wrinkle curves over four age groups on top and an example of generating the wrinkles image from the wrinkle curves on the bottom.

Patterson et al. [31] presented a framework for aging synthesis based on a face-model including landmarks for shape, and AAMs for both shape and texture. They learned age-related AAM parameters from a training set annotated with landmarks using support-vector regression (SVR). The learned AAM parameters were used to generate feasible random faces along with their age estimated by SVR. In the final step, these simulated faces were used to generate a table of 'representative age parameters' which then manipulated the AAM parameters in the

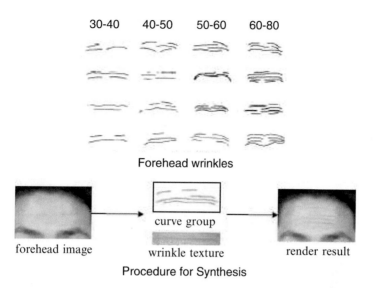

Fig. 7 Generation of wrinkle curves for different age patterns and synthesis of a wrinkle pattern over aged image (reproduced from [39])

Fig. 8 The *top row* shows original images of an individual. The *bottom row* shows synthetic aged images where each image is synthesized at approximately the same age as that in the image above (reproduced from [31])

feature space. The manipulated AAM parameters thus obtained were used to age-progress or regress a given face image. Figure 8 shows synthesized aged images vs. original images for a subject using their AAM-SVR face model.

Ramanathan and Chellappa [33] proposed a shape variation model and a texture variation model towards modeling of facial aging in adults. Attributing facial shape variations during adulthood to the changing elastic properties of the underlying facial muscles, the shape variation model was formulated by means of physical models that characterized the functionalities of different facial muscles. Facial feature drifts were modeled as linear combinations of the drifts observed on individual facial

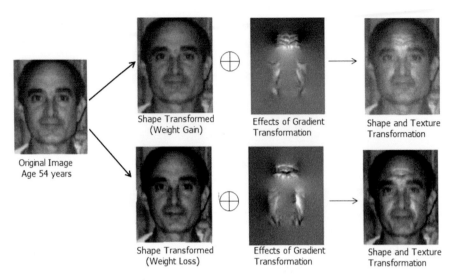

Fig. 9 Facial shape variations induced for the cases of weight-gain/loss in [33]. Further, the effects of gradient transformations in inducing textural variations using Poisson image editing are illustrated as well (reproduced from [33])

muscles. The aging texture variation model was designed specifically to characterize facial wrinkles in predesignated facial regions such as the forehead, nasolabial region, etc. To synthesize aging texture, they proposed a texture variation model by means of image gradient transformation functions. The transformation functions for a specific age gap and wrinkle severity class (subtle/moderate/strong) were learnt from the training set. Given a test image, the transformed image according to an age group and wrinkles severity was then obtained by solving the Poisson equation of image reconstruction from gradient fields. Figure 9 illustrates the process of transforming facial appearances with increase in age in their work.

Fu and Zheng introduced a novel framework for appearance-based photorealistic facial modeling called Merging Face (M-Face) [13]. They introduced 'merging ratio images' which were defined to be as the seamless blending of individual expression ratio images, aging ratio images, and illumination ratio images. Thus the aging skin texture was also represented as a ratio image. Derived from the average face, the caricatured shape was obtained by accentuating an average face by exaggerating individual distinctiveness of the subject while the texture ratio image was rendered during the caricaturing. This way, the expression morphing, chronological aging or rejuvenating, and illumination variance could be merged seamlessly in a photorealistic way on desired view-rotated faces yielded by view morphing. Figure 10 shows an example image and the corresponding rendered images for different facial attributes in their work.

As regards with aging in M-Face framework, the 'age space' including both shape and aging ratio images (ARI), was assumed to be a low-dimensional manifold of the image space where the origin of the manifold represented the shape and

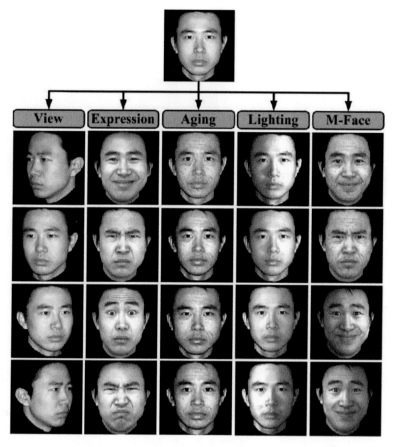

Fig. 10 An example image with photorealistically rendered images for different attributes (reproduced from [13])

texture of the average face of a young face set. Each point in the manifold denoted a specific image with distinctive shape and ARI features. Facial attributes of a given image lay on this manifold, at some point P. Points at a farther distance from the origin than that of the original image represented aging and those closer to the origin represented rejuvenating. Different aged and rejuvenated faces were rendered by using features belonging to the points on this manifold by processing along the line from the origin to the point P (Fig. 11).

Following a similar approach based on ratio images, Liu et al. [23] presented a framework to map subtle changes in illumination and appearance corresponding to facial creases and wrinkles in the context of facial expressions instead of facial aging. Their work was an attempt to complement traditional expression mapping techniques which focused mostly on the analysis of facial feature motions and ignored details in illumination changes due to expression wrinkles/creases.

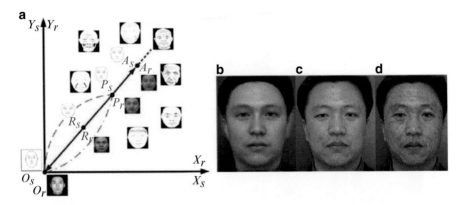

Fig. 11 (**a**) Age space for aging and rejuvenating. The origin is the average face of a young face set. (**b**) Rejuvenation of an adult male face. (**c**) Original face image. (**d**) Aging of the face (reproduced from [13])

In a generative framework, they proposed 'expression ratio images (ERI)' which captured illumination changes of a person's expressions as we describe next. Under the Lambertian model, ERI is defined in terms of the changes in the illumination of skin surface due to the skin folds. For any point P on a surface, let $\mathbf{n}$ denote its normal and assume m point light sources. Let $\mathbf{l}_i$, $1 \leq i \leq m$ denote the light direction from point P to the ith light source, and I_i its intensity. Assuming a diffuse surface let ρ be its reflectance coefficient at P. Under the Lambertian model, the intensity at P is:

$$I_P = \rho \sum_{i=1}^{m} \mathbf{l}_i \mathbf{n} \cdot I_i. \tag{13}$$

With the deformation of skin due to wrinkles, the surface normals and light intensity change. Consequently, new intensity value at P is calculated as:

$$I'_P = \rho \sum_{i=1}^{m} \mathbf{l}_i \mathbf{n}' \cdot I'_i. \tag{14}$$

The ratio image, ERI, is defined to be the ratio of the two images:

$$ERI = \frac{I'_P}{I_P}. \tag{15}$$

The ERIs obtained in this way, corresponding to one person's expression, were mapped to another person's face image along with geometric warping to generate similar, and sometimes more 'expressive', expressions. Figure 12 depicts an example of synthesis of more expressive faces using this method.

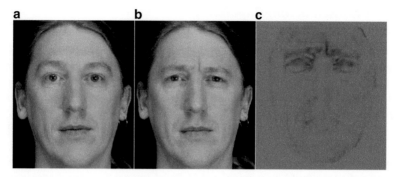

Fig. 12 An expression used to map to other subjects' facial images. (**a**) Neutral face. (**b**) Result from ERI and geometric warping. (**c**) The ERI used in (**b**) and obtained from another person's face (wrinkles due to expressions are prominent—reproduced from [23])

3.2 Age Estimation

Shape changes are prominent in facial aging during younger years, while wrinkles and other textural pattern variations are more prominent during older years. Hence, age estimation methods try to learn patterns in both shape and textural variations using appropriate image features for specific age intervals and then infer the age of a test face image using the learned classifiers. Some of the popular image features to learn age-related changes have been Gabor features, AAM features, LBP features, LTP features or a combination of them.

Luu et al. [24] proposed an age estimation technique combining holistic and local features where AAM features were used as holistic features and local features were extracted using LTP features. These combined features from training set were then used to train age classifiers based on PCA and Support Vector Machines (SVM). The classifiers were then used to classify faces into one of two age groups—pre-adult (youth) and adult.

Chen et al. [7] conducted thorough experiments on facial age estimation using 39 possible combination of four feature normalization methods, two simple feature fusion methods, two feature selection methods, and three face representation methods as Gabor, AAM and LBP features. LBP encoded the local texture appearance while the Gabor features encoded facial shape and appearance information over a range of coarser scales. They systematically compared single feature types vs. all possible fusion combinations of AAM and LBP, AAM and Gabor, and, LBP and Gabor. Feature fusion was performed using feature selection schemes such as Least Angle Regression (LAR) and sequential selection. They concluded that Gabor feature outperformed LBP and even AAM as single feature type. Furthermore, feature fusion based on local feature of Gabor or LBP with global feature AAM achieved better accuracy than each type of features independently.

3.3   Facial Retouching/Inpainting

Facial retouching is widely used in media and entertainment industry and consists of changing facial features such as removing imperfections, enhancing skin fairness, skin tanning, applying make-up, etc. A few attempts that detect and manipulate facial wrinkles and other marks for such retouching application are described here.

In their work Mukaida and Ando emphasized the importance of wrinkles and spots for understanding and synthesizing facial images with different ages [26]. A method based on local analysis of shape properties and pixel distributions was proposed for extracting wrinkles and spots. It was also demonstrated that extracted wrinkles and spots could be manipulated in facial images for visual perception of aging. The morphological processing of the luminance channel was used to divide resulting binary images in regions of wrinkles and dark spots. The extracted regions were then used to increase/decrease the luminance of the source facial image thus giving an impression of aging/rejuvenating. Figure 13 shows an example of facial image and the extracted binary template. The template is then used to manipulate the original facial image to give a perception of aging/rejuvenating.

Batool and Chellappa [3] presented an approach for facial retouching application based on the semi-supervised detection and inpainting of facial wrinkles and imperfections due to moles, brown spots, acne and scars. In their work, the detection of wrinkles/imperfections allowed those skin features to be processed differently than the surrounding skin without much user interaction. Hence, the algorithm resulted in better visual results of skin imperfection removal than contemporary algorithms. For detection, Gabor filter responses along with texture orientation field were used as image features. A bimodal Gaussian mixture model (GMM) represented distributions of Gabor features of normal skin vs. skin imperfections. Then a Markov random field model (MRF) was used to incorporate spatial relationships among neighboring pixels for their GMM distributions and texture orientations. An Expectation-Maximization (EM) algorithm was used to

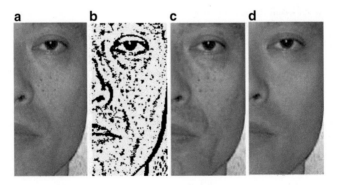

Fig. 13 Manipulation of facial skin marks. (**a**) Original image. (**b**) Binary image. (**c**) Strengthening. (**d**) Weakening (reproduced from [26])

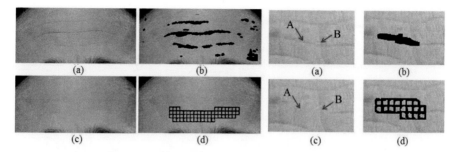

Fig. 14 (*Left*) Wrinkle removal. (**a**) Original image. (**b**) Wrinkled areas detected by GMM-MRF. (**c**) Inpainted image with wrinkles removed. (**d**) Patches from regular grid fitted on the gap which were included in texture synthesis. (*Right*) (**a**) Original image. (**b**) Wrinkled areas detected by GMM-MRF. (**c**) Inpainted image with wrinkles removed; note that wrinkle 'A' has been removed since it was included in the gap whereas a part of wrinkle 'B' is not removed. (**d**) Stitching of skin patches to fill the gap (reproduced from [3])

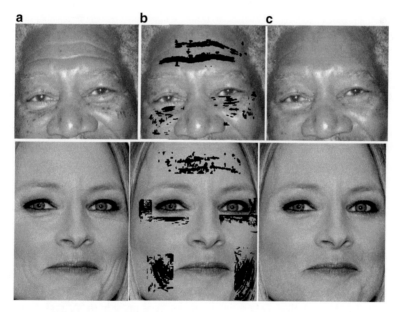

Fig. 15 Results of wrinkle detection and removal for a subject. (**a**) Original image. (**b**) Detected wrinkled areas. (**c**) Image after wrinkle removal (reproduced from [3])

classify skin vs. skin wrinkles/imperfections. Once detected automatically, wrinkles/imperfections were removed completely instead of being blended or blurred. For inpainting, they proposed extensions to current exemplar-based constrained texture synthesis algorithms to inpaint irregularly shaped gaps left by the removal of detected wrinkles/imperfections. Figures 14, 15 and 16 show some results of detection and removal of wrinkles and other imperfections using their algorithms.

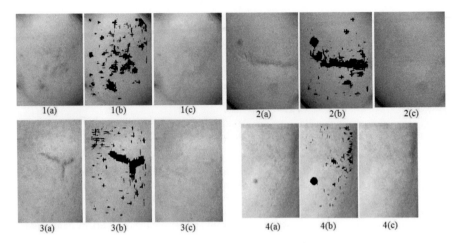

Fig. 16 Results of detection and removal of skin imperfections including wound scars, acne, brown spots and moles. (**a**) Original images. (**b**) Detected imperfections. (**c**) Images after inpainting (reproduced from [3])

4 Applications Incorporating Wrinkles as Curvilinear Objects

In this section, we present applications incorporating facial wrinkles as curvilinear objects instead of image texture features. Curvilinear objects are detected or hand-drawn in images and then analyzed for the specific application. In this section, we first describe work aimed at accurate localization of wrinkles in images.

4.1 Detection/Localization of Facial Wrinkles

Localization techniques can be grouped in two categories: stochastic and deterministic modeling techniques where Markov point process has been the main stochastic model of choice. Deterministic techniques include modeling of wrinkles as deformable curves (snakelets) and image morphology.

Localization Using Stochastic Modeling Batool and Chellappa [1, 2] were the first to propose a generative stochastic model for wrinkles using Marked Point Processes (MPP). In their proposed model wrinkles were considered as stochastic spatial arrangements of sequences of line segments, and detected in an image by proper placement of line segments. Under Bayesian framework, a prior probability model dictated more probable geometric properties and spatial interactions of line segments. A data likelihood term, based on intensity gradients caused by wrinkles and highlighted by Laplacian of Gaussian (LoG) filter responses, indicated more

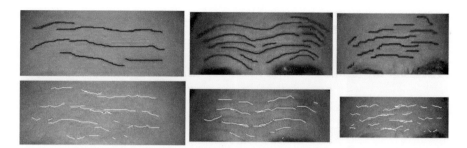

Fig. 17 Localization of wrinkles in three FG-NET images using MPP model in [1]. (*Top*) Ground Truth. (*Bottom*) Localization results (reproduced from [1])

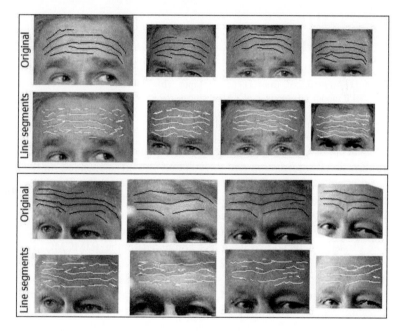

Fig. 18 Localization of wrinkles as line segments for eight images of two subjects (reproduced from [2])

probable locations for the line segments. Wrinkles were localized by sampling MPP posterior probability using the Reversible Jump Markov Chain Monte Carlo (RJMCMC) algorithm. They proposed two MPP models in their work, [1, 2], where the latter MPP model produced better localization results by introducing different movements in RJMCMC algorithm and data likelihood term. They also presented an evaluation setup to quantitatively measure the performance of the proposed model in terms of detection and false alarm rates in [2]. They demonstrated localization results on a variety of images obtained from the Internet. Figures 17 and 18 show examples of wrinkle localization from the two MPP models in [1, 2] respectively.

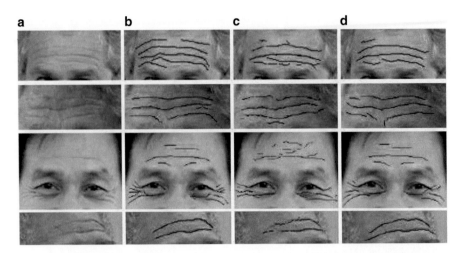

Fig. 19 Localization of wrinkles using Jeong et al.'s MPP model [17] vs. Batool and Chellappa's MPP model [1] (reproduced from [17]). (**a**) Input. (**b**) Manually labelled. (**c**) Batool and Chellappa's MPP model. (**d**) Jeong et al.'s MPP model

The Laplacian of Gaussian filter used by Batool and Chellappa [1, 2] could not measure directional information and the solution strongly depended on the initial condition determined by the placement of first few line segments. To address these shortcomings, Jeong et al. [17] proposed a different MPP model. To incorporate directional information, they employed steerable filters at several orientations and used second derivative of Gaussian functions as the basis filter to extract linear structures caused by facial wrinkles. As compared to the RJMCMC algorithm used by Batool and Chellappa [1, 2], their RJMCMC algorithm included two extensions: affine movements of line segments in addition to birth and deletion as well as 'delayed' rejection/deletion of line segments. Figure 19 shows comparison of localization results using MPP models of Jeong et al. [17] and Batool and Chellappa [1]. However, they reported results on fewer test images as compared to those in [1, 2].

Several parameters are required in an MPP model to interpret the spatial distribution of curvilinear objects i.e. modeling parameters for the geometric shape of objects and hyper-parameters to weigh data likelihood and prior energy terms. To bypass the computationally demanding estimation of such large number of parameters, in further work, Jeong et al. presented a generic MPP framework to localize curvilinear objects including wrinkles in images [18]. They introduced a novel optimization technique consisting of two steps to bypass the selection of hyper-parameters. In the first step, an RJMCMC sampler with delayed rejection [17] was employed to collect several line configurations with different hyper-parameter values. In the second step, the consensus among line detection results was max-imized by combining the whole set of line candidates to reconstruct the most

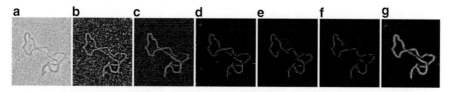

Fig. 20 Localization of a DNA strand using different hyperparameter values in [18]. (**a**) Original image. (**b**) Gradient magnitude. (**c**) Mathematical morphology operator, path opening. (**d**)–(**f**) Line configurations associated with different hyperparameter vectors. (**g**) Final composition result (reproduced from [18])

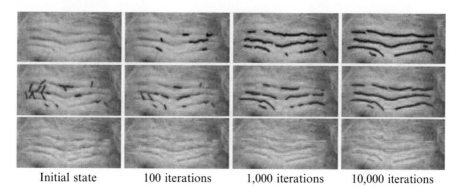

Initial state   100 iterations   1,000 iterations   10,000 iterations

Fig. 21 Localization of wrinkles using different initial conditions; *every image row* represents a different initial condition (reproduced from [18])

plausible curvilinear structures. Figure 20 shows an example of combining linear structures using different hyper-parameter values for a DNA image. Figure 21 shows localization results for a wrinkle image using different initial conditions in RJMCMC algorithm. Thus their optimization scheme made the RJMCMC algorithm almost independent of the initial conditions.

Localization Using Deterministic Modeling The MPP model, despite its promising localization results, requires a large number of iterations in the RJMCMC algorithm to reach global minimum resulting in considerable computation time. To avoid such long computation times for larger images, Batool and Chellappa [4] proposed a deterministic approach based on image morphology for fast localization of facial wrinkles. They used image features based on Gabor filter bank to highlight subtle curvilinear discontinuities in skin texture caused by wrinkles. Image morphology was used to incorporate geometric constraints to localize curvilinear shapes at image sites of large Gabor filter responses. In this work, they reported experiments on much larger set of high resolution images. The localization results showed that not only the proposed deterministic algorithm was significantly faster than MPP modeling but also provided visually better results.

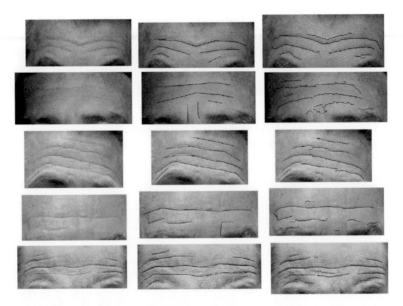

Fig. 22 A few examples of images with detection rate greater than 70 %. (*Left*) Original. (*Middle*) Hand-drawn. (*Right*) Automatically localized (reproduced from [4])

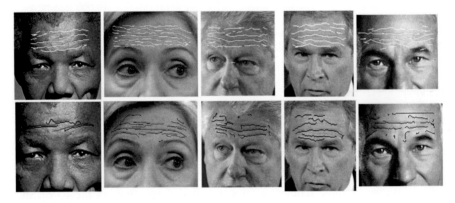

Fig. 23 Comparison of localization results using MPP modeling (*top row*) and deterministic algorithm proposed by Batool and Chellappa (*bottom row*) (reproduced from [4])

Figure 22 includes some examples of localization with high detection rate using their deterministic algorithm and Fig. 23 presents comparison of localization results between their proposed MPP modeling [2] and deterministic algorithm [4].

For the localization of wrinkles, Ng et al. assumed facial wrinkles to be ridge-like features instead of edges [27]. They introduced a measure of ridge-likeliness obtained on the basis of all eigenvalues of the Hessian matrix (Sect. 2). The eigenvalues of the Hessian matrix were analyzed at different scales to locate ridge-like features in images. A few post-processing steps followed by a curve fitting

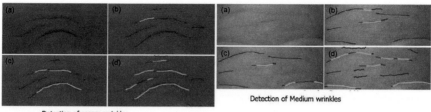

Detection of Medium wrinkles

Detection of coarse wrinkles

Fig. 24 Automatic detection of coarse wrinkles. (**a**) Original image. (**b**), (**c**) and (**d**) are the wrinkle detection by two other methods and Ng et al.'s method respectively. *Red*: ground truth, *green*: true positive, *blue*: false positive (reproduced from [27])

step were then used to place wrinkle curves at image sites of ridge-like features. Figure 24 presents an example of wrinkle localization. Although, their localization results were compared with earlier methods, no comparison results were reported with those of MPP modeling [1, 2].

4.2 Age Estimation Using Localized Wrinkles

One of the initial efforts related to age estimation from digital images of face and those also using detection of facial wrinkles as curvilinear features was reported by Kwon and da Vitoria Lobo [21, 22]. They used 47 high resolution facial images for classification into one of three age groups: babies, young adults or senior adults. Their approach was based on geometric ratios of so-called primary face features (eyes, nose, mouth, chin, virtual-top of the head and the sides of the face) based on cranio-facial development theory and wrinkle analysis. In secondary feature analysis, a wrinkle geography map was used to guide the detection and measurement of wrinkles. A wrinkle index was defined based on detected wrinkles which was sufficient to distinguish seniors/aged adults from young adults and babies. A combination rule for the face ratios and the wrinkle index allowed the categorization of a face into one of the above-mentioned three classes.

In their 2-step wrinkle detection algorithm, first snakelets were dropped in random orientations in the input image in user-provided regions of potential wrinkles around eyes and forehead. The snakelets were directed according to the directional derivatives of image intensity taken orthogonal to the snakelet curves. The snakelets that had found shallow image intensity valleys were eliminated based on the assumption that only the deep intensity valleys corresponded to narrow and deep wrinkles. In the second step, a spatial analysis of the orientations of the stabilized snakelets determined wrinkle snakelets from non-wrinkle snakelets. Figure 25a1, b1 shows the stabilized snakelets on an aged adult face and young adult face respectively. It can be seen in Fig. 25a2, b2 that a large number of stabilized snakelets correspond to wrinkles in an aged face. Figure 26 shows two examples of final results of detection of wrinkles from initial random snakelets.

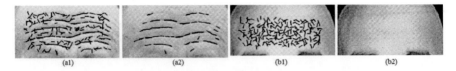

Fig. 25 (**a1, b1**) Stabilized snakelets. (**a2, b2**) Snakelets passing the spatial orientation test and corresponding to wrinkles (reproduced from [22])

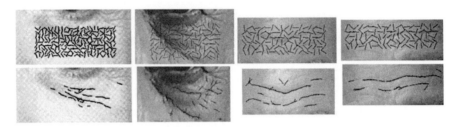

Fig. 26 Examples of detection of wrinkles using snakelets. (*Top*) Initial randomly distributed snakelets. (*Bottom*) Snakelets representing detected wrinkles (reproduced from [22])

4.3 Localized Wrinkles as Soft Biometrics

Recently, due to the availability of high resolution images, a new area of research in face recognition has focused on analysis of facial marks such as scars, freckles, moles, facial shape, skin color, etc. as biometric traits. For example, facial freckles, moles and scars were used in conjunction with a commercial face recognition system for face recognition under occlusion and pose variation in [16, 30]. Another interesting application presented in [20, 32] was the recognition between identical twins using proximity analysis of manually annotated facial marks along with other typical facial features. Where the uniqueness of the location of facial marks is obvious, the same uniqueness of wrinkles is not that obvious. Batool and Chellappa [5] investigated the use of a group of hand-drawn or automatically detected wrinkle curves as soft biometrics. First, they presented an algorithm to fit curves to automatically detected wrinkles which were localized as line segments using MPP modeling in their previous work. Figure 27 includes an example of curves fitted to the detected line segments using their algorithm.

Then they used the hand-drawn and automatically detected wrinkle curves on subjects' foreheads as curve patterns. Identification of subjects was then done based on how closely wrinkle curve patterns of those subjects matched. The matching of curve patterns was achieved in three steps. First, possible correspondences were determined between curves from two different patterns using a simple bipartite graph matching algorithm. Second, several metrics were introduced to quantify the similarity between two curve patterns. The metrics were based on the Hausdorff distance and the determined curve-to-curve correspondences. Third, the nearest neighborhood algorithm was used to rate curve patterns in the gallery in terms of

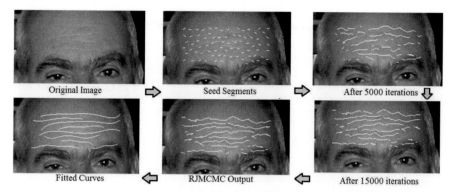

Fig. 27 Fitting of curves to detected wrinkles as line segments using MPP modeling (reproduced from [5])

similarity to that of the probe pattern using their defined metrics. The recognition rate in their experiments was reported to exceed 65 % at rank 1 and 90 % at rank 4 using matching of curve patterns only.

4.4 Applications in Skin Research

Cula et al. [9, 10] proposed digital imaging as a non-invasive, less expensive tool for the assessment of the degree of facial wrinkling to establish an objective baseline and for the assessment of benefits to facial appearance due to various dermatological treatments. They used finely tuned oriented Gabor filters at specific frequencies and adaptive thresholding for localization of wrinkles in forehead images acquired in controlled settings. They introduced a wrinkle measure, referred to as wrinkle index, as the product of both wrinkle depth and wrinkle length to score the severity of wrinkling. The wrinkle index was calculated from Gabor responses and the length of localized wrinkles. The calculated wrinkle indices were then validated using 100 clinically graded facial images. Figure 28 shows examples of localization of wrinkles with different severity in images acquired in controlled setting along with a plot of clinical vs. computer generated scores given in their work.

Jiang et al. [19] also proposed an image based method named 'SWIRL' based on different geometric characteristics of localized wrinkles to score the severity of wrinkles. However, they used a proprietary software tool to localize wrinkles in images which were taken in controlled lighting settings. The goal was to quantitatively assess the effectiveness of dermatological/cosmetic products and procedures on wrinkles. In their controlled illumination settings, the so-called raking light optical profilometry, lighting was cast at a scant angle to the face of the subject casting wrinkles as dark shadows. The resulting high-resolution digital images were analyzed for the length, width, area, and relative depth of automatically localized

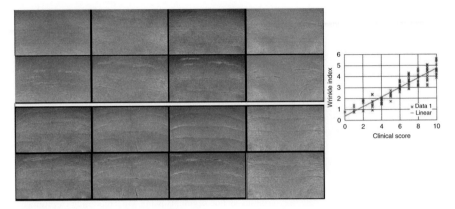

Fig. 28 (*Left*) Localization of wrinkles with varying severity. (*Right*) Plot of clinical scores vs. computer generated scores for 100 images (reproduced from [10])

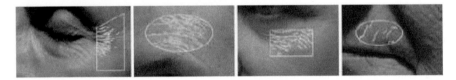

Fig. 29 Localization of wrinkles in different facial regions using a proprietary software used in [19] (reproduced from [19])

wrinkles. The parameters were shown to be correlated well with clinical grading scores. Furthermore, the proposed assessment method was also sensitive enough to detect improvement in facial wrinkles after 8 weeks of product application. Figure 29 shows few images from different facial regions with localized wrinkles using a proprietary software tool used in their work.

4.5 Facial Expression Analysis

The conventional methods on the analysis of facial expression are usually based on Facial Action Coding System (FACS) in which a facial expression is specified in terms of Action Units (AU). Each AU is based on the actions of a single muscle or a cluster of muscles. On the other hand little investigation has been conducted on wrinkle texture analysis for facial expression recognition. In this section we present research work in expression analysis which incorporates facial wrinkles. Facial wrinkles deepen, change or appear due to expressions and can be an important clue to recognizing expressions. Hence, the following approaches treat changes in facial wrinkles due to expressions as transient or temporary facial features.

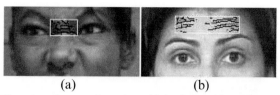

(a)   (b)

Transient feature detection; (a) horizontal wrinkles
between eyes; (b) horizontal wrinkles on the forehead

the example results of nasolabial fold detection

Fig. 30 Examples of detection of transient wrinkles during expressions in different facial regions
in [45]

Zang and Ji [45] presented a 3-layer probabilistic Bayesian Network (BN) to
classify expressions from videos in terms of probability. The BN model consisted
of three primary layers: classification layer, facial AU layer and sensory information layer. Transient features e.g. wrinkles and folds were part of the sensory
information and were modeled in the sensory information layer containing other
visual information variables, such as brows, lips, lip corners, eyelids, cheeks, chin
and mouth. The static BN model for static images was then extended to dynamic
BN to express temporal dependencies in image sequences by interconnecting time
slices of static BNs using Hidden Markov modeling. In their work the presence of
furrows and wrinkles was determined by edge feature analysis in the areas where
transient features appear i.e. forehead, on the nose bed/between eyes and around
mouth (nasolabial area). Figure 30 shows examples of transient feature detection in
three regions. The contraction and extension of facial muscles due to expressions
result in wrinkles/folds in particular shapes detected by edge detectors. The shape
of wrinkles was approximated by fitting quadratic forms passing through a set of
detected edge points in a least-square sense. The coefficients in the quadratic forms
then signified the curvature of the folds and indicated presence of particular facial
AUs.

Tian et al. [42] proposed a system to analyze facial expressions incorporating
facial wrinkles/furrows in addition to commonly studied facial features of mouth,
eyes and brows. Facial wrinkles/furrows appearing or deepening during a facial
expression were termed as 'transient' features and were detected in pre-defined
three regions of a face namely around eyes, nasal root/bed or around mouth. The
Canny edge detector was used to analyze frames of a video to determine if wrinkles
appeared or deepened in later frames of a video. The presence/absence of wrinkles

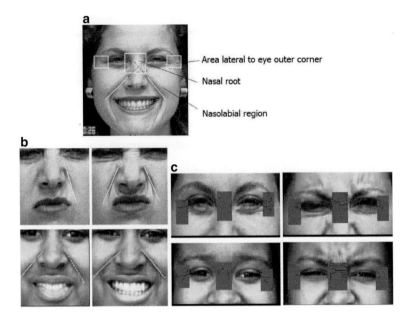

Fig. 31 (**a**) Three pre-defined areas of interest for detection of transient features (wrinkles/furrows). (**b**) Detection of orientation of expressive wrinkles. (**c**) Example of detection of wrinkles around eyes (reproduced from [42])

in three facial regions of interest as well as the orientations of the detected wrinkles were incorporated as an indication to the presence of specific AUs in their expression analysis system. Figure 31 shows three examples of the detection of the orientation of wrinkles around mouth for a certain expression.

Yin et al. [44] explored changing facial wrinkle textures exclusively in videos for recognizing facial expressions. They assumed that facial texture consisted of static and active parts where the active part of texture was changed with an expression due to muscle movements. Hence they presented a method based on the extraction of active part of texture and its analysis for expression recognition where the wrinkle textures were analyzed in four regions of face as shown in Fig. 32a. In their method the correlation between wrinkles texture in the neutral expression and the active expression was determined using Gaussian blurring. The two textures were correlated several times as they gradually lost detail due to blurring. The rate-of-change of correlation values reflected the dissimilarity of the two textures in four facial regions of interest and was used as a clue to the determination of six universal expressions.

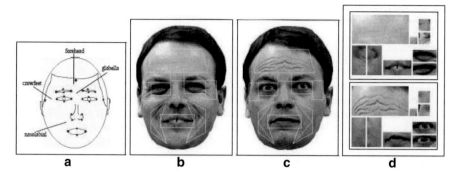

Fig. 32 Example of wrinkle textures extracted from two expressions (smile and surprise). (**a**) Facial regions of interest. (**b, c**) Example of textures extracted from smile and surprise expressions. (**d**) Normalized textures (reproduced from [44])

5 Summary and Future Work

In this chapter, we presented a review of the research in computer vision focusing on the analysis of facial wrinkles as image texture or curvilinear objects with several applications. Facial wrinkles are important features in terms of facial aging/expressions and can be a cue to several aspects of a person's identity and lifestyle. Image-based analysis of facial wrinkles can improve existing algorithms on facial analysis as well as pave way to new applications. For example, patterns of personalized aging can be deduced from the spatio-temporal analysis of changes in facial wrinkles. A person's smoking habits, facial expression and sun exposure history can be inferred from the severity of wrinkling. The specific patterns of wrinkles appearing on different facial regions can be added to facial soft biometrics or to the analysis of facial expressions. Furthermore, analysis of subtle changes in facial wrinkles can quantify the effects of different dermatological treatments. However, the first step in any of these applications would be the accurate and fast localization of facial wrinkles in high resolution images.

References

1. N. Batool, R. Chellappa, A Markov point process model for wrinkles in human faces, in *19th IEEE International Conference on Image Processing, ICIP 2012*, Lake Buena Vista, Orlando, 30 September–3 October 2012, pp. 1809–1812. doi:10.1109/ICIP.2012.6467233
2. N. Batool, R. Chellappa, Modeling and detection of wrinkles in aging human faces using marked point processes, in *ECCV Workshops (2)* (2012), pp. 178–188
3. N. Batool, R. Chellappa, Detection and inpainting of facial wrinkles using texture orientation fields and Markov random field modeling. IEEE Trans. Image Process. **23**(9), 3773–3788 (2014). doi:10.1109/TIP.2014.2332401

4. N. Batool, R. Chellappa, Fast detection of facial wrinkles based on gabor features using image morphology and geometric constraints. Pattern Recogn. **48**(3), 642–658 (2015). doi:10.1016/j.patcog.2014.08.003

5. N. Batool, S. Taheri, R. Chellappa, Assessment of facial wrinkles as a soft biometrics, in *10th IEEE International Conference and Workshops on Automatic Face and Gesture Recognition, FG 2013*, Shanghai, 22–26 April 2013, pp. 1–7. doi:10.1109/FG.2013.6553719

6. L. Boissieux, G. Kiss, N. Thalmann, P. Kalra, Simulation of skin aging and wrinkles with cosmetics insight, in *Computer Animation and Simulation 2000, Eurographics*, ed. by N. Magnenat-Thalmann, D. Thalmann, B. Arnaldi (Springer, Vienna, 2000), pp. 15–27

7. C. Chen, W. Yang, Y. Wang, K. Ricanek, K. Luu, Facial feature fusion and model selection for age estimation, in *2011 IEEE International Conference on Automatic Face Gesture Recognition and Workshops (FG 2011)* (2011), pp. 200–205

8. T.F. Cootes, G.J. Edwards, C.J. Taylor, Active appearance models, in *Proceedings of the 5th European Conference on Computer Vision - ECCV'98*, Freiburg, vol. II 2–6 June 1998, pp. 484–498

9. G.O. Cula, P.R. Bargo, N. Kollias, Assessing facial wrinkles: automatic detection and quantification (2009). doi:10.1117/12.811608

10. G.O. Cula, P.R. Bargo, A. Nkengne, N. Kollias, Assessing facial wrinkles: automatic detection and quantification. Skin Res. Technol. **19**(1), e243–e251 (2013). doi:10.1111/j.1600-0846.2012.00635.x

11. O.G. Cula, K.J. Dana, F.P. Murphy, B.K. Rao, Skin texture modeling. Int. J. Comput. Vision **62**(1–2), 97–119 (2005). doi:10.1007/s11263-005-4637-2

12. W. Freeman, E. Adelson, The design and use of steerable filters. IEEE Trans. Pattern Anal. Mach. Intell. **13**(9), 891–906 (1991)

13. Y. Fu, N. Zheng, M-face: an appearance-based photorealistic model for multiple facial attributes rendering. IEEE Trans. Circuits Syst. Video Technol. **16**(7), 830–842 (2006)

14. Y. Fu, G. Guo, T.S. Huang, Age synthesis and estimation via faces: a survey. IEEE Trans. Pattern Anal. Mach. Intell. **32**(11), 1955–1976 (2010). doi:http://doi.ieeecomputersociety.org/10.1109/TPAMI.2010.36

15. U. Hess Jr., R.B. Adams, A. Simard, M.T. Stevenson, R.E. Kleck, Smiling and sad wrinkles: age-related changes in the face and the perception of emotions and intentions. J. Exp. Soc. Psychol. **48**(6), 1377–1380 (2012)

16. A. Jain, U. Park, Facial marks: Soft biometric for face recognition, in *2009 16th IEEE International Conference on Image Processing (ICIP)* (2009), pp. 37–40

17. S. Jeong, Y. Tarabalka, J. Zerubia, Marked point process model for facial wrinkle detection, in *2014 IEEE International Conference on Image Processing (ICIP)* (2014), pp. 1391–1394. doi:10.1109/ICIP.2014.7025278

18. S. Jeong, Y. Tarabalka, J. Zerubia, Marked point process model for curvilinear structures extraction, in *Energy Minimization Methods in Computer Vision and Pattern Recognition - Proceedings of the 10th International Conference, EMMCVPR 2015*, Hong Kong, 13–16 January 2015, pp. 436–449

19. L.I. Jiang, T.J. Stephens, R. Goodman, SWIRL, a clinically validated, objective, and quantitative method for facial wrinkle assessment. Skin Res. Technol. **19**, 492–498 (2013). doi:10.1111/srt.12073

20. B. Klare, A.A. Paulino, A.K. Jain, Analysis of facial features in identical twins, in *Proceedings of the 2011 International Joint Conference on Biometrics, IJCB '11* (2011), pp. 1–8

21. Y.H. Kwon, N. da Vitoria Lobo, Age classification from facial images, in *1994 IEEE Computer Society Conference on Computer Vision and Pattern Recognition, 1994. Proceedings CVPR '94* (1994), pp. 762–767

22. Y.H. Kwon, N. da Vitoria Lobo, Age classification from facial images. Comput. Vis. Image Underst. **74**(1), 1–21 (1999). doi:10.1006/cviu.1997.0549

23. Z. Liu, Y. Shan, Z. Zhang, Expressive expression mapping with ratio images, in *Proceedings of the 28th Annual Conference on Computer Graphics and Interactive Techniques, SIGGRAPH '01* (ACM, New York, 2001), pp. 271–276. doi:10.1145/383259.383289

24. K. Luu, T.D. Bui, C. Suen, K. Ricanek, Combined local and holistic facial features for age-determination, in *2010 11th International Conference on Control Automation Robotics Vision (ICARCV)* (2010), pp. 900–904

25. N. Magnenat-Thalmann, P. Kalra, J. Luc Leveque, R. Bazin, D. Batisse, B. Querleux, A computational skin model: fold and wrinkle formation. IEEE Trans. Inf. Technol. Biomed. **6**(4), 317–323 (2002). doi:10.1109/TITB.2002.806097

26. S. Mukaida, H. Ando, Extraction and manipulation of wrinkles and spots for facial image synthesis, in *Proceedings of the Sixth IEEE International Conference on Automatic Face and Gesture Recognition* (2004)

27. C.C. Ng, M. Yap, N. Costen, B. Li, Automatic wrinkle detection using hybrid hessian filter, in *2014 Proceedings of the Asian Conference on Computer Vision ACCV*, Singapore (2014)

28. T. Ojala, M. Pietikainen, T. Maenpaa, Multiresolution gray-scale and rotation invariant texture classification with local binary patterns. IEEE Trans. Pattern Anal. Mach. Intell. **24**(7), 971–987 (2002)

29. H.C. Okada, B. Alleyne, K. Varghai, K. Kinder, B. Guyuron, Facial changes caused by smoking: a comparison between smoking and nonsmoking identical twins. Plast Reconstr. Surg. **132**(5), 1085–1092 (2013)

30. U. Park, A. Jain, Face matching and retrieval using soft biometrics. Inf. Forensics Secur. **5**(3), 406–415 (2010)

31. E. Patterson, A. Sethuram, K. Ricanek, F. Bingham, Improvements in active appearance model based synthetic age progression for adult aging, in *Proceedings of the 3rd IEEE International Conference on Biometrics: Theory, Applications and Systems, BTAS'09* (2009), pp. 104–108

32. P. Phillips, P. Flynn, K. Bowyer, R. Bruegge, P. Grother, G. Quinn, M. Pruitt, Distinguishing identical twins by face recognition, in *2011 IEEE International Conference on Automatic Face Gesture Recognition and Workshops (FG 2011)* (2011), pp. 185–192

33. N. Ramanathan, R. Chellappa, Modeling shape and textural variations in aging faces, in *8th IEEE International Conference on Automatic Face Gesture Recognition, 2008 (FG '08)* (2008), pp. 1–8

34. N. Ramanathan, R. Chellappa, S. Biswas, Computational methods for modeling facial aging: A survey. J. Vis. Lang. Comput. **20**(3), 131–144 (2009)

35. C. Robert, M. Bonnet, S. Marques, M. Numa, O. Doucet, Low to moderate doses of infrared a irradiation impair extracellular matrix homeostasis of the skin and contribute to skin photodamage. Skin Pharmacol. Physiol. **28**(4), 196–204 (2015)

36. A. Sethuram, E. Patterson, K. Ricanek, A. Rawls, Improvements and performance evaluation concerning synthetic age progression and face recognition affected by adult aging, in *Advances in Biometrics*, ed. by M. Tistarelli, M. Nixon. Lecture Notes in Computer Science, vol. 5558 (Springer, Berlin/Heidelberg, 2009), pp. 62–71

37. J. Suo, F. Min, S. Zhu, S. Shan, X. Chen, A multi-resolution dynamic model for face aging simulation, in *IEEE Conference on Computer Vision and Pattern Recognition, 2007 (CVPR '07)* (2007), pp. 1–8

38. J. Suo, X. Chen, S. Shan, W. Gao, Learning long term face aging patterns from partially dense aging databases, in *2009 IEEE 12th International Conference on Computer Vision* (2009), pp. 622–629

39. J. Suo, S.C. Zhu, S. Shan, X. Chen, A compositional and dynamic model for face aging. IEEE Trans. Pattern Anal. Mach. Intell. **32**(3), 385–401 (2010)

40. J. Suo, X. Chen, S. Shan, W. Gao, Q. Dai, A concatenational graph evolution aging model. IEEE Trans. Pattern Anal. Mach. Intell. **34**(11), 2083–2096 (2012)

41. X. Tan, Triggs, B.: Enhanced local texture feature sets for face recognition under difficult lighting conditions. IEEE Trans. Image Process. **19**(6), 1635–1650 (2010)

42. Y.L. Tian, T. Kanade, J. Cohn, Recognizing action units for facial expression analysis. IEEE Trans. Pattern Anal. Mach. Intell. **23**(2), 97–115 (2001). doi:10.1109/34.908962

43. Y. Wu, N. Magnenat-Thalmann, D. Thalmann, A plastic-visco-elastic model for wrinkles in facial animation and skin aging, in *Proceedings of the Second Pacific Conference on Fundamentals of Computer Graphics, Pacific Graphics '94* (World Scientific, River Edge, 1994), pp. 201–213
44. L. Yin, S. Royt, M. Yourst, A. Basu, Recognizing facial expressions using active textures with wrinkles, in *Proceedings of the 2003 International Conference on Multimedia and Expo, 2003 (ICME '03)*, vol. 1 (2003), pp. 177–180. doi:10.1109/ICME.2003.1220883
45. Y. Zhang, Q. Ji, Facial expression understanding in image sequences using dynamic and active visual information fusion, in *Proceedings of the Ninth IEEE International Conference on Computer Vision, 2003*, vol. 2 (2003), pp. 1297–1304. doi:10.1109/ICCV.2003.1238640

Communication-Aid System Using Eye-Gaze and Blink Information

Kiyohiko Abe, Hironobu Sato, Shogo Matsuno, Shoichi Ohi, and Minoru Ohyama

Abstract Recently, a novel human-machine interface, the eye-gaze input system, has been reported. This system is operated solely through the user's eye movements. Using this system, many communication-aid systems have been developed for people suffering from severe physical disabilities, such as amyotrophic lateral sclerosis (ALS). We observed that many such people can perform only very limited head movements. Therefore, we designed an eye-gaze input system that requires no special tracing devices to track the user's head movement. The proposed system involves the use of a personal computer (PC) and home video camera to detect the users' eye gaze through image analysis under natural light. Eye-gaze detection methods that use natural light require only daily-life devices, such as home video cameras and PCs. However, the accuracy of these systems is frequently low, and therefore, they are capable of classifying only a few indicators. In contrast, our proposed system can detect eye gaze with high-level accuracy and confidence; that is, users can easily move the mouse cursor to their gazing point. In addition, we developed a classification method for eye blink types using the system's feature parameters. This method allows the detection of voluntary (conscious) blinks. Thus, users can determine their input by performing voluntary blinks that represent mouse clicking. In this chapter, we present our eye-gaze and blink detection methods. We also discuss the communication-aid systems in which our proposed methods are applied.

K. Abe (✉) • H. Sato
Kanto Gakuin University, 1-50-1, Mutsuura-higashi, Kanazawa-ku, Yokohama, Kanagawa 236-8501, Japan
e-mail: abe@kanto-gakuin.ac.jp; hsato@kanto-gakuin.ac.jp

S. Matsuno
The University of Electro-Communications, 1-5-1, Chofugaoka, Chofu, Tokyo, Japan
e-mail: m1440004@edu.cc.uec.ac.jp

S. Ohi • M. Ohyama
Tokyo Denki University, 2-1200, Muzaigakuendai, Inzai, Chiba, Japan
e-mail: ohi@mail.dendai.ac.jp; ohyama@mail.dendai.ac.jp

© Springer International Publishing Switzerland 2016
M. Kawulok et al. (eds.), *Advances in Face Detection and Facial Image Analysis*,
DOI 10.1007/978-3-319-25958-1_12

1 Introduction

Recently, eye-gaze input systems have been proposed as novel human-machine interfaces [1–10]. Users can employ these systems to input characters or commands in personal computers (PCs) using only their eye movements. In other words, the operation of these systems involves the detection of the user's eye-gaze. Therefore, these systems are used to develop communication-aid systems for people with severe physical disabilities, such as amyotrophic lateral sclerosis (ALS). In addition, eye blink input systems have also been proposed to support communication for people with severe physical disabilities [11–13]. Eye blink input systems detect the occurrence of user's voluntary blinks, and users can use such systems to input characters or commands in PCs as well as with eye-gaze input systems. The detection method for eye blink is simpler than for eye-gaze detection. However, it is difficult for many eye blink detection methods to classify eye blink types as voluntary (conscious) or involuntary (unconscious).

We developed an eye-gaze input system [8–10] that utilizes a PC and home video camera to detect eye-gaze by image analysis. Our eye-gaze input system can be used under natural light (and artificial light sources such as fluorescent or LED lamps). The traditional eye-gaze input systems that operate under natural light often have low accuracy. Therefore, they are capable of classifying only a few indicators [4, 5]. To solve this problem, a system that uses multi-cameras is proposed [6]; consequently, our proposed eye-gaze input system employs multi-indicators (27 indicators in three rows and nine columns) [8], in addition to the PC and home video camera. In other words, the system is not only inexpensive, but also user friendly; therefore, it is suitable for personal use, such as for welfare device applications. In addition, we developed an application system for our eye-gaze input system that supports PCs (Microsoft Windows), English and Japanese text input, etc. [9].

Our proposed system can detect horizontal and vertical eye-gaze using image analysis based on the limbus tracking method [8–10], which arranges its detection area on the open-eye area of the eye image. The user's eye-gaze is estimated by the integral values of the light intensity on the detection area. In addition, we developed a new method for estimating detection area automatically [10]. Our proposed eye-gaze detection method can estimate the coordinate data of the user's gazing point with high accuracy; that is, users can easily move the mouse cursor to their gazing point [10].

In general eye-gaze input systems, users select the icons displayed on the PC monitor by gazing at them [1–4, 8, 9]. These icons are called "indicators" and are assigned to characters or functions of the application program. If the application programs are oriented to eye-gaze input, the programs can recognize the gazed indicator easily. Using eye-gaze input systems, the interface for mouse operation can move the cursor to the user's gazing point; however, operating system of PC does not recognize the gazed indicator. Therefore, it is difficult for users using eye-gaze input systems to operate general application programs.

To resolve this problem fundamentally, we developed an interface that utilizes information on eye-gaze and blink. We call this type of interface, "eye-gaze and blink input." We confirmed that there is a large difference in the duration of voluntary and involuntary blinks [14–16]. In addition, we also confirmed that the duration of eye blink varies widely among individuals. Using these results, we developed a new input interface that utilizes the information of eye-gaze and voluntary blink occurrence. That is, users can determine their input by performing voluntary blinks that represent mouse clicking. Using this eye-gaze and blink input system, users can operate general application software on Windows. In this chapter, we propose our detection method for eye-gaze and blink through image analysis and the input interface using the information obtained from eye-gaze and blink.

2 Current Situation of Eye-Gaze and Blink Input

Many systems and devices have been developed as communication aids for people with severe physical disabilities, such as ALS patients. For example, the E-tran (eye transfer) frame is used for communication between disabled users and others [17]. The E-tran frame is a conventional device with a simple structure: a transparent plastic board with characters, such as the alphabet, or simple words printed on it. To use the E-tran frame, a communication partner (helper) holds such device over the user's face. In particular, the user gazes at the place where the character or word that the user wishes to communicate is positioned. The helper moves the E-tran frame until the eye gaze of the user corresponds with that of the helper. Therefore, the helper can determine the character from the user's eye gaze. A user who can gaze at the characters on the E-tran frame can also communicate with others. In addition, the E-tran frame does not require power supply, and is therefore, highly portable. However, considerable skill is required to use the E-tran frame.

The row-column scanning system is also used to aid the communication of people with severe physical disabilities. This system can be operated with one switch. In other words, the user can input characters or operate a PC using their physical residual function. Considerable time is required to input using the row-column scanning system because this system operates by scanning the rows and columns of keyboards using only one switch. To improve upon this situation, a new method for row-column scanning has been reported [18]. This method optimizes the speed of row-column scanning using a Bayesian network for machine learning.These methods are frequently for people with severe physical disabilities to support their communication. However, a proposal for a new method is expected because the operation of these methods is difficult and requires considerable time.

The eye-gaze input system mitigates these weaknesses. In a general eye-gaze input system, the icons displayed on the PC monitor are selected by the user gazing at them, as shown in Fig. 1. These icons are assigned to characters or functions of the application program. The eye-gaze input has to detect the user's gaze in order

Fig. 1 Overview of eye-gaze
input

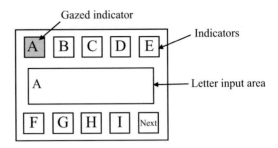

to ascertain the selected icon. In general, if a user gazes an icon of eye-gaze input system, the icon switches color. In other words, the icon indicates the gazed position. Therefore, the icon is called an "indicator."

Many eye-gaze detection methods have been developed in the past. Several systems use the electro-oculogram (EOG) method for eye-gaze detection [7], which detects eye-gaze by the difference in the electrical potential between the cornea and the retina. However, many systems detect eye-gaze using non-contact methods to enhance usability. In particular, the user's gaze is detected by analyzing images of the eye (and its surrounding skin) captured by a video camera. To classify the many indicators, most conventional systems use special devices, such as infrared light or multiple cameras [1–3, 6]. Nevertheless, in order to be suitable for personal use, the system should be inexpensive and user-friendly. Therefore, a simple system that uses a single camera in natural light is desirable. However, natural-light systems often have low accuracy and are capable of classifying only a few indicators [4, 5]. This makes it difficult for users to perform a task that requires many functions, such as text input. To solve these problems, a simple eye-gaze input system that can classify many indicators is required.

In addition, input decision requires the detection of not only the location of the user's gaze, but also the user's selection command. Selection can be performed either with an eye blink or using eye fixations (measuring how long the eye fixates on a target such as an indicator). These two methods are effective for the input decision of eye-gaze input. However, the usability of the interface that uses eye blink is higher than the method that uses eye fixations.

3   Eye-Gaze Detection by Image Analysis

The aim of the eye-gaze input system is to detect the eye-gaze of the user. Several eye-gaze detection methods have been studied. To detect eye-gaze, these methods analyze eye images captured by a video camera [1–6]. This method of tracking the iris by image analysis is the most popular detection method used under natural light. However, it is difficult to distinguish the iris from the sclera through image analysis because of the smooth transition of the luminance between the iris and the sclera. In some users, the iris is hidden by the eyelids, which makes it difficult to estimate the location of the iris by elliptical approximation.

We propose a new eye-gaze detection method that involves image analysis using the limbus tracking method [8–10], which does not estimate the edge of the iris. Here, we describe the eye-gaze input detection method in detail. To support reader's understanding, the location and size of detection area is fixed, as described in Sect. 3.1.

3.1 Eye-Gaze Detection by Image Analysis Based on Limbus Tracking Method

The limbus tracking method is well known as a practical method for eye-gaze detection [19]. This method detects eye-gaze using the light reflected from the eyeball. This reflected light is measured by a photo sensor, such as a phototransistor. An infrared LED is often used as the light source. If the subject's eyeball moves, the reflection value changes. Therefore, eye-gaze can be estimated from this change value. We apply the limbus tracking method to our new eye-gaze detection method by image analysis [8–10].

Eye outlines are shown in Fig. 2, and an overview of the proposed horizontal eye-gaze detection method is shown in Fig. 2a. The difference in the reflectance between the iris and the sclera is used to detect the horizontal eye-gaze. In other words, the horizontal eye-gaze is estimated using the difference between the integral values of the light intensity on areas A and B, as shown in Fig. 2a. We define this differential value as the horizontal eye-gaze value.

An overview of the vertical eye-gaze detection method is shown in Fig. 2b. We estimate the vertical eye-gaze using the integral value of the light intensity on area C, as shown in Fig. 2b. We define this integral value as the vertical eye-gaze value. If the eye-gaze input system is calibrated using the relationship between these eye-gaze values and the angle of sight, we can estimate the horizontal and vertical eye-gaze of the user. Details of the calibration method are described in Sect. 3.4.

Fig. 2 Overview of horizontal and vertical eye-gaze detection: (**a**) horizontal detection, (**b**) vertical detection

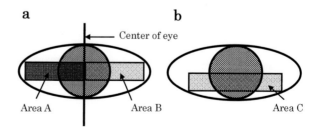

3.2   *Automatic Arrangement of Detection Area*

We developed an eye-gaze input system as described in the previous sections. Using this system, users can select icons on the computer screen or control the mouse cursor by eye-gaze. In general, if the user's eye-gaze moves, the shape of the eyelids (upper or lower) changes. We observed that the error of detection increases with a change in the shapes of the eyelids because the shape of the detection area also changes. To resolve this problem, we developed a new method for estimating the detection area. This method extracts the exposed eye area from an eye image by analyzing the part of the image where the exposed eye area is not hidden by the eyelids [10]. The detection area can be used to detect both the horizontal and vertical eye-gaze.

We can estimate the exposed eye area by merging the eye shape images (open-eye areas). The merging method uses a bitwise OR operation when the pixel value on the open-eye area is "0;" otherwise, it is "1." The open-eye areas are extracted from eye images when the user directs his or her gaze at indicators 1–4, shown in Fig. 3. The horizontal and vertical distances of the indicators are 24 deg. and 18 deg., respectively. These degrees indicate vision angle ("deg" in Fig. 3 indicates vision angle).These open-eye areas are extracted by binarization using the color information of the skin [8]. We use the exposed eye area estimated from these open-eye areas to extract the detection area for the horizontal and vertical eye-gaze. An overview of the exposed eye area extraction is shown in Fig. 4. Details of the binarization that uses color information are provided in Sect. 3.3.

To utilize the image analysis based on the limbus tracking method, we have to focus the detection area on the pixels whose light intensity changes. Therefore, we estimate the detection area of the horizontal and vertical eye-gaze using the eye images when a user looks at indicator 1 (up) and 4 (down), or indicator 2 (left) and 3 (right), respectively. First, we create an image that consists of the differences in the images captured when a user looks at indicators 1 and 4. We estimate the detection area of the vertical eye-gaze from the created image. In practical terms, the detection area for the vertical eye-gaze is arranged on the pixels with positive difference values. This process is executed inside the exposed eye area, as shown in Fig. 4.

Fig. 3 Indicators for eye expose area extraction

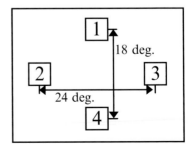

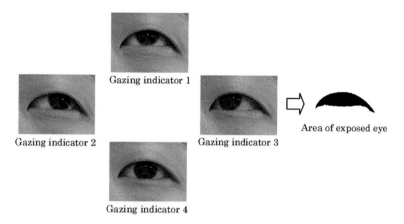

Fig. 4 Extraction of exposed eye area

Fig. 5 A sample of detection area

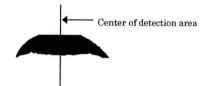

   Next, we estimate the centerline of the detection area using the horizontal eye-gaze detection method shown in Fig. 2a. In practical terms, this centerline is located where the horizontal eye-gaze value is at a maximum. We estimate the horizontal eye-gaze from the eye images when a user looks at indicators 2 and 3. This process is executed inside the detection area of the vertical eye-gaze. Figure 5 shows a sample of the detection area estimated by our proposed method. We estimate the vertical eye-gaze using the integral value of the light intensity on the total detection area. We estimate the horizontal eye-gaze from the difference between the integral values of the light intensity on the two detection areas divided by the centerline.

3.3   Compensation for Head Movement

In our eye-gaze detection method, a video camera records an image of the user's eye from a distant location (the distance between the user and camera is approximately 70 cm), and then this image is enlarged. The user's head movement induces a large error in the measurements. Therefore, the system must compensate for such movement, which we accomplish by tracing the location of the inner corner of the eye. This tracking method is based on image analysis and is executed in real time [8–10]. The open-eye area can be estimated from the eye image. If the inner-corner point of the open-eye area is extracted, the location of the inner-corner of the eye is determined.

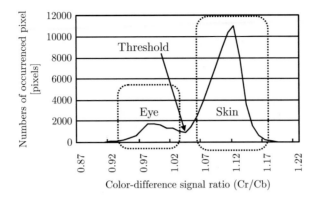

Fig. 6 A histogram of color difference signal ratio [8]

In our method, the open-eye area is estimated by binarization using the facial skin color. The skin color threshold is determined by a histogram of the color-difference signal ratio of each pixel (Cr/Cb) calculated from the YCbCr image transformed from the RGB image. This histogram has two peaks that indicate the skin area and open-eye area [8]. The minimum Cr/Cb value between the two peaks is designated as the threshold for open-eye area extraction. A sample histogram of the color-difference signal ratio is shown in Fig. 6.

Our method can extract the open-eye area almost completely. However, the results sometimes leave deficits around the corner of the eye because for some subjects, the Cr/Cb value around the corner of the eye is similar to the value on the skin. To solve this problem, we developed a method for open-eye area extraction without deficit by combining two extraction results. One is based on a binarized image that uses color information. The other is based on a binarized image that uses light-intensity information, which includes, in the extraction result, the area around the corner of the eye. These two binarized images are combined by a bitwise AND operation when the pixel value on the open-eye area is "0;" otherwise, it is "1."

We extract the location of the inner corner of the eye using the open-eye area image described above, and an eye image that enhances the edge of the inner corner of the eye. The enhanced image is processed through a special differential filter. Using these two images, the open-eye area and the enhanced, this method almost completely extracts the location of the inner corner of the eye. Hence, user head movement is compensated in the eye-gaze detection method [8]. The operator for the differential filter is shown in Fig. 7, where the noticed pixel appears gray. Figure 8 shows an example of the extraction result for the open-eye area, an eye image enhanced at the inner corner of the eye, and an extraction of the inner corner of the eye.

Fig. 7 Operator for
differential filter

1	0	0	0	0
0	3	0	0	0
1	3	0	−3	−1
0	0	0	−3	0
0	0	0	0	−1

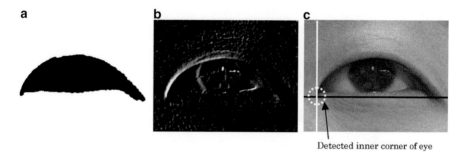

Detected inner corner of eye

Fig. 8 Detection of inner corner of left eye: (**a**) open-eye area, (**b**) eye image enhanced at the inner corner of the eye, (**c**) extraction of the inner corner of the eye

Fig. 9 Indictors for
calibration

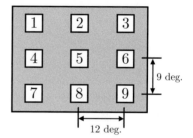

3.4   Calibration Method for Eye-Gaze Detection

The characteristics of eye-gaze detection are different for different users. Therefore, the eye-gaze input system needs to be calibrated for eye-gaze detection. For typical calibration, the user gazes at the indicators arranged at regular intervals on the computer screen, as shown in Fig. 9, and the results obtained are used to calibrate the system. The characteristics of the horizontal and vertical eye-gaze are shown in Fig. 10 as a scatter plot, where the x-axis and y-axis indicate the horizontal and vertical eye-gaze value, respectively. In addition, the indicator numbers (Ind. 1–9) are displayed near the plot points. These correspond to the indicator numbers shown in Fig. 9.

From Fig. 10, it is evident that if the eye-gaze of the user moves only in a horizontal direction, the vertical eye-gaze values change as well. For example, indicators 1–3 are arranged on the same horizontal line; however, the estimated

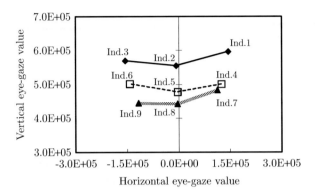

Fig. 10 A characteristic of eye-gaze [10]

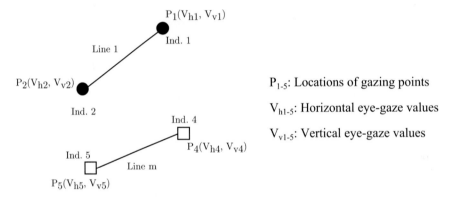

Fig. 11 An overview of our calibration method for eye-gaze detection

vertical eye-gaze values are different. Similarly, it is also evident that if the user's eye-gaze moves in a vertical direction only, the horizontal eye-gaze values change as well. In our method, there is a dependency relationship between the characteristics of the horizontal and vertical eye-gaze. Considering this, we calibrate our eye-gaze input system using four indicator groups, "indicators 1, 2, 4, 5 (upper left group)," "indicators 2, 3, 5, 6 (upper right group)," "indicators 4, 5, 7, 8 (lower left group)," and "indicators 5, 6, 8, 9 (lower-right group)." The eye-gaze input system is calibrated using each indicator group [10]. An overview of the calibration method using indicators 1, 2, 4, 5 is shown in Fig. 11.

First, we note the characteristics of the horizontal eye-gaze. The characteristics from indicators 1 and 2 and indicators 4 and 5 are defined as Line l and Line m, respectively. In Fig. 11, we see that the gradient of Line l is greater than Line m. If the user's eye-gaze moves down, the gradient of the horizontal eye-gaze characteristic decreases. The calibration method for horizontal eye-gaze detection calculates the change in the parameters of these calibration functions in order to calculate their gradients. We assume that the changes in the horizontal parameters are proportional to the vertical eye-gaze movement. Then, we estimate the calibration function for the vertical eye-gaze detection using a similar method. If we calculate the calibration functions using the horizontal and vertical eye-gaze values, ($V_{h1,2,4,5}$, $V_{v1,2,4,5}$) while the user looks at the indicators (Ind.1, 2, 4, and 5), we can estimate where the user is looking. In other words, the coordinate of the user's gazing point can be estimated by our calibration method.

Using this calibration method, the user's gazing point can be estimated with a high degree of accuracy. In the results of an evaluation experiment with five subjects, we confirmed that the average errors of vertical and horizontal eye-gaze detection are 1.09 deg and 0.56 deg, respectively [10]. These results indicate that our proposed method can detect vertical and horizontal eye-gaze to a high degree of accuracy, and performs as well as the detection method that uses infrared light [20].

4 Eye Blink Detection by Image Analysis

Human eye blinks can be classified into three types: involuntary, voluntary, and reflex. Involuntary blinks occur unconsciously, whereas voluntary blinks are generated consciously by a cue. Reflex blinks are induced by stimuli, such as loud sounds or bright lights. In this chapter, we note two blink types, voluntary and involuntary; the former is used as cue for input decisions, and the latter is treated as noise on the input process.

If voluntary blinks could be detected automatically, it would be possible to develop a user-friendly interface. We confirmed the feature parameters of involuntary and voluntary eye blinks, which can be used to distinguish between these two types of blinks. We present the eye blink detection method and the feature parameters of voluntary and involuntary blinks.

4.1 Eye Blink Measurement by Frame-Splitting Method

If the entire process of an eye blink is captured, the wave pattern of the eye blink can be generated. We developed a new method for measuring the wave pattern of an eye blink. This method can be used under natural light, and it can measure the wave pattern automatically [14]. In this method, eye blinks are detected by measuring the pixels of the open-eye area [14]. The extraction method for open-eye area is discussed in Sect. 3.3. In other words, the processes for eye blink and eye-gaze detection can be executed simultaneously using our proposed methods. This point is an advantage for developing a practical eye-gaze and blink input system.

Commonly used NTSC or 1080i Hi-vision video cameras output interlaced images. If one interlaced image is divided by scanning even and odd-numbered lines separately, two field images are generated. Thus, the time resolution for the motion images doubles, but the amount of information in the vertical direction decreases by

Fig. 12 Overview of frame
splitting method

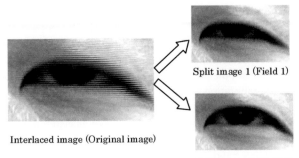

Interlaced image (Original image)

Split image 1 (Field 1)

Split image 2 (Field 2)

half. These field images are captured at 60 fields/s, and the NTSC or 1080i Hi-vision
interlaced moving images are captured at 30 fps; therefore, this method yields a time
resolution double that available in the NTSC or 1080i Hi-vision format.

We propose a new method for measuring the eye blink that uses split-interlaced
images generated from an interlaced image. We define this new method as the
frame-splitting method. In general, it is difficult for NTSC or 1080i Hi-vision
cameras to measure the detailed temporal change that occurs during the process
of eye blinking because eye blinks occur relatively fast (within a few hundred
milliseconds). However, eye blink detail can be measured using the frame-splitting
method. In addition, the feature parameters of the eye blink can be estimated from its
wave pattern using this method. In this section, we describe the feature parameters
of the eye blink, which are measured with the frame splitting method. However,
the practical input system that uses eye-gaze and blink does not require the frame-
splitting method. The details of the practical system are described in Sects. 5 and 6.

Examples of interlaced images during eye blinking and the split-interlaced
images are shown in Fig. 12. The two eye images shown on the right are enlarged in
the vertical direction, and were generated from the interlaced image shown on the
left. The interlace image has comb-like noise. To describe this phenomenon most
clearly, Fig. 12 was captured at low resolution (145×80 pixels).

4.2 Automatic Extraction for Wave Pattern of Eye Blink

Captured moving images of the eye include images of the blinking eye and
those of the open eye. Thus, the contiguous data of the estimated open-eye area
include samples with these mixed situations. If we want to classify the eye blink
types, we need to extract data only for the blinking eye. Then, the parameters
of eye blinks are estimated in order to classify the different types of eye blinks.
To estimate the threshold for the extraction of the wave pattern of eye blinks,
we utilize the difference wave pattern of eye blinks [14]. A sample of the wave
pattern of an eye blink and its difference wave pattern are shown in Figs. 13 and

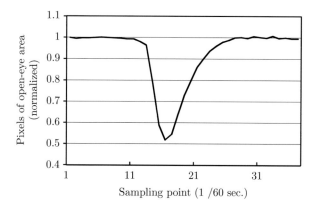

Fig. 13 A wave pattern of eye blink [14]

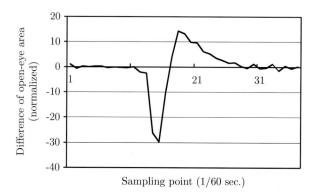

Fig. 14 A differential wave pattern of eye blink [14]

14, respectively. In these figures, the x-axis indicates the sampling point (interval = 1/60 s), and the y-axis indicates the open-eye area pixels and its difference value, respectively. These plots are normalized relative to the pixels in the open-eye area in the first field image.

From Fig. 14, it is evident that the difference value for the blinking eye is greater than that for the open eye. In other words, the fluctuation in the difference value for the open eye is smaller than for the blinking eye. Thus, we can use statistical information to distinguish between data for the open eye and the blinking eye. In particular, we estimate the average and standard deviation of the difference value of the open-eye area when the eye is open. Using these values, the threshold is estimated automatically. The thresholds for the discrimination of data when the eye is open, Th_1 and Th_2, are determined by (1) and (2) [14].

$$Th_1 = E_{avg} + 2\sigma \qquad (1)$$

$$Th_2 = E_{avg} - 2\sigma \qquad (2)$$

where E_{avg} is the average difference value of the open-eye area when the eye is open, and sigma is its standard deviation. If the sample of the open-eye area is greater than Th_1, it indicates that the eye is opening. Similarly, if the sample of the open-eye area is smaller than Th_2, it indicates that the eye is closing. The part between these two results is the wave pattern for the blinking eye.

4.3   Feature Parameters for Classification of Eye Blink Types

If voluntary blinks could be detected automatically, it would be possible to develop a user-friendly interface. Thus, users could employ this interface to input commands in their PC consciously. Many input interfaces that utilize the information of voluntary blinks have been developed that users can use to input text in PCs. However, these interfaces utilize specific patterns of eye blinks. For example, they use a method that classifies voluntary blinks based on the occurrence of multiple blinks [11] or on fixed duration [12, 13]. To relax these conditions, we developed a new interface. If users close their eyes firmly, the new interface captures an input command [14, 15]. In other words, the constraint present when users employ the interface by means of an eye blink is alleviated.

To classify the eye blink types, we need to scrutinize the feature parameters of an eye blink. Eye blink patterns can be extracted automatically from detection results using the method described in Sect. 4.1. The feature parameters can be estimated from the eye blink pattern. We reported that there is a large difference in the duration of voluntary and involuntary blinks [14]. We conducted metering experiments to confirm the parameters, and the outline of an eye blink pattern is shown in Fig. 15. We define the feature parameters as the duration (Td) and maximum amplitude (Am) of an eye blink.

We confirmed the feature parameters, duration Td, and maximum amplitude Am of the eye blink types through the metering experiment, which involved five subjects. Figure 16 shows samples of the voluntary and involuntary blinks patterns.

Fig. 15 An outline of eye blink wave pattern and its feature parameters

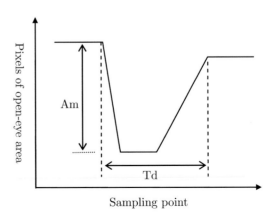

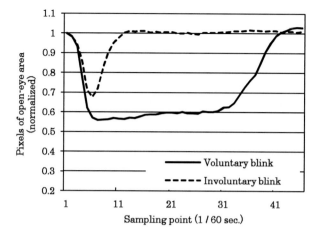

Fig. 16 Wave patterns of voluntary and involuntary blink

In Fig. 16, the x-axis indicates the sampling point (interval = 1/60 s), and the y-axis indicates the open-eye area pixels. The two wave patterns show voluntary and involuntary blinks, respectively. This plot is normalized relative to the pixels in the open-eye area of the first field image.

From Fig. 16, it is evident that there are large differences in the duration and amplitude of voluntary and involuntary blinks [14, 15]. Through the metering experiments with five subjects, we confirmed that the average duration of involuntary and voluntary blinks is 253.4 and 715.0 ms, respectively [14]. In addition, we confirmed that the average of the normalized maximum amplitudes of involuntary and voluntary blinks is 0.283 and 0.402, respectively [14]. In other words, it is evident that the duration and maximum amplitude of involuntary blinks are shorter than those of voluntary blinks.

5   Communication-Aid System Using Eye-Gaze and Blink Information

Using the method shown in Sects. 3 and 4, we developed an eye-gaze and blink input system to aid in communication. Users can move the mouse cursor on the screen of PCs through eye-gaze using this interface. In addition, users can determine their input through voluntary blinks. In this section, we present our proposed input interface for the eye-gaze and blink input system. We also describe the input method for this interface and the problem points on practical use.

5.1   Input Decision Using Detection of Gaze State

In the early phase of our research, we developed a text input system with eye-gaze only [9]. For this system, we designed indicators for text input through eye-gaze, considering the success rate of gaze selection for our proposed system. There are 12 indicators (two rows, six columns). However, approximately 60 indicators are required to input Japanese text (our native language). A total of 12 indicators are also insufficient for English text input because the English language contains uppercase and lowercase letters and symbols. To solve this problem, we designed a new interface where users can select any characters (English or Japanese) by choosing the indicator group. An overview of the interface is shown in Fig. 17.

This interface requires two selections: one for character input and another for character group selection, for example, "group A–E." Fig. 17 shows the interface. Alphabets and symbols ("etc." in Fig. 17) require two selections; however, commonly used characters (for example, "space") require only one selection. To input character "C," the user first selects the indicator for "group A–E," then the indicator for "C." Japanese can be input in the same way. In the system shown in Fig. 17, input is determined through the information on the user's eye-movement history. The eye-movement history of a user is the gaze-state history, and the gaze state has two settings: initial and continuous. If the user gazes at an indicator for more than a defined period of time, the gaze state changes to the initial state. When the initial state is complete, the state changes to continuous. An overview of this method is shown in Fig. 18.

This method measures the gaze-state history from the two gaze states (initial and continuous). The indicators assigned the majority of the time in each state are extracted as candidate inputs. If these two values are equal, the candidate input is determined. The user's gaze has noise such as blinking and involuntary eye movement. However, the user's input decision is extracted correctly. Users can input text at a rate of 16.2 characters/min through eye-gaze. This rate is comparable to that of a conventional infrared eye-gaze input system without requiring word prediction

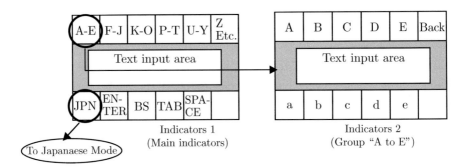

Fig. 17 Indicators for text input

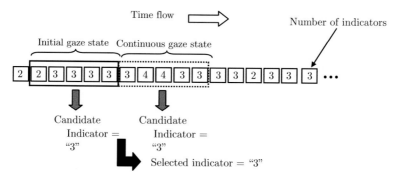

Fig. 18 Overview of input decision using gaze-state history

or special devices [9]. Using this type of interface, users can determine input by gazing at an indicator because the application program for eye-gaze input classifies such indicator.

5.2 Input Decision by Voluntary Blink

Using the input interface that classifies the gaze state of users described in Sect. 5.1, users can operate PCs for communication-aid through eye-gaze only. However, this type of interface system requires special application programs for eye-gaze input. In order to provide additional improvements for the Quality of Life (QoL) of users, a more versatile environment for eye-gaze input is required. For example, some users might want to explore Web services, such as Facebook and Twitter. Developing new software for these users individually is difficult. To solve this problem, we need to improve our interface for mouse operation through eye-gaze. In other words, the eye-gaze and blink input system can operate all the functions of generic Windows application software.

The eye blink feature parameters are described in Sect. 4.3. We confirmed that there is a large difference in the duration of voluntary and involuntary blinks. Therefore, we classify eye blink types using eye blink duration as a feature parameter [14]. Our method is simple and robust. Methods for classifying voluntary blinks based on duration have been proposed [12, 13], and these methods use a fixed threshold. However, we confirmed that the threshold duration varies significantly and that it depends on the subject and experiment conditions [14]. To classify eye blink types with certainty, we focused on these points and developed a new method for estimating the threshold automatically. The threshold Th_c is calculated using (3)

$$Th_c = \frac{Tdv_{avg} - Tdiv_{avg}}{2} + Tdiv_{avg} \qquad (3)$$

where Tdv$_{avg}$ and Tdiv$_{avg}$ are the average duration of voluntary and involuntary blinks, respectively. If the duration of an eye blink is greater than the threshold Th$_c$, this eye blink is classified as voluntary; otherwise, the eye blink is classified as involuntary. Using our proposed method, the average rate of successful classification of voluntary eye blinks is approximately 95 % for the experiment sample of five subjects [14]. This rate is sufficient to input by voluntary blink. Therefore, our proposed method determines input through voluntary blink using eye blink duration as the feature parameter.

5.3 Feature Parameters of Eye Blinks when the Sampling Rate is Changed

To develop an eye blink input interface suitable for practical use, the interface system uses a standard video camera, such as an NTSC-based model. In particular, this system requires feature parameters that can be estimated by a standard video camera. In addition, if other application programs are executed while the eye blink detection program is in use, the sampling rate of video capture may be decreased. For example, the sampling rate of our proposed eye-gaze input system is approximately 10 fps. In other words, when eye blink and gaze are estimated simultaneously by our system, the sampling rate is decreased, compared with the method described in Sect. 4. To confirm the changes in these feature parameters when the sampling rate is decreased, we conducted experiments that utilize downsampled moving images [16].

The typical wave patterns of voluntary and involuntary blinks are displayed in Figs. 19, 20, and 21. Figure 19 displays wave patterns of voluntary blinks, whereas both Figs. 20 and 21 display wave patterns of involuntary blinks. The samples displayed in Figs. 20 and 21 indicate different trends. In addition, we estimated the data for sampling intervals of 16, 32, and 128 ms. These were estimated based on the moving images captured by a high-speed camera (sampling rate: 500 Hz, sampling interval: 2 ms). The sampling intervals of 16 and 32 ms approximately correspond to frame rates of 60 and 30 fps, respectively. These frame rates can be captured by a standard video camera. Furthermore, we assumed that the sampling interval of 128 ms was utilized for the eye-gaze input system, with the interface using the eye blink information. In these figures, the x-axis indicates the sampling point (interval = 2 ms), and the y-axis indicates the open-eye area pixels. These plots are normalized relative to the pixels in the open-eye area in the first field image.

From Figs. 20 and 21, we confirmed that the timing of eye blink occurrences affects the estimation of the wave patterns in an adverse way. At low sampling rates, the data plot errors in Fig. 21 are larger than in Fig. 20. In particular, there is a significant error at the bottom of the curve in Fig. 21. Therefore, the amplitude of this wave pattern changes significantly. In other words, it is difficult for estimation at low sampling rates to classify the eye blink type using its maximum amplitude.

Fig. 19 Wave pattern of
voluntary blink [16]

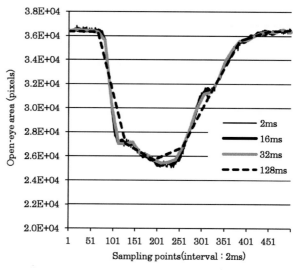

Fig. 20 Wave pattern of
involuntary blink 1 [16]

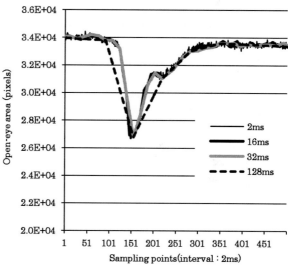

On the other hand, we confirmed that the eye blink types can be classified using the
duration at low sampling rates, such as 128 ms.

Considering this point, it is deemed desirable for the system to utilize eye blink
duration at low sampling rates. In other words, the method that uses eye blink
duration is suitable for the eye gaze input system using the information of voluntary
blink occurrence.

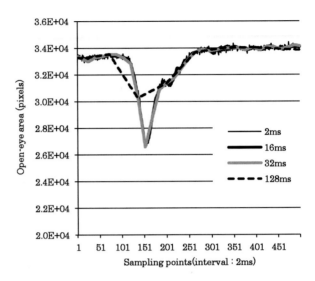

Fig. 21 Wave pattern of involuntary blink 2 [16]

6 Eye-Gaze and Blink Input System for Practical Use

Using the methods described in Sects. 3–5, we developed an eye-gaze and blink input system for practical use under natural light. Users can move the mouse cursor to a gazing point, and determine input through voluntary blinks using this system. In this section, we describe our eye-gaze and blink input system.

6.1 Detection of Voluntary Blink Through Eye-Gaze and Blink Input System for Practical Use

To detect the occurrence of voluntary blinks, we need to detect the occurrence of eye blink and measure its duration. As described in Sect. 4, if the detail of an eye blink can be estimated using the frame-splitting method, its duration can be measured easily. However, it is difficult for the measuring system to measure the detail of eye blinks at low sampling rates. In other words, if our proposed eye-gaze input system is used for this purpose, there will be a new estimation method for the feature parameter of the eye blink because the sampling rate of our proposed eye-gaze input system is 10 fps, which a low sampling rate.

To estimate the feature parameters of eye blinks, such as duration, we need to detect the start and end point of eye blinking. To detect these feature points, we proposed two methods, one that uses the differential wave pattern of the eye blink [14, 15], as described in Sect. 4.2, and another that uses the time when the open-eye area is lower than a threshold value (for example, this threshold is 90 % of the open-eye area when the eye is open completely) [16]. However, if the measuring

Fig. 22 An outline of eye
blink and its new feature
parameter

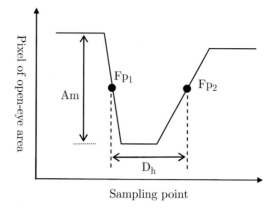

sampling rate decreases, the value of the differential wave pattern of the eye blink changes significantly. Thus, it is difficult for our proposed method to detect the start and end points of an eye blink. In addition, if the wave pattern of the eye blink is measured for a prolonged time, it is difficult for our proposed method to detect the 90 % value of the open-eye area because the base line of the wave pattern of the eye blink changes with a prolonged measurement.

To resolve these points, we proposed a new feature parameter for the eye blink that can be detected easily. In particular, we use the half value of the maximum amplitude of the eye blink wave pattern. This half-value width is defined as the eye blink duration. Such duration is the new feature parameter for the eye blink, and it is shown in Fig. 22.

In Fig. 22, Fp_1 and Fp_2 show the half-value of the maximum amplitude Am. D_h shows the half-value width of the eye blink wave pattern. Feature parameter D_h varies widely among individuals. In addition, we confirmed that there is a large difference between the D_h of voluntary and involuntary blinks. In other words, when we implement the function of voluntary blink detection to the practical eye-gaze input system, calibration is required. To realize a simple calibration method, we use the feature parameter D_h estimated from one voluntary blink. Before using the eye-gaze input system, the user voluntarily blinks one time, and this is used for calibration. D_h is estimated from the wave pattern of this voluntary blink. The maximum amplitudes of voluntary blinks vary narrowly. Therefore, the system can calibrate using one sample voluntary blink.

In order to detect the voluntary blink, we estimate the duration threshold Th_c, which is estimated through (3) as shown in Sect. 5.2. In (3), Tdv_{avg} is the half-value width of the eye blink wave pattern D_h estimated from the calibration, and $Tdiv_{avg}$ is 200 ms (two samples of our eye-gaze input system). The value $Tdiv_{avg}$ is selected empirically. If the duration of an eye blink is greater than Th_c, this eye blink is classified as a voluntary blink. Using this method, a user can determine input using voluntary blinks at lower sampling rates.

Fig. 23 Hardware
configuration of our eye-gaze
and blink input system

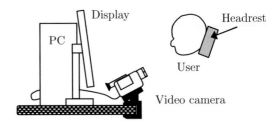

Fig. 23 Hardware
configuration of our eye-gaze
and blink input system

6.2   Practical Eye-Gaze and Blink Input System
and Its Evaluation

Our eye-gaze and blink input system is comprised of a PC for image analysis and
a home video camera for capturing the eye image. The camera, for example, a
DV camera, can be connected easily to the PC with the IEEE1394 interface. The
images captured by the camera can be analyzed in real time using the DirectShow
library from Microsoft. The system detects both horizontal and vertical eye-gaze
positions through a simple image analysis, and does not require special image-
processing units or sensors. Therefore, this system is cost effective and versatile
and it can be easily customized. The system can also be used in natural light, and
therefore, it is suitable for personal use. The hardware configuration of our system is
shown in Fig. 23. As described in Sects. 3–5, eye-gaze and blink characteristics vary
widely among individuals. Therefore, an eye-gaze and blink input system requires
calibration for detect eye-gaze and blink.

Users must calibrate the system for eye-gaze detection before using it for
tasks. After the camera location is adjusted, calibration starts. The indicators for
calibration are displayed on the PC screen, as shown in Fig. 9 and described in
Sect. 3.4. While calibration is performed, users gaze at each indicator when its color
switches to red. In addition, the user gazes at the center indicator (indicator 5) at
the start of calibration. At that time, the PC emits a "beep" sound. To calibrate
for voluntary blink detection, the user blinks once voluntarily upon hearing this
sound. In this calibration for voluntary blink detection, the system utilizes the simple
calibration method described in Sect. 6.1. After calibration, users can control the
mouse cursor through their eye-gaze, and command the signal for mouse clicking
to the PC operating system (Windows) through voluntary blinks.

We evaluated our system by conducting an experiment that involved eight
subjects. In this experiment, the subject selected a circle indicator (diameter: 4 deg)
on the PC screen by gazing, as shown in Fig. 24. After selecting the indicator, the
subject determined the input through voluntary blinks. The location of the indicator
was changed ten times randomly per experiment. The results of the experiment
show that the average time for indicator selection and input determination is 4.77 s
(standard deviation: 1.53 s) [21]. The results also show that all subjects can operate
our proposed system. Thus, we think that QoL for people with severe physical
disabilities, such as ALS patients, can increase with our proposed system.

Fig. 24 Circle indicator for evaluation experiment

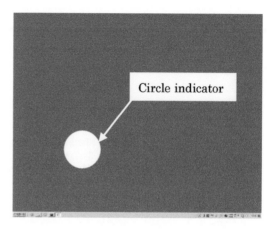

Involuntary blinks were detected in this evaluation experiment. However, input errors by involuntary blink were not detected. In addition that, some subjects could not decide input by voluntary blink in some samples, after selecting indicators by eye-gaze. We think that there are two causes of failed inputs. One of the causes is the error of voluntary blink detection. Another is the measurement error of eye-gaze detection. We will confirm the cause of failed input in future experiments.

The size of the circle indicator is larger than general GUI icons, which is approximately 1 deg. It is difficult for the present method for eye-gaze detection to select GUI icons using eye-gaze because the measuring accuracy of eye-gaze detection is approximately 1 deg, and involuntary eye movements occur with eye fixation. Therefore, users find it difficult to select GUI icons using eye-gaze. In the future, we have to develop a new interface for selecting small indicators, such as GUI icons, using eye-gaze.

7   Conclusion

We developed an eye-gaze input system for people with severe physical disabilities. This system detects the horizontal and vertical eye-gaze components of users under natural light, in addition to incandescent, fluorescent, and LED lights. In addition, we developed a voluntary blink detection method using the eye image. These eye-gaze and blink detection methods can be executed simultaneously. We developed a new input system through eye-gaze and blinks using these two methods. Users can move the mouse cursor and determine input using our new system.

Our proposed eye-gaze and blink input system can operate general application software for Windows without special customization. In addition, we developed special text input software that works through eye-gaze. The function of this software increases QoL for people with severe physical disabilities, such as ALS patients. In future works, we plan to develop a new input interface that uses eye-gaze and blink. Using this new system, uses can operate software with small icons with certainty using eye-gaze and blink.

References

1. T.E. Huchinson, P. White, Jr, W.N. Martin, C. Reichert, L.A. Frey, Human-computer interaction using eye-gaze input. IEEE Trans. Syst. Man Cybern. **19**(7), 1527–1534 (1989)
2. D.J. Ward, D.J.C. MacKay, Fast Hands-free Writing by Gaze Direction. Nature **418**, 838 (2002)
3. J.P. Hansen, K. Torning, A.S. Johansen, K. Itoh, H. Aoki, Gaze typing compared with input by head and hand, in *Proceedings of Eye Tracking Research and Applications Symposium on Eye Tracking Research and Applications*, San Antonio (2004), pp. 131–138
4. F. Corno, L. Farinetti, I. Signorile, A cost-effective solution for eye-gaze assistive technology, in *Proceedings of IEEE International Conference on Multimedia and Expo*, Lausanne, vol. 2 (2002), pp. 433–436
5. K.N. Kim, R.S. Ramakrishna, Vision-based eye-gaze tracking for human computer interface, in *Proceedings of IEEE International Conference on Systems, Man and Cybernetics*, Tokyo, vol. 2 (1999), pp. 324–329
6. J.G. Wang, E. Sung, Study on eye-gaze estimation. IEEE Trans. Syst. Man Cybern. **32**(3), 332–350 (2002)
7. J. Gips, P. DiMattia, F.X. Curran, P. Olivieri, Using EagleEyes - an electrodes based device for controlling the computer with your eyes - to help people with special needs, in *Proceedings of 5th International Conference on Computers Helping People with Special Needs, Part 1* (1996), pp. 77–83
8. K. Abe, M. Ohyama, S. Ohi, Eye-gaze input system with multi-indicators based on image analysis under natural light. J. Inst. Image Inf. Telev. Eng. **58**(11), 1656–1664 (in Japanese) (2004)
9. K. Abe, S. Ohi, M. Ohyama, An eye-gaze input system using information on eye movement history, in *Universal Access in Human-Computer Interaction. Ambient Interaction Lecture Notes in Computer Science*, vol. 4555 (Springer, New York, 2007), pp. 721–729
10. K. Abe, S. Ohi, M. Ohyama, Eye-gaze detection by image analysis under natural light, in *Human-Computer Interaction. Interaction Techniques and Environments*. Lecture Notes in Computer Science, vol. 6762, pp. 19–26 (2011)
11. D.O. Gorodnichy, Second order change detection, and its application to blink-controlled perceptual interfaces, in *Proceedings of the International Association of Science and Technology for Development Conference on Visualization, Imaging and Image Processing*, Benalmadena (2003), pp. 140–145
12. A. Krolak, P. Strumillo, Vision-based eye blink monitoring system for human-computer interfacing, in *Proceedings on Human System Interaction (HIS2008), Krakow* (2008), pp. 994–998
13. I.S. MacKenzie, B. Ashitani, BlinkWrite: efficient text entry using eye blinks. Univ. Access Inf. Soc. **10**, 69–80 (2011)
14. K. Abe, H. Sato, S. Matsuno, S. Ohi, M. Ohyama, Automatic classification of eye blink types using a frame-splitting method, in *Engineering Psychology and Cognitive Ergonomics*. Understanding Human Cognition Lecture Notes in Computer Science, vol. 8019 (2013), pp. 117–124
15. S. Matsuno, M. Ohyama, K. Abe, H. Sato, S. Ohi, Automatic discrimination of voluntary and spontaneous eyeblinks -Use of the blink as a switch interface, in *Proceedings of The Sixth International Conference on Advances in Computer-Human Interactions*, Nice (2013), pp. 433–439

16. K. Abe, H. Sato, S. Ohi, M. Ohyama, Feature parameters of eye blinks when the sampling rate is changed, in *Proceedings of TENCON 2014–2014 IEEE Region 10 Conference*, Bangkok, Thailand (2014), pp. 1–6
17. COGAIN - Communication by Gaze Interaction, Eye Gaze Communication Board. http://wiki.cogain.org/index.php/Eye_Gaze_Communication_Board (2015). Accessed 1 March 2015
18. R.C. Simpson, H.H. Koester, Adaptive one-switch row-column scanning. IEEE Trans. Rehabil. Eng. **7**(4), 464–73 (1999)
19. L. Stark, G. Vossius, L.R. Young, Predictive control of eye tracking movements. IRE Trans. Hum. Factors Electron. **3**, 52–57 (1962)
20. Z. Ramdane-Cherif, A. Nait-Ali, An adaptive algorithm for eye-gaze-tracking-device calibration. IEEE Trans. Instrum. Meas. **57**(4), 716–723 (2008)
21. K. Abe, H. Sato, S. Matsuno, S. Ohi, M. Ohyama, Input interface using eye-gaze and blink information, in *HCI International 2015 Poster Short Papers Proceedings (Part I)*, CCIS528, Los Angeles, vol. 27 (2015), pp. 463–467

The Utility of Facial Analysis Algorithms in Detecting Melancholia

Matthew P. Hyett, Abhinav Dhall, and Gordon B. Parker

Abstract Facial expressions reliably reflect an individual's internal emotional state and form an important part of effective social interaction and communication. In clinical psychiatry, facial affect is routinely assessed, and any identified deviations from normal affective range and reactivity may signal the presence of a potential psychiatric disorder. An example is melancholic depression or 'melancholia' where facial immobility and non-reactivity are viewed as sensitive diagnostic indicators of the illness. However, affect in depressive disorders such as melancholia, and indeed psychiatric conditions more broadly, is largely assessed by clinicians, without biological or computational quantification. While such clinical assessment provides useful qualitative descriptors of illness features, the inherent subjectivity of this approach raises concerns regarding diagnostic reliability, and may hinder communication between clinicians. Methodological advances and algorithm development in the field of affective computing have the potential to overcome such limitations through objective characterization of facial features. Among these methods are implicit face analysis techniques, which are based on local spatio-temporal descriptors such as the space-time interest points and Bag-of-Words framework, and explicit face analysis techniques based on deformable model fitting methods such as Constrained Local Models and Active Appearance Models. In this chapter we overview these approaches and discuss their application toward detection and diagnosis of depressive disorders, in particular their capacity to delineate melancholia from the residual non-melancholic conditions.

M.P. Hyett (✉) • G.B. Parker
School of Psychiatry, University of New South Wales, Sydney, NSW, Australia 2052

Black Dog Institute, Prince of Wales Hospital, Randwick, NSW, Australia 2031
e-mail: m.hyett@unsw.edu.au

A. Dhall
Human-Centred Technology Research Centre, University of Canberra, University Drive, Bruce, Canberra, ACT, Australia 2617

Research School of Computer Science, Australian National University, Acton ACT, Australia 0200

© Springer International Publishing Switzerland 2016
M. Kawulok et al. (eds.), *Advances in Face Detection and Facial Image Analysis*,
DOI 10.1007/978-3-319-25958-1_13

359

1 Introduction

Clinical psychiatry has long utilized so-called mental state signs to assist in the diagnosis of mental disorders. In assessing the presence of a psychiatric disorder, particular attention is given to appearance, behavior, speech, thought, perception, insight into illness and, of key relevance to this chapter, mood and facial affect [1]. The latter contributes to delineating depressive illnesses such as melancholia from non-melancholic depression, and is typically assessed by clinicians (e.g., psychiatrists). However, such clinical observation is, by virtue of its non-standardized nature, subjective and thus may not reliably capture the presence or absence of specific disorders. As more objective means of quantifying facial affect and emotion emerge, driven largely by advances in affective computing, the diagnosis of disorders characterized by specific affective signs—such as melancholia—will be made more valid. Prior to examining the potential utility of these emerging technologies in depressive illness, we broadly overview conceptual models of emotion and depression classification, with an emphasis on the relevance of affect in such disorders.

2 Emotion and Affect

The subjective affective experience of an individual, variably referred to as one's 'passions', 'emotions', 'feelings', or 'moods' have, to this day, largely eluded definition. Over a century ago, James [2] suggested that emotions evoke a certain "phenomenal quality"; referring to the notion that emotions are "*sensation-like* mental states" [3], and were in effect seen as 'intuitions' in response to emotion-eliciting events. Others (e.g., [4]) challenged James's notion that emotions were reflex-like in nature (e.g., fear *in response to* a bear in the woods), and instead positioned emotion as an appraisal-based process (e.g., feeling happy or sad towards an object). Throughout the 1960s, appraisal theories of emotion dominated [5], but even to this day there continues to be an ongoing philosophical debate over what constitutes an emotional state [6]. The fundamental biological and psychological processes underlying the above theories of emotion generation and/or appraisal are referred to as the 'evolutionary core' [3]—that is, a set of discrete emotion mechanisms aligned to one's "basic" emotions.

There have been several dominant theories in the field of psychology that offer insight into such basic emotions. Beginning with McDougall [7], emotion was defined as a small set of adaptive processes, so-called emotional *instincts*, and included the behaviorally well-defined emotions of flight, repulsion, curiosity, pugnacity, self-abasement, self-assertion, and the parental instinct [8]. Most modern variants of discrete emotion theory correspond at least partially to McDougall's model. For instance, Tomkins [9–11], who is credited with the development of "affect theory", considered there to be nine independent affects: two positive (enjoyment/joy and interest/excitement), one neutral (surprise/startle), and six negative affects (anger/rage, disgust, dissmell [similar to distaste], distress/anguish, fear/terror, and shame/humiliation). Likewise, Izard [12] considered fear, anger, shame, contempt, disgust, guilt, distress, interest, surprise, and joy to be primary emotions, and with various combinations of these giving rise to tertiary emotions (e.g., such as affection). While the basic emotions were initially believed to be largely subjective states of mind, Ekman [13] postulates that seven basic emotions correspond near-universally to observable facial expressions, namely anger, disgust, fear, happiness, sadness, surprise, and contempt. Such categorical definitions of emotion have received substantial empirical support, but there is also a question as to whether affect can be conceptualized dimensionally [14]; that is that emotions exist along interrelated continuums of, for example, arousal and valence [15].

It has been suggested that discrete (i.e., categorical) emotion theory provides a more accurate index of the current, momentary experience of emotion, while dimensional perspectives may be most relevant to temporal emotional experience, such as mood states [16]. Rather than being mutually exclusive, however, it is likely that both viewpoints are pertinent to affective states such as depression. The work of Ekman in particular [17], which highlights the importance of emotional expressions, is especially relevant for quantifying affect. Indeed, 'affect display' refers to the externally displayed (i.e., observable) affect of an individual, through facial, vocal or gestural means. When affect is in line with the subjective mood state of an individual, it is termed 'congruent affect', but when subjective states and affect is misaligned it is referred to as 'incongruent affect'. As will become apparent throughout this chapter, affect in depression is typically congruent with the individual's self-reported, subjective mood state; for instance, subjective low mood or sadness is often identifiable in the depressed patient through observation of a flat and/or non-reactive affect.

2.1 What Precisely Is an Affective Disorder?

Historical descriptions of disordered affective states date back over 2000 years. Hippocrates described the existence of "melancholia" as a disease-like state arising from an excess of black bile (this being one of four 'humors', or bodily fluids, that were thought to directly influence health and temperament). Current day formulations of melancholia position it as the prototypical depressive disease, principally of biological and genetic origin [18], which presents with characteristic clinical features such as psychomotor slowing (i.e., physical slowing, concentration impairment), anergia, anhedonia, diurnal mood variation (mood worse in the morning), early morning wakening, and appetite and weight loss. Despite evidence for the existence of melancholia as a distinct condition, it was largely abandoned in psychiatric circles in 1980 with the introduction of the Diagnostic and Statistical Manual of Mental Disorders (DSM), which popularized 'symptom-centric' models of affective illness. Its third edition (DSM-III) [19] brought three psychiatric disorders under the diagnostic umbrella of the 'affective disorders', including major depression (subsuming melancholia), bipolar disorder (depression and/or hypomania or mania) and dysthymia (chronic low mood, defined as having fewer symptoms than major depression). Under the DSM framework, symptom reports by patients guide diagnostic decisions, which in isolation are thought to contribute to misdiagnosis, thus impacting on appropriate management [18]. Hence, while the affective disorders began with melancholia as observable illness states, they are now characterized by the severity of symptoms. It is argued that this has contributed to an era of insipidity in depression research, and is of limited clinical import. Here, we argue that there is scope for departure away from dimensional accounts of depression (e.g., major depression), back to more refined categorical depressive subtypes (e.g., melancholic vs. non-melancholic conditions).

3   The Case for Objectively Informed Classification of Depression Subtypes

The classification of depression has long been contentious, with numerous typologies being proposed since the beginning of the twentieth century. So-called 'simple typologies' conceptualized depression as comprising anywhere between one and five categories. The father of modern psychiatry, Emil Kraepelin, distinguished between *dementia praecox* (now schizophrenia) and manic-depressive insanity [20]. All major affective disturbances, namely mania and depression, were thought by Kraepelin to be part of the same illness (manic-depressive insanity). In 1926, British psychiatrist Edward Mapother at the Maudsley Hospital in London claimed [21] that there was only one form of depression, and did not distinguish between depressive subtypes—and hence saw depression as varying along a continuum. The influential theories of Kraepelin and Mapother laid the foundations for a move toward unitary models of depression. Sir Aubrey Lewis, an eminent psychiatrist at the same institute as Mapother, also viewed all depressions as essentially the same (and hence proposed a single category, "depressive illness"), but conceded that there are likely differences between individuals regarding its causation: specifically depressions that are more hereditary versus those caused by environmental factors [22].

The "separatists" opposed the unitary view of depression from the mid-twentieth century, arguing that depression could be classified into different types [23]. Such categorical views principally saw the diagnosis and classification of depression according to a "binary" model—differentiating those of 'constitutional origin' (i.e., caused from within the individual) versus those that were reactions to environmental stressors. There have been several examples in the literature that support the existence of different depressive subtypes, which is typically achieved by identifying differences between groups on some metric (e.g., clinical or self-report ratings of an illness variable). Kiloh and Garside [24] quantified the independence of neurotic and endogenous (synonymous with melancholic) depressives through analysis of reported symptoms and clinical variables. Features such as psychomotor retardation and concentration difficulties correlated with the diagnosis of endogenous depression, whereas 'reactivity of depression' (by which it is assumed to mean reactivity of affect), irritability and variability of illness were correlated with neurotic depression. Similarly, a 'point of rarity' was identified between endogenous and neurotic depressives in a series of papers from the Newcastle school [25, 26]. Here, clinical ratings of depressive symptoms and signs, along with personality and anxiety, were shown to differentiate endogenous and neurotic depressive subgroups. Again, psychomotor change was more prevalent in the endogenous group than the neurotic group. Despite strong support for the notion that endogenous depression is a categorically distinct entity (one either has it or does not), and that neurotic depression consisted of symptoms varying dimensionally [23], it was abandoned under the DSM system in favor of a dimensional approach to diagnosis.

'Melancholia' was retained in DSM-III (and subsequent iterations of DSM), albeit in much-diluted form, as a 'specifier' diagnosis [19]. The retaining of a melancholic specifier allows for diagnosis of major depression with melancholic features, but has been criticized for its focus on symptom expression (a severity framework), while also explicitly disregarding differing aetiological contributions [18]. In the DSM, melancholia is diagnosed by the presence of additional symptoms of anhedonia (reduced interest or reactivity to previously pleasurable events) and psychomotor slowing, amongst others, which are present in almost all individuals with clinically significant depression [27]. Observable (not just reported) psychomotor disturbance has been proposed as a specific diagnostic marker of melancholia, which aligns with some of the earliest definitions of the disorder, from classical antiquity, where "symptoms ... were not part of the concept" ([6] p. 298). We therefore developed a clinician-rated scale (the CORE) to measure psychomotor disturbances in depressed patients [28]. The tool allows rating of 18 clinical signs (observable features) across the domains of non-interactiveness (including features such as emotional non-reactivity and inattentiveness), retardation (including slowed motor movements and facial immobility), and agitation (including facial apprehension and motor agitation). Clinically diagnosed melancholia (still arguably the "gold standard" in diagnosing the condition) was associated with "substantial" CORE scores. In these studies a score of >8 defined "substantial" psychomotor disturbances [27], with such scores being representative of those with melancholia but not those with non-melancholic depression. Despite the high sensitivity in detecting melancholia, such systems—much like clinical diagnosis—require extensive training and exposure to appropriate clinical populations (i.e., in services where those with melancholia are likely to present) to be of any benefit. Several investigators, including our own research team, have thus sought to clarify biological correlates of melancholia in the hope that any identified perturbations will eventually assist in its diagnosis.

Throughout the late 1970s and early 1980s there were many investigations into disturbances of hypothalamic-pituitary-adrenal (HPA) axis function in depressive illness [29]. The HPA axis plays a central role in regulating homeostasis of many physical systems, including the metabolic, reproductive, immune, and central nervous systems [30]. Insights into the function of this system in depression has been achieved by challenging patients with the synthetic corticosteroid, dexamethasone (DEX), and then observing fluctuations in plasma cortisol—referred to as the DEX suppression test or DST. It was seen as a watershed for psychiatry when Carroll and colleagues [31] reported that 48 % of those with primary depression were DEX 'non-suppressors', compared to only 2 % of other psychiatric patients. Carroll et al. [32] subsequently demonstrated that the DST had utility in detecting melancholic depression, with sensitivity of 67 % and specificity of 96 %. This measure hence appeared to be highly informative in depressive subtyping (i.e., nearly all non-melancholic patients were correctly classified as not having melancholia). Despite such promising early findings, the DST lost favor as a diagnostic tool in the mid-1980s after the DSM-III was introduced—given it lacked sensitivity in later studies using broader diagnostic criteria [33]. Since then, disruptions across other cognitive and biological systems have been identified with specificity to the melancholic phenotype, and include working memory impairments and disturbances in sleep architecture [34]. Our team also recently identified a neurobiological signature for melancholia [35]—which was not observed in non-melancholic depression—involving disrupted integration of brain regions supporting interoception and attention.

While such research has been of key importance in understanding the underlying causes of melancholia, their use as diagnostic tools will continue to be limited given their invasiveness, relatively high cost, and difficulty of access (e.g., in rural areas). Facial imaging has the potential to overcome these barriers and become an important tool in the diagnosis of melancholia. In the following sections we overview methodological advances in facial imaging research, and highlight the utility of the methods in contributing to an objectively-informed diagnostic tool for depressive disorders.

4 Methodological Considerations for Quantifying Facial Affect in Depression

Significant advances have been made for inferring affect using facial imaging over the past two decades [36–40]. In this section we discuss the general methodological approaches of affect recognition based on facial analysis. A typical facial analysis system has the following main components: face and fiducial points (facial parts location) detection, feature extraction, and classification. Figure 1 depicts the main components of such a system. Once the image has been captured, face detection algorithms, such as the popular Viola-Jones (VJ) detector [41], are used to locate the face. Next, fiducial points are inferred using parametric models such as the widely used Active Appearance Models (AAM) [42]. Once the location of facial parts such as the eyes and mouth are known, facial features are computed. Features can be extracted either on a holistic level or on individual parts of the face. The

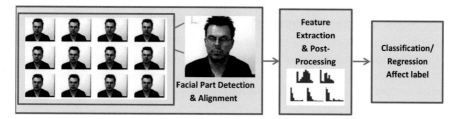

Fig. 1 A typical FAR system

individual features computed in the latter case are concatenated to construct the final feature vector. Further, dimensionality reduction methods such as Principal Component Analysis (PCA) can be applied to the feature vector. This provides a compact and less noisy feature representation capable of higher discrimination (e.g., between features). The classifier then categorizes a given face into differing affective states (such as depressed/non-depressed, various emotion types, and continuous valence/arousal labels etc). The components of a facial affect recognition (FAR) system are discussed in detail in the following sections.

4.1 Face and Fiducial Points Detection

One of the most widely used face detectors is the classic VJ face detector [41]. It is based on a cascade-boosting framework [43] in which haar-like features are extracted using an integral image. The use of an integral image leads to near real-time execution of the face detection method. The cascade classifier scans an image using a sub-window at different scales and localizes the regions that are labeled as faces by the classifier. The Adaboost algorithm is also applied for selecting discriminative haar-features during the training phase of the classifier. The cascade classifier consists of various weak classifiers, which together act as a strong classifier. The use of weak classifiers early on allows fast rejection of patches that do not resemble faces. The open source computer vision library, OpenCV, contains an implementation of the VJ object detector. For multi-view face detection, multiple models are learned for different head poses [44]. These select features using forward feature selection before training the cascade classifier.

There are several other facial detection methods, including energy-based models which infer the face location and head pose simultaneously [45], and vector boosting-based methods where a tree representation is proposed for dividing the face space into smaller subspaces [46]. The power of non-rigid deformable models stems from the low-dimensional representation of the shape and texture of a face they provide. One of the earliest deformable model methods is the Active Contour Model [47]. The Active Shape Model (ASM) algorithm [48] models the shape of an object and has been used extensively in face tracking. During the training process, the landmark points of all input samples are aligned into a common co-ordinate frame using Procrustes analysis. Following this, the model is computed by applying PCA over the shapes. The shape of the model can be controlled/changed via parameters of deformation. Then, fitting of the model can be performed on a new image using an iterative method, which calculates the best match for the model boundary and hence decides the new location for the model points.

4.2   Active Appearance Models

The Active Appearance Models (AAM) are an extension of the ASM. However, they not only model the shape but also consider the grey-level appearance. During the training process the grey-level appearance is modeled by warping each training sample image using a triangulation algorithm that aligns it to the mean shape. The grey-level information is then sampled and PCA is applied to the samples. Hence, this model can then be fitted to a new image using an optimization algorithm that uses the difference between intensities of the learnt model and the reference image. There are a number of AAM fitting algorithms that can be broadly classified into two classes: generative and discriminative fitting. In generative fitting (*Fixed Jacobian* [42], *Project Out Inverse Compositional* [49], *Simultaneous Inverse Compositional* [50]), minimization/maximization of some measure of fitness between the model's texture and warped image region is applied to the image. In discriminative fitting (*Iterative Error Bound Minimization* [51], *Haar-like Feature Based Iterative Discriminative Method* [52]) a relationship is learned between the features and the parameters, by using the features extracted from parameter settings, which are perturbed from their optimal setting in each image. The disadvantage of AAM is their limited generalizability to unobserved subjects. AAM's can be classified as subject-dependent or subject-independent. Subject-dependent AAM is best suited to scenarios when the train and test images have the same subjects. Subject-independent modeling is for situations where the subjects in the train and test set are different. In one of the earliest works in automatic depression analysis, McIntyre et al. [53] and Cohn et al. [54] used subject-dependent AAM for facial part detection. In such studies, the facial points were used as feature descriptors for learning a classifier.

Constrained Local Models (CLM) [55] are an extension of the AAM algorithms. The texture is divided into blocks. This helps in generalization and better subject-independent performance. Subject-dependent AAM methods perform better than subject-independent CLM [56]. However, the current state-of-art descriptors compensate for small errors introduced by subject-independent CLM [56]. Another limitation of both AAM and CLM is their requirement of large volumes of labeled data representing different scenarios, such as illumination, pose and expression during training. However, despite the advantages, labeling fiducial points is a manually laborious and erroneous task.

4.3 Pictorial Structure

In the Pictorial Structure (PS) framework [57], an object is represented as a graph with n vertices $V = \{v_1, \ldots, v_n\}$ for the parts and a set of edges E, where each $(v_i, v_j) \in E$ pair encodes the spatial relationship between parts i and j. For a given image I, PS learns two models. The first one learns the evidence of each part as an *appearance model*, where each part is parameterized by its location (x, y), orientation θ, scale s, and foreshortening. All of these parameters (together referred to as D) are learned from exemplars and produce a likelihood model for I. The second model learns the kinematic constraints between each pair of parts in a prior *configuration model*. L is the parts configuration. Given the two models, the posterior distribution over the whole set of part locations is:

$$p\left(L \big| I, D\right) \alpha \, p(I \big| L, D)$$

(1)

where $p(I|L, D)$ measures the likelihood of representing I in a particular configuration and $p(L|D)$ is the kinematic prior configuration. A major problem of this framework is the low contribution of the occluded parts, resulting in either erroneous or missed detection of these parts, leading to inaccurate pose estimation. Everingham and colleagues [58] proposed a PS based fiducial point detector, which is initialized using the VJ face detector. Recently, Zhu and Ramanan [59] proposed an extension to the PS framework by adding mixtures representing different face poses. This latter study performed face and fiducial point detection and head pose inference in the one framework. The face detector performs better than the VJ face detector. The disadvantage of the PS based method as used by Everingham et al. [58] is that it requires initialization from a face detector like VJ. However, Zhu and Ramanan [59] overcome this limitation by using multiple pose as mixture detectors.

Selecting the appropriate face and fiducial point detector is problem driven. For example, in the case of affect analysis, it is desirable that the system should generalize over subjects. Joshi et al. [60] used the PS framework of Everingham and colleagues [58] for fiducial points detection. Even though the Mixture of Pictorial Structures (MoPS) framework performs better than the PS method, our own research group [60] prefer the use of PS. We argue that the inference time for MoPS is substantially longer than that obtained in the Everingham et al. [58] study, which matters when analyzing long duration depression video clips. Furthermore, such work can utilize spatio-temporal descriptors (Local Binary Pattern in Three Orthogonal Planes (LBP-TOP) [61]) that compensate for algorithm error. On the other hand, applications such as facial performance transfer [62] require accurate fiducial points. Asthana and colleagues [62] use subject dependent AAM models as they are more accurate as compared to CLM and subject-independent AAM.

4.4   Facial Descriptors

Once the face and facial parts location is identified and aligned, feature descriptors are computed for extracting information for learning classifiers. FAR techniques can be segregated on the basis of the type of descriptors used, broadly classified as either geometric- or appearance-based features. Geometric features [63–65] correspond to facial points and to the location of different facial parts. Appearance features generally correspond to face texture information [61, 66, 67]. Furthermore, facial feature descriptors can be divided on the basis of temporal information (i.e., spatio-temporal descriptors [61] and frame-based descriptors [67]). A popular method for modeling geometric features is based on Facial Animation Parameters (FAP), defined in the MPEG-4 video-coding standard. Lavagetto and Pockaj [68] present a method for synthesizing facial animations using FAP and Facial Definition Parameters (FDP). Similarly, Sebe and colleagues [69] use the Piecewise Beizier Volume Deformation (PBVD) tracker for tracking facial parts. The motion information between two consecutive frames is measured using template matching.

Various classifiers can also be compared, allowing insight to be gained as to their utility in depression classification. Asthana et al. [70] compute geometric features by fitting AAM models on input faces. They compared various AAM fitting techniques and experimented on a Cohn-Kanabe (CK+) database. One of the limitations of this work is that it required manual initialization of facial parts. Dhall and colleagues [63] propose the use of the geometric descriptor algorithm, "Emotion Image" (EI). This feature constructs a visual map based on an undirected map derived from a facial points detector. EIs of two faces (images) is compared using the Structural Similarity Index Metric (SSIM) of Wang et al. [71], allowing computation of their similarity, and is applied to the problem of expression based album creation. The discriminative ability of EI is dependent on fiducial point detection quality, which may introduce some errors when the fiducial points detection is not very accurate. Valstar and Pantic [72] showed that the performance of geometric features is similar to that of the appearance features. However, the limitation of geometric features comes from their dependence on accurate facial parts location information. Facial parts detection is relatively accurate on lab-controlled scenario data; however, it is still an open problem for images in real-world conditions. If there is an error in the facial parts detection, the error generally propagates in the geometric feature representation. Chew et al. [56] argue that appearance descriptors are able to compensate error produced by facial parts detectors to some extent. Popular appearance descriptors are described below.

4.4.1   Local Binary Patterns (LBP)

The LBP family of descriptors has been extensively used in computer vision for texture and face analysis [59, 73, 74]. The LBP descriptor assigns binary labels to pixels by thresholding the neighborhood pixels with the central value. Therefore, for a center pixel p of an image I and its neighboring pixels N_i, a decimal value d is assigned to it:

$$d = \sum_{i=1}^{k} 2^{i-1} I(p, Ni) \tag{2}$$

$$\text{where } I(p, Ni) = \begin{cases} 1 \text{ if } c < N_i \\ 0 \text{ otherwise} \end{cases}$$

4.4.2 Local Binary Pattern-Three Orthogonal Planes (LBP-TOP)

Local Binary Pattern-Three Orthogonal Planes (LBP-TOP) [61] is a popular descriptor in computer vision. It considers patterns in three orthogonal planes: *XY*, *XT* and *YT*, and concatenates the pattern co-occurrences in these three directions. The local binary pattern (LBP-TOP) descriptor assigns binary labels to pixels by thresholding the neighborhood pixels with the central value. Therefore, for a center O_p of an orthogonal plane O and its neighboring pixels N_i, a decimal value d is assigned to it:

$$d = \sum_{O}^{XY,XT,YT} \sum_{p} \sum_{i=1}^{k} 2^{i-1} I(O_p, N_i) \tag{3}$$

Joshi et al. [75] proposed a LBP-TOP based framework for analyzing depression data. The video clips were divided into temporal slices and LBP-TOP was computed on each time slice. Temporal slicing helps in encoding spatio-temporal changes. Further, a Bag-of-Words (BoW) representation was learnt with LBP-TOP from each temporal slice from an interview-based video (a 'document'). BoW-based representations come from the domain of document processing. A BoW feature represents a document (image/video) as an unordered set of frequencies of words. Li and colleagues [76] were the first to use a BoW for FAR—they fused PHOG- and BoW-based histograms constructed from a dictionary based on Scale Invariant Feature Transform (SIFT). Even though BoW-based vectors can represent the frequency of different stages of an expression, the temporal sequencing information is still missing. To overcome this problem, a data-driven technique was recently proposed to explicitly encode the temporal information using n-grams. Bettadapura et al. [77] performed experiments on human action recognition and activity analysis and showed that adding temporal sequencing information based on their method increases the accuracy of the BoW-based techniques.

The facial descriptor representation modeled on computing features based on the output of the facial parts detection module can be referred to as explicit modeling of affect. Here, an explicit model (face model) is used to localize the facial parts. In a different approach (implicit modeling), Joshi et al. [60] proposed the computation of Space Time Interest Points (STIP) [78] on the upper body of a subject in the video frame in depression. STIP are widely used in the computer vision community

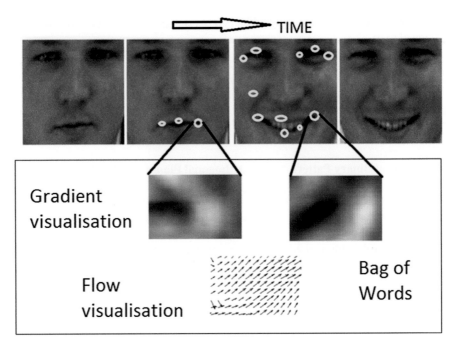

Fig. 2 The figure describes the STIP computation on a video from FEEDTUM database [42]. The blocks represent the gradient information [41] around the generated interest points. The HOG and HOF features represent the local spatio-temporal movements

for human action recognition. STIP are salient spatio-temporal locations in a video, where a change has occurred. These are based on a 3D extension of the 2D Harris interest point detector. Once an interest point is detected, Histogram of Gradients (HOG) and Histogram of Flow (HOF) is computed around the interest point. Joshi et al. [75] computed STIP on the upper body and compared the performance with face area only STIP in depressed subjects. Further, a BOW is learnt on HOG and HOF features generated around the interest points. Figure 2 describes the STIP computation on a sample from the FEEDTUM database [79].

Note that the yellow ellipses around the eyes and mouth are the interest points generated. The regions of interest around the interest points are scaled up, and gradient information is shown (as demonstrated previously [80]). It is easy to notice that the two gradient blocks around the same region of interest (i.e., right tip of the mouth) show different gradients due to changes in facial expression. The flow information here describes the motion change around the interest point on the right tip point of the mouth. Hence, STIP captures local movements on a holistic level (full-face), and information is captured implicitly without having to use any output from a facial parts localization method.

4.5 Objective Identification of Melancholia

Many of the above methods have recently been used to facilitate better identification of those with depression. Williamson and colleagues [81] showed that depression was associated with changes in the coordination and movement of facial gestures using facial action units. When used in multivariate modeling, such facial features were predictive of self-reported depressive symptomatology, highlighting the utility of the approach in quantifying the depressive syndrome. There is also evidence to suggest that facial landmark fitting (66 points to the face on each frame) allows robust prediction of affective dimensions (i.e., valence/arousal) and global depression state (measured through self-rated depression scores) [82]. In addition, the predictive capacity of facial imaging in detecting depression appears to increase with the addition of other 'affective computing' data modalities. Pérez-Espinosa and colleagues [83] demonstrated that fusing affective dimensions and audiovisual features (facial imaging) allowed for accurate construction of depression recognition models. Whilst no studies have directly examined the performance of the methods above in classifying different types of depression such as melancholia, we propose that they will likely be of significant benefit in future studies. Indeed, our own research team recently completed recruitment for a large facial imaging study of those with melancholic depression, those with a non-melancholic depression, and healthy controls, to determine whether the above methods could be used to accurately classify these groups. Preliminary analyses using data from this study have been completed [84], and analyses are now underway to determine whether melancholia can be detected on the basis of its unique, and quantifiable, facial features.

5 Summary and Future Trends

In this chapter we have reviewed the role of facial imaging technologies in detecting affective states, and specifically the potential of emerging methods in classifying depressive disorders. Methodological advances and continuing algorithm development in the field of affective computing have the potential to offer unique insights regards depression detection. Objective characterization of facial expressions with specificity to melancholia will be the next step in determining the overall clinical utility of these methods. Based on the literature to date, however, facial imaging technologies are well positioned to contribute to the assessment and monitoring of depressive disorders.

References

1. S. Bloch, B.S. Singh, *Foundations of Clinical Psychiatry*, 2nd edn. (Melbourne University Publishing, Melbourne, 2001)
2. W. James, What is an emotion? Mind **9**, 188–205 (1894)
3. R. Reisenzein, A short history of psychological perspectives on emotion, in *The Oxford Handbook of Affective Computing*, ed. by R. Calvo, S. D'Mello, J. Gratch, A. Kappas (Oxford University Press, Oxford, 2015)
4. M.B. Arnold, *Emotion and Personality* (Columbia University Press, New York, 1960)
5. R.S. Lazarus, *Psychological Stress and the Coping Process* (McGraw-Hill, New York, 1966)
6. G.E. Berrios, *The History of Mental Symptoms: Descriptive Psychopathology Since the Nineteenth Century* (Cambridge University Press, Cambridge, UK, 1996)
7. W. McDougall, *An Introduction to Social Psychology* (Methuen, London, 1908/1960)
8. K.L. Davis, J. Panksepp, The brain's emotional foundations of human personality and the affective neuroscience personality scales. Neurosci. Biobehav. Rev. **35**, 1946–1958 (2011)
9. S.S. Tomkins, *Affect Imagery Consciousness: Volume I: The Positive Affects* (Springer Publishing Company, New York, 1962)
10. S.S. Tomkins, *Affect Imagery Consciousness: Volume II: The Negative Affects* (Springer Publishing Company, New York, 1963)
11. S.S. Tomkins, *Affect Imagery Consciousness: Volume III: Anger and Fear* (Springer Publishing Company, New York, 1991)
12. C.E. Izard, *The Face of Emotion* (Appleton-Century Crofts, New York, 1971)
13. P. Ekman, Facial expression and emotion. Am. Psychol. **48**, 384 (1993)
14. C.E. Izard, Basic emotions, natural kinds, emotion schemas, and a new paradigm. Perspect. Psychol. Sci. **2**, 260–280 (2007)
15. J.A. Russell, A circumplex model of affect. J. Pers. Soc. Psychol. **39**, 1161–1178 (1980)
16. D. Keltner, P. Ekman, Facial expressions of emotions, in *Handbook of Emotions*, ed. by M. Lewis, J. Haviland-Jones, 2nd edn. (Guildford Press, New York, 2000)
17. P. Ekman, W.V. Friesen, *The Facial Action Coding System: A Technique for the Measurement of Facial Movement* (Consulting Psychologists, Palo Alto, 1978)
18. G. Parker, Classifying depression: should paradigms lost be regained? Am. J. Psychiatry **157**, 1195–1203 (2000)
19. APA, *Diagnostic and Statistical Manual of Mental Disorders (DSM-III)*, 3rd edn. (American Psychiatric Association, Washington, DC, 1980)
20. E. Kraepelin, *Manic-Depressive Insanity and Paranoia* (E. & S. Livingstone, Edinburgh, 1921)
21. E. Mapother, Discussion on manic-depressive psychosis. Br. Med. J. **ii**, 872–879 (1926)
22. A. Lewis, Melancholia: a clinical survey of depressive states. J. Ment. Sci. **80**, 277–378 (1934)
23. G. Parker, *The Bonds of Depression* (Angus & Robertson Publishers, Sydney, 1978)
24. L.G. Kiloh, R.F. Garside, The independence of neurotic depression and endogenous depression. Br. J. Psychiatry **109**, 451–463 (1963)
25. M.W.P. Carney, M. Roth, R.F. Garside, The diagnosis of depressive syndromes and the prediction of ECT response. Br. J. Psychiatry **111**, 659–674 (1965)
26. C. Gurney, M. Roth, R.F. Garside, T.A. Kerr, K. Schapira, Studies in the classification of affective disorders. The relationship between anxiety states and depressive illnesses. II. Br. J. Psychiatry **121**, 162–166 (1972)
27. G. Parker, P.D. Hadzi, *Melancholia: A Disorder of Movement and Mood* (Cambridge University Press, New York, 1996)
28. G. Parker, D. Hadzi-Pavlovic, K. Wilhelm et al., Defining melancholia: properties of a refined sign-based measure. Br. J. Psychiatry **164**, 316–326 (1994)
29. C.M. Pariante, S.L. Lightman, The HPA axis in major depression: classical theories and new developments. Trends Neurosci. **31**, 464–468 (2008)
30. J.P. Herman, W.E. Cullinan, Neurocircuitry of stress: central control of the hypothalamo-pituitary-adrenocortical axis. Trends Neurosci. **20**, 78–84 (1997)

31. B.J. Carroll, G.C. Curtis, J. Mendels, Neuroendocrine regulation in depression. II. Discrimination of depressed from nondepressed patients. Arch. Gen. Psychiatry **33**, 1051–1058 (1976)
32. B.J. Carroll, M. Feinberg, J.F. Greden et al., A specific laboratory test for the diagnosis of melancholia. Standardization, validation, and clinical utility. Arch. Gen. Psychiatry **38**, 15–22 (1981)
33. E. Shorter, M. Fink, *Endocrine Psychiatry: Solving the Riddle of Melancholia* (Oxford University Press, New York, 2010)
34. G. Parker, M. Fink, E. Shorter et al., Issues for DSM-5: whither melancholia? The case for its classification as a distinct mood disorder. Am. J. Psychiatry **167**, 745–747 (2010)
35. M.P. Hyett, M.J. Breakspear, K.J. Friston, C.C. Guo, G.B. Parker, Disrupted effective connectivity of cortical systems supporting attention and interoception in melancholia. JAMA Psychiatry **72**, 350–358 (2015)
36. I. Cohen, N. Sebe, L. Chen, A. Garg, T.S. Huang, Facial expression recognition from video sequences: temporal and static modelling. Comput. Vis. Image Underst. **91**, 160–187 (2003)
37. B. Fasel, J. Luettin, Automatic facial expression analysis: a survey. Pattern Recogn. **36**, 259–275 (2003)
38. M. Pantic, L.J.M. Rothkrantz, Facial action recognition for facial expression analysis from static face images. IEEE Trans. Syst. Man Cybern. B Cybern. **34**, 1449–1461 (2004)
39. E. Sariyanidi, H. Gunes, A. Cavallaro, Automatic analysis of facial affect: a survey of registration, representation and recognition. IEEE Trans. Pattern Anal. Mach. Intell. **37**, 39–58 (2015)
40. Z. Zeng, M. Pantic, G.I. Roisman, T.S. Huang, A survey of affect recognition methods: audio, visual, and spontaneous expressions. IEEE Trans. Pattern Anal. Mach. Intell. **31**, 39–58 (2009)
41. P.A. Viola, M.J. Jones, Rapid object detection using a boosted cascade of simple features, in *Proceedings of the IEEE Conference on Computer Vision and Pattern Recognition (CVPR)* (IEEE, 2001) pp. 511–518
42. T.F. Cootes, G.J. Edwards, C.J. Taylor, Active appearance models, in *Proceedings of the European Conference on Computer Vision (ECCV)* (Springer,1998), pp. 484–498
43. Y. Freund, R.E. Schapire, A decision-theoretic generalization of on-line learning and an application to boosting, in *Computational Learning Theory*, 23–27, Springer (Springer, 1995), pp. 23–27
44. M. Jones, P. Viola, Fast multi-view face detection, in *Mitsubishi Electric Research Lab TR-20003-96*, vol. 3 (MERL, 2003), p. 14
45. M. Osadchy, Y.L. Cun, M.L. Miller, Synergistic face detection and pose estimation with energy-based models. J. Mach. Learn. Res. **8**, 1197–1215 (2007)
46. C. Huang, H. Ai, Y. Li, S. Lao, Vector boosting for rotation invariant multi-view face detection, in *Proceedings of the IEEE International Conference on Computer Vision (ICCV)* 446–453. (IEEE, 2005), pp. 446–453
47. M. Kass, A. Witkin, D. Terzopoulos, Snakes: active contour models. Int. J. Comput. Vis. **1**, 321–331 (1988)
48. T.F. Cootes, C.J. Taylor, D.H. Cooper, J. Graham, Active shape models-their training and application. Comput. Vis. Image Underst. **61**, 38–59 (1995)
49. S. Baker, I. Matthews, Equivalence and efficiency of image alignment algorithms, in *Proceedings of the IEEE Conference on Computer Vision and Pattern Recognition (CVPR)* (IEEE, 2001), pp. 1090–1097
50. S. Baker, R. Gross, I. Matthews, *Lucas-Kanade 20 Years on: A Unifying Framework: Part 3* (Carnegie Mellon University, RI, USA, 2003)
51. J. Saragih, R. Göcke, Iterative error bound minimisation for AAM alignment, in *Proceedings of the International Conference on Pattern Recognition (ICPR)* (IEEE, 2006), pp. 1192–1195
52. J. Saragih, R. Göcke, A Nonlinear discriminative approach to AAM fitting, in *Proceedings of the IEEE International Conference on Computer Vision (ICCV)* (IEEE, 2007), pp. 1–8
53. G. McIntyre, R. Goecke, M. Hyett, M. Green, M. Breakspear, An approach for automatically measuring facial activity in depressed subjects, in *Proceedings of Affective Computing and Intelligent Interaction (ACII)* (Springer, 2009), pp. 1–8

54. J.F. Cohn, T.S. Kreuz, I. Matthews, et al. Detecting depression from facial actions and vocal prosody, in *Proceedings of Affective Computing and Intelligent Interaction (ACII)*, (Springer, 2009), pp. 1–7
55. J.M. Saragih, S. Lucey, J. Cohn, Face alignment through subspace constrained mean-shifts, in *Proceedings of the IEEE International Conference of Computer Vision (ICCV)*, September (IEEE, 2009), pp. 1034–1041
56. S.W. Chew, P. Lucey, S. Lucey et al., In the pursuit of effective affective computing: the relationship between features and registration. IEEE Trans. Syst. Man Cybern. B Cybern. **42**, 1006–1016 (2012)
57. P.F. Felzenszwalb, D.P. Huttenlocher, Pictorial structures for object recognition. Int. J. Comput. Vis. **61**, 55–79 (2005)
58. M. Everingham, J. Sivic, A. Zisserman, Hello! My name is: Buffy" – automatic naming of characters in TV Video, in *Proceedings of the British Machine and Vision Conference (BMVC)* (Springer, 2006), pp. 899–908
59. X. Zhu, D. Ramanan, Face detection, pose estimation, and landmark localization in the wild, in *Proceedings of the IEEE Conference on Computer Vision and Pattern Recognition (CVPR)* (IEEE, 2012), pp. 2879–2886
60. J. Joshi, A. Dhall, R. Goecke, M. Breakspear, G. Parker, Neural-net classification for spatio-temporal descriptor based depression analysis, in *Proceedings of the International Conference on Pattern Recognition (ICPR)* (IAPR, 2012), pp. 2634–2638
61. G. Zhao, M. Pietikainen, Dynamic texture recognition using local binary patterns with an application to facial expressions. IEEE Trans. Pattern Anal. Mach. Intell. **29**, 915–928 (2007)
62. A. Asthana, M. de la Hunty, A. Dhall, R. Goecke, Facial performance transfer via deformable models and parametric correspondence. *IEEE Trans. Vis. Comput. Graph.* **18**, 1511–1519 (2012)
63. A. Dhall, A. Asthana, R. Goecke, Facial expression based automatic album creation, in *Proceedings of the Neural Information Processing. Models and Applications (ICONIP)*, (Springer, 2010), pp. 485–492
64. M. Pantic, L. Rothkrantz, Expert system for automatic analysis of facial expression. Image Vision Comput. J. **18**, 881–905 (2000)
65. L. Zhang, D. Tjondronegoro, V. Chandran, Evaluation of texture and geometry for dimensional facial expression recognition, in *Proceedings of the International Conference on Digital Image Computing Techniques and Applications (DICTA)* (IEEE, 2011), pp. 620–626
66. A. Dhall, A. Asthana, R. Goecke, T. Gedeon, Emotion recognition using PHOG and LPQ features, in *Proceedings of the IEEE Conference Automatic Faces & Gesture Recognition workshop FERA* (IEEE, 2011), pp. 878–883
67. K. Sikka, T. Wu, J. Susskind, M. Bartlett, Exploring bag of words architectures in the facial expression domain, in *Proceedings of the European Conference on Computer Vision and Workshops (ECCVW)* (Springer, 2012), pp. 250–259
68. F. Lavagetto, R. Pockaj, The facial animation engine: towards a high-level interface for the design of MPEG-4 compliant animated faces. IEEE Trans. Circuits Syst. Video Technol. **9**, 277–289 (1999)
69. N. Sebe, I. Cohen, T. Gevers, T.S. Huang, Emotion recognition based on joint visual and audio cues, in *Proceedings of the International Conference on Pattern Recognition (ICPR)*, 2006
70. A. Asthana, J. Saragih, M. Wagner, R. Goecke, Evaluating AAM fitting methods for facial expression recognition, in *Proceedings of the IEEE International Conference on Affective Computing and Intelligent Interaction (ACII)* (Springer, 2009), pp. 598–605
71. Z. Wang, A.C. Bovik, H.R. Sheikh, S. Member, E.P. Simoncelli, S. Member, Image quality assessment: from error visibility to structural similarity. IEEE Trans. Image Process. **13**, 600–612 (2004)
72. M. Valstar, M. Pantic, Fully automatic facial action unit detection and temporal analysis, in *Proceedings of the International Conference on Computer Vision and Pattern Recognition Workshop (CVPRW)*, (IEEE, 2006), pp. 149–149
73. T. Ojala, M. Pietikinen, T. Menp, Multiresolution gray-scale and rotation invariant texture classification with local binary patterns. IEEE Trans. Pattern Anal. Mach. Intell. **24**, 971–987 (2002)

74. V. Ojansivu, J. Heikkilä, Blur insensitive texture classification using local phase quantization, in *Proceedings of the Image and Signal Processing (ICISP)* (Springer, 2008), pp. 236–243
75. J. Joshi, R. Goecke, M. Breakspear, G. Parker, Can body expressions contribute to automatic depression analysis? in *Proceedings of the International Conference on Automatic Face and Gesture Recognition (FG)* (IEEE, 2013), pp. 1–7
76. Z. Li, J.-I. Imai, M. Kaneko, Facial-component-based bag of words and PHOG descriptor for facial expression recognition, in *Proceedings of the IEEE International Conference on Systems, Man and Cybernetics (SMC)* (IEEE, 2009), pp. 1353–1358
77. V. Bettadapura, G. Schindler, T. Plötz, I. Essa, Augmenting bag-of-words: data-driven discovery of temporal and structural information for activity recognition, in *Proceedings of the IEEE Conference on Computer Vision and Pattern Recognition (CVPR)* (IEEE, 2013), pp. 2619–2626
78. I. Laptev, T. Lindeberg, Space-time interest points, in *International Conference on Computer Vision (ICCV)* (IEEE, 2003), pp. 432–439
79. F. Wallhoff, *Facial Expressions and Emotion Database*, Technical Report (2006)
80. C. Vondrick, A. Khosla, T. Malisiewicz, A. Torralba, HOGgles: Visualizing object detection features, in *Proceedings of the IEEE International Conference on Computer Vision (ICCV)* (IEEE, 2013), pp. 1–8
81. J.R. Williamson, T.F. Quatieri, B.S. Helfer, G. Ciccarelli, D.D. Mehta, Vocal and facial biomarkers of depression based on motor incoordination and timing, in *Proceedings of the 4th International Workshop on Audio/Visual Emotion Challenge (AVEC)* (ACM, 2014), pp. 65–72
82. R. Gupta, N. Malandrakis, B. Xiao, et al. Multimodal prediction of affective dimensions and depression in human-computer interactions, in *Proceedings of the 4th International Workshop on Audio/Visual Emotion Challenge (AVEC)* (ACM, 2014), pp. 33–40
83. H. Pérez-Espinosa, H.J. Escalante, L. Villaseñor-Pineda, M. Montes-y-Gómez, D. Pinto Avedaño, V. Reyez-Meza, Fusing affective dimensions and audio-visual features from segmented video for depression recognition, in *Proceedings of the 4th International Workshop on Audio/Visual Emotion Challenge* (ACM, 2014), pp. 49–55
84. J. Joshi, R. Goecke, S. Alghowinem et al., Multimodal assistive technologies for depression diagnosis and monitoring. J. Multimodel User Interfaces **7**, 217–228 (2013)

Visual Speech Feature Representations: Recent Advances

Chao Sui, Mohammed Bennamoun, and Roberto Togneri

Abstract Exploiting the relevant speech information that is embedded in facial images has been a significant research topic in recent years, because it has provided complementary information to acoustic signals for a wide range of automatic speech recognition (ASR) tasks. Visual information is particularly important in many real applications where acoustic signals are corrupted by environmental noises. This chapter reviews the most recent advances in feature extraction and representation for Visual Speech Recognition (VSR). In comparison with other surveys published in the past decade, this chapter presents a more up-to-date survey and highlights the strengths of two newly developed approaches (i.e., graph-based learning and deep learning) for VSR. In particular, we summarise the methods of using these two techniques to overcome one of the most challenging difficulties in this area-that is, how to automatically learn good visual feature representations from facial images to replace the widely used handcrafted features. This chapter concludes by discussing potential visual feature representation solutions that may overcome the remaining challenges in this domain.

1 Introduction

Given that speech is widely acknowledged to be one of the most effective means of communication between humans, researchers in the automatic speech recognition (ASR) community have made great efforts to provide users with a natural way to communicate using intelligent devices. This is particularly important for disabled people, who may be incapable of using a keyboard, mouse or joystick. As a result of the great achievements made by the ASR community in recent years in terms of the

C. Sui (✉) • M. Bennamoun
School of Computer Science and Software Engineering, University of
Western Australia, Perth, WA, Australia
e-mail: chao.sui@uwa.edu.au; mohammed.bennamoun@uwa.edu.au

R. Togneri
School of Electrical, Electronic and Computer Engineering, University of Western
Australia, Perth, WA, Australia
e-mail: roberto.togneri@uwa.edu.au

© Springer International Publishing Switzerland 2016   377
M. Kawulok et al. (eds.), *Advances in Face Detection and Facial Image Analysis*,
DOI 10.1007/978-3-319-25958-1_14

Fig. 1 Possible application scenarios of VSR. In an acoustically noisy environment, using an intelligent handset to capture and extract visual features is an effective solution for ASR

application of novel techniques such as deep learning [26], people generally believe that we are getting closer to talking naturally and freely to our computers[12].

Although a number of ASR systems have been commercialised and have entered our daily lives (e.g., Apples Siri and Microsofts Cortana), several limitations still exist in this area. One major limitation is that ASR systems are still prone to environmental noises, thereby limiting their applications. Given ASR's vulnerability, research in the area of Visual Speech Recognition (VSR) has emerged to provide an alternative solution to improve speech recognition performance. Further, VSR systems have a wider range of applications compared to their acoustic-only speech recognition counterparts. For example, as shown in Fig. 1, in many practical applications where speech recognition systems are exposed to noisy environments, acoustic signals are almost unusable for speech recognition. Conversely, with the availability of front and rear cameras on most intelligent mobile devices, users can easily record facial movements to perform VSR. In extremely noisy environments, visual information basically becomes the only source that ASR systems can use for speech recognition.

Moreover, inspired by bimodal human speech production and perception even in clean and moderate noise conditions, where good-quality acoustic signals are available for speech recognition visual information can provide complementary information for ASR [54, 55]. Therefore, research on VSR is of particular importance, because once an adequate VSR result is obtained, speech recognition performance can be boosted through the fusion of audio and visual modalities.

Despite the wide range of applications of VSR systems, there are two main limitations related to this area: the development of appropriate dynamic audio-visual fusion and the development of appropriate visual feature representations. Regarding dynamic audio-visual fusion, although several high-quality works on this topic have been published recently [15, 49, 61, 64], a similar fusion framework was used in most of the cases. More specifically, in these works, the quality of both the audio and visual signals was evaluated using different criteria, such as signal-to-noise ratio, dispersion and entropy. Weights were dynamically assigned to the audio and visual streams according to the quality of the audio and visual signals. However, compared with the audio-visual fusion method, visual feature representation techniques are more controversial. The goal of visual feature representation is to embed spatio-temporal visual information into a compact visual feature vector. This is the most fundamental problem for VSR, because it directly affects the final recognition performance. Hence, in this survey, we mainly focus on the most recent advances in the area of visual feature representation, and we discuss potential solutions and future research directions.

Regarding audio feature representation, Mel-Frequency Cepstral Coefficients (MFCCs) are generally acknowledged to be the most widely used acoustic features for speech recognition. However, unlike audio feature extraction, there is no universally accepted visual feature extraction technique that can achieve promising results for different speakers and different speech tasks, as three fundamental issues remain unresolved [79]: (1) how to extract visual features with constant quality from videos with different head and pose positions; (2) how to remove speech-irrelevant information from the visual data; (3) how to encode temporal information into the visual features. This chapter will summarise recent research that has examined solutions to these issues, and it will provide an insight into the relationships between these methods.

This chapter is organised as follows. Section 2 introduces handcrafted visual feature extraction methods, which are still the most widely used techniques for visual feature representation, and they are sometimes used in pre-processing steps for automatic feature learning. Sections 3 and 4 respectively describe graph-based feature learning and deep learning-based feature learning methods. Finally, Sect. 5 provides insights into potential solutions for the remaining challenges and possible future research directions in this area.

2 Hand Crafted Feature Extraction

Before introducing visual feature learning techniques, this section describes some of the handcrafted visual features that still play a dominant role in VSR. In addition, handcrafted feature extraction methods can be used in the pre-processing steps of many visual feature learning frameworks. In terms of the type of information embedded in the features, visual features can be categorised into two classes: appearance-based and geometric-based features [7].

For appearance-based visual features, the entire ROIs (e.g., mouth, lower face or even the face area) are considered informative regions in terms of VSR. However, it is infeasible to use all the pixels of ROIs because the dimensions of the features are too large for the classifiers to process. Hence, appropriate transformations of the ROIs are used to map the images to a much lower-dimensional feature space. More specifically, given the original image $\mathbf{I}$ in the feature space $\mathbb{R}^D$ (where D is the feature dimension), appearance-based feature extraction methods seek to transform matrix **P** to map **I** to a lower feature space $\mathbb{R}^d$ ($d << D$), such that the transformed feature vector contains the most speech-relevant information with a much smaller feature dimension.

The Discrete Cosine Transform (DCT) [54, 55] is among the most commonly used appearance-based visual feature extraction methods. It can be formulated as:

$$Y(i,j) = \sum_{j=1}^{N}\sum_{i=1}^{N} I(i,j)\cos\left(\frac{\pi(2j+1)j}{2N}\right)\cos\left(\frac{\pi(2i+1)i}{2N}\right), \qquad (1)$$

for $i, j = 1, 2, \ldots, N$, where N is the width and height of the mouth ROI, the value of N is a power of two and $I(i,j)$ is the grey-level intensity value of the ROI. To avoid the curse of dimensionality, low-frequency coefficients are selected and used as the static components of the visual feature. To encode the temporal information, the first and second derivatives of the DCT coefficients are used along with the static coefficients of the DCT ($Y(i,j)$) as the dynamic components of the visual feature. Other appearance-based techniques can also be used to extract appearance-based visual features [55], such as Principle Component Analysis (PCA) [13], Hadamard and Haar transform [60] and Discrete Wavelet Transform (DWT) [53].

In addition to the methods described above, other appearance-based visual feature extraction methods have been proposed. More specifically, instead of seeking a global transformation that can be used on the entire ROI, other methods use a feature descriptor to describe a small region centred at each pixel in the ROI, and to count the descriptors response occurrence in the ROI. Typical methods in this category include Local Binary Pattern (LBP) [45] and Histogram of Oriented Gradients (HOG) [9]. However, these methods are incapable of extracting temporal dynamic information from the ROIs. Hence, a number of variants have been proposed. For example, Zhao et al. [74] proposed a local spatio-temporal visual feature descriptor for automatic lipreading. This visual feature descriptor can be viewed as an extension of the basic LBP [45]. More specifically, to encode the temporal information into the visual feature vector, Zhao et al. [74] extracted LBP features from Three Orthogonal Planes (LBP-TOP), which contain the spatial axes of the images (X and Y) and the time axis (T), as shown in Fig. 2. Although the LBP-TOP feature contains rich visual speech-relevant information, the dimensionality of the original LBP-TOP feature is too large to be used directly for VSR. Hence, in [74], AdaBoost was used to select the most informative components from the original LBP-TOP feature for VSR.

Numerous works [2, 77, 78] have used LBP-TOP for VSR, and variations of the original LBP-TOP feature have also been proposed. Pei et al. [51] used Active Appearance Models (AAM) to track keypoints on the lips. For each small patch centred around the keypoints of the lips, LBP-TOP and HOG were used to extract the texture features. In addition to the texture features, the difference between the patch positions in the adjacent frames was used as a shape feature. Given that rich speech-relevant information is embedded in LBP-TOP features, a number of feature reduction techniques have been introduced to extract a more compact visual feature

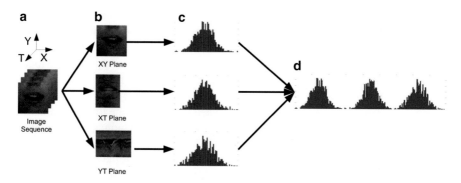

Fig. 2 Lip spatio-temporal feature extraction using LBP-TOP feature extraction. (**a**) Lip block volumes; (**b**) lip images from three orthogonal planes; (**c**) LBP features from three orthogonal planes; (**d**) concatenated features for one block volume with appearance and motion

from the original LBP-TOP feature. In addition to the LBP-TOP feature, a number of other appearance-based feature descriptors have been used to extract temporal dynamic information, such as LPQ-TOP [30], LBP-HF [76], and LGBP-TOP [1].

Although both DCT and LBP-TOP are widely used for VSR, they are quite different because they represent visual information from different perspectives. More specifically, as shown in (1), each component ($Y(i,j)$) of the DCT feature is a representation of the entire mouth region at a particular frequency. Hence, DCT is a global feature representation method. Conversely, the LBP-TOP feature uses a descriptor to represent the local information in a small neighbourhood; therefore, the LBP-TOP is a local feature representation method. Hence, the development of a method that can combine both global and local information using a compact feature vector would be expected to boost visual speech accuracy. Although Zhao et al. [75] showed that combining different types of visual features (LBP-TOP and EdgeMap [17]) can improve recognition accuracy, finding an effective way to combine DCT and LBP-TOP features is still an undeveloped area.

Although the dimensionality of the appearance-based visual features is much smaller compared to the number of pixels in the ROI, it still makes the system succumb to the curse of dimensionality. Hence, a feature dimension reduction process is essential as a prior step to VSR. Among the feature reduction methods, LDA and PCA are the most widely used [54, 55]. In addition, Gurban et al. [20] presented a Mutual Information Feature Selector (MIFS)-based scheme to select an informative visual feature component subset and thus reduce the dimensionality of the visual feature vector. Unlike feature reduction schemes such as PCA and LDA, MIFS analyses each feature component in the visual feature vector and selects the most informative components using the greedy algorithm. In addition, Gurban et al. [20] proposed that penalizing features for their redundancy is essential to yield a more informative visual feature vector.

Geometric visual features explicitly model the shape of the mouth and are potentially more powerful than appearance-based features. However, they are sensitive to lighting conditions and image quality. Geometric-based features include Deformable Template (DT), Active Shape Model (ASM), Active Appearance Model (AAM) and Active Contour Model (ACM). DT [36] is a method that uses a parametric lip template to partition an input image into a lip region and a non-lip region. However, this approach is degraded when the shape of the lip is irregular or the mouth is opened wide [31]. The ASM [42] uses a set of landmarks to describe the lip model. The AAM approach [38] can be viewed as an extension of ASM that incorporates grey-level information into the model. However, as the landmarks need to be manually labelled during training, it is very laborious and time-consuming to train the ASM and AAM for lip extraction.

In terms of ACM-based lip extraction, there are two main categories, namely edge-based and region-based. With respect to the edge-based extraction approach, the image gradients are calculated to locate the lip potential boundary [11]. Unfortunately, given that the intensity contrast between the lip and the face region is usually not large enough, the edge-based ACM is likely to achieve incorrect extraction results. Moreover, this method has been confirmed to be prone to image noise, and it is highly dependent on the initial parameters of the ACM [31]. In terms of region-based techniques, the foreground is segmented from the background by finding the optimum intensity energy in the images. Compared to its edge-based counterpart, this method has been shown to be robust with respect to the initial curve selection and the influence of noise [31]. In contrast, because of the appearance of the teeth and tongue, intensity values inside the lips are usually different. In this situation, a Global region-based ACM (GACM) can fail because all of the pixels inside the lips are taken into consideration. However, with a Localised region-based ACM (LACM), only the pixels around the objects contour are taken into account. This method can therefore successfully avoid the influence of the appearance of the teeth and the tongue [8].

However, provided that the LACM is used solely for lip extraction and the initial contour is far away from the actual lip contour, the curvature may converge to a local minima without finding the correct lip boundary. Therefore, the initial contour needs to be specified near the lip boundary as a priori. The common method for specifying the initial contour is to detect several lip corners [10, 35] and to construct an ellipse surrounding the lip. Unfortunately, this approach is either sensitive to the image noise and illuminations or needs a complex training process. In order to effectively solve this problem, Sui et al. [69] presented a new extraction framework that synthesises the advantages of both the global and localised region-based ACMs.

Although the geometric feature can explicitly model the shape of the lips, it is difficult to derive an accurate model that can describe the dynamic movement of the mouth. Hence, appearance-based features remain the most widely used features in the VSR community.

3 Graph Based Visual Feature Representations

In most cases, the dimensions of the visual features that are extracted using handcrafted feature extraction methods are usually too large for the classifiers. Graph-based learning methods that non-linearly map the original visual features to a more compact and discriminatory feature space have also been used in recent years.

Initially, graph-based methods were commonly used in human activity recognition [52]. Given that both human activity and speech recognition deal with the analysis of spatial and temporal information, graph-based feature learning can therefore also be used for VSR. The idea behind graph-based learning is that visual features can be represented as the elements of a unified feature space, and the temporal evolution of lip movements can be viewed as the trajectory connecting these elements in the feature space. Hence, after the extracted feature sequences from the videos have been correctly mapped to the corresponding trajectories, the speech can be correctly recognised. In addition, it is generally believed that the dimension of the underlying structure of the visual speech information should be significantly smaller than the dimension of the corresponding observed videos. Based on the above assumptions, numerous papers have proposed different frameworks to parameterise the original high-dimensional visual features to the trajectories to extract lower-dimensional features. An illustration of the concept behind graph-based feature representation methods is shown in Fig. 3.

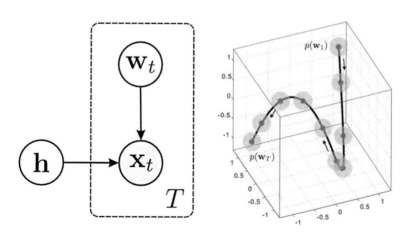

Fig. 3 The idea behind graph-based feature representation methods is to project the original high-dimensional spatio-temporal visual features to a trajectory in a lower-dimensional feature space, thereby reducing the feature dimension to boost the performance of speech recognition. Each point ($p(\mathbf{w}_T)$) of the projected trajectory represents a frame in the corresponding video. This figure appeared in [78]. In this work, each image $\mathbf{x}_i$ of the T-frame video is assumed to be generated by the latent speaker variable h and the latent utterance variable $\mathbf{w}_i$

Zhou et al. [77] proposed a path graph based method to map the image sequence of a given utterance to a low-dimensional curve. Their experimental results showed that the recognition rate of this method is 20 % higher than the recognition rate reported in [74] on the OuluVS data corpus. Based on this work, the visual feature sequence of a speakers mouth when talking is further assumed to be generated from a speaker-dependent Latent Speaker Variable (LSV) and a sequence of speaker-independent Latent Utterance Variables (LUV). Hence, Zhou et al. [78] presented a Latent Variable Model (LVM) that separately represents the video by LSV and LUV, and the LUV is further used for VSR. Given an image sequence of length T, $\mathbf{X} = \{\mathbf{x}_t\}_{t=1}^{T}$, the LVM of an image $\mathbf{x}_t$ which is generated from the inter-speaker variations h (LSV) and dynamic changes of the mouth $\mathbf{w}_t$, can be formulated by (2):

$$\mathbf{x}_t = \mu + \mathbf{F}\mathbf{h} + \mathbf{G}\mathbf{w}_t + \epsilon_t, \tag{2}$$

where μ is the global mean, $\mathbf{F}$ is a factor matrix whose columns span the inter-speaker space, $\mathbf{G}$ is the bias matrix that describes the uttering variations and ϵ_t is the noise term. The model described in (2) is a compact representation of high-dimensional visual features. Compared with the 885-dimensional raw LBP-TOP feature, the six-dimensional LUV feature is very compact and can yield better accuracy than other features, such as PCA [4], DCT [18], AF[57] and AAM [38].

Pei et al. [51] presented a method based on the concept of unsupervised random forest manifold alignment. In this work, both appearance and geometric visual features were extracted from the lip videos, and the affinity of the patch trajectories in the lip videos was estimated by a density random forest. A multidimensional scaling algorithm was then used to embed the original data into a low-dimensional feature space. Their experimental results showed that this method was capable of handling large datasets and low-resolution videos effectively. Moreover, the exploitation of depth information for VSR was also discussed in this paper.

Unlike the unsupervised manifold alignment approach proposed by Pei et al. [51], Bakry and Elgammal [2] presented a supervised visual feature learning framework where each video was first mapped to a manifold by the manifold parametrisation [14], and then kernel partial least squares was used in the manifold parameterisation space to yield a latent low-dimensional manifold parameterisation space.

It is well known that different people speak at different rates, even when they are uttering the same word. The varying rates of speech result in random parameterisations of the same trajectory, which leads to a failure in speech recognition. Hence, a temporal alignment is essential for VSR to remove any temporal variabilities caused by different speech rates. Su et al. [65] applied a statistical framework (introduced in [66]) and proposed a rate-invariant manifold alignment method for VSR. In this method, each trajectory α of the video sequence in the trajectory set $\mathbb{M}$ is represented by a Transported Square-Root Vector Field (TSRVF) to a reference point c:

$$h_\alpha(t) = \frac{\dot{\alpha}(t)_{\alpha(t) \to c}}{\sqrt{|\dot{\alpha}(t)|}}, \tag{3}$$

where $h_\alpha(t)$ is the TSRVF of trajectory α at time t, $\dot{\alpha}(t)$ it the velocity vector of $\alpha(t)$, and $|\cdot|$ is defined as the Riemannian metric on the Riemannian manifold. Given the TSRVFs of two smooth trajectories α_1 and α_2, these two trajectories can be aligned according to:

$$\gamma^* = \underset{\gamma \in \Gamma}{\arg\min} \sqrt{\int_0^1 |h_{\alpha_1}(t) - h_{\alpha_2}(\gamma(t))\sqrt{\dot{\gamma}(t)}|^2 dt}, \tag{4}$$

where Γ is the set of all diffeomorphisms of $[0, 1]$: $\Gamma = \{\gamma : [0, 1] \to [0, 1] | \gamma(0) = 0, \gamma(1) = 1, \gamma$ is a diffeomorphism$\}$. The minimization over Γ in (4) can be solved using dynamic programming. After the trajectories have been registered, the mean of the multiple trajectories can be used as a template for visual speech classification. Although the method introduced in [65] did not produce superior performance over other recent graph-based methods [2, 51, 77, 78], and although only speech-dependent recognition was reported, this work provided a general mathematical speech-rate-invariant framework for the registration of trajectories and for comparison.

Despite graph-based methods have shown promising recognition performance compared to conventional feature reduction methods [79] such as LDA and PCA, it should be noted that none of the above graph-based methods were tested on continuous speech recognition. Even though Zhou et al. [78] reported that their method achieved promising results on classifying visemes, which are generic images that can be used to describe a particular sound, it is still unclear whether their graph-based method can be used for continuous speech recognition.

4   Visual Feature Learning Using Deep Learning

Section 3 introduced various graph-based methods that can map high-dimensional visual features to non-linear feature spaces. However, the use of graph-based methods for VSR requires prior extraction of the visual features, and the classification performance largely depends on the quality of the extracted visual features. In this section, we introduce deep feature learning-based methods, which can directly learn visual features from videos. These techniques offer the potential to replace handcrafted features with deep learned features for the VSR task.

Deep learning techniques were first proposed by Hinton et al. [25], who used the greedy, unsupervised, layer-wise pre-training scheme to train a Restricted Boltzmann Machine (RBM) to model each layer of a Deep Belief Network (DBN), which effectively solved the difficulty of training multiple hidden-layer neural networks. Later works showed that a similar pre-training scheme could also be

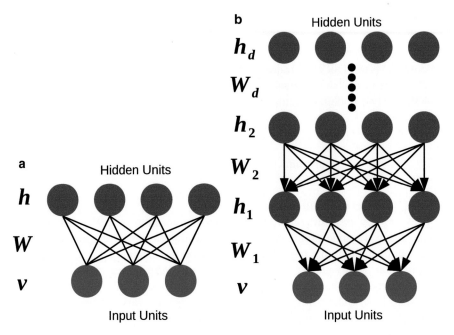

Fig. 4 Two RBM-based deep models. *Blue* circles represent input units and *red* units represent hidden units. (**a**): An RBM. (**b**): A Stacked RBM-based Auto-Encoder

used by stacked auto-encoders [3] and Convolutional Neural Networks (CNN) [56]. These techniques achieved great success in various classification tasks, such as acoustic speech recognition and image set classification [21, 22].

After deep learning techniques had been successfully applied to a single modality for the task of feature learning, Ngiam et al. [41] used it for a bimodal (i.e., audio and video) task. This was the first deep learning work in the domain of VSR and Audio-Visual Speech Recognition (AVSR). Since then, a number of other methods have been proposed that employed deep learning techniques to learn visual features for visual speech classification. Deep learning techniques used for VSR and AVSR can be categorised into three types: RBM-based deep models, stacked denoising auto-encoder-based methods and CNN-based methods.

The RBM is a particular type of Markov random field with hidden variables $\mathbf{h}$ and visible variables $\mathbf{v}$ (Fig. 4a). The connections W_{ij} between the visible and hidden variables are symmetrical, but there are no connections within the hidden and visible variables. The model defines the probability distribution $P(\mathbf{v}, \mathbf{h})$ over $\mathbf{v}$ and $\mathbf{h}$ via an energy function, which can be formulated by (5). The log-likelihood of $P(\mathbf{v}, \mathbf{h})$ can be maximised by minimising the energy function in (5):

$$E(\mathbf{v}, \mathbf{h}; \theta) = -\sum_{i=1}^{m}\sum_{j=1}^{n} W_{ij} v_i h_j - \sum_{i=1}^{m} b_i v_i - \sum_{j=1}^{n} a_j h_j, \qquad (5)$$

where a_j and b_i are the biases of the hidden units and visible unit respectively, m and n are the numbers of hidden units and visible units, and θ includes the parameters of the model. As the computation of the gradient of the log-likelihood is intractable, the parameters of the model have usually been learned using contrastive divergence [24]. With the proper configurations of the RBM, the visual feature is fed to the first layer of the RBM, the posteriors of the hidden variables (given the visible variables) are obtained using $p(h_j|\mathbf{v}) = \text{sigmoid}(b_j + W_j^T\mathbf{v})$, and $p(h_j|\mathbf{v})$ can be used as the new training data for the successive layers of the RBM-based deep networks. This process is repeated until the subsequent layers are all pre-trained.

In Ngiam et al.'s work [41], the deep auto-encoder, which consisted of multiple layers of sparsity RBMs [34], was used to learn a shared representation of the audio and visual modalities for speech recognition. The authors discussed two learning architectures in their paper. The first model investigated was cross-modality learning, where the model learned to reconstruct both the audio and video modalities, while only the video was used as an input during the training and testing stage. The second model was used for the training of the multimodal deep auto-encoder with both audio and video data. However, two-thirds of the used data had zero values in one of the input modalities (e.g., video), and the original values were used in the other input modality (e.g., audio). Experimental results in [41] showed an improvement over previous handcrafted visual features [20, 38, 74]. However, their bimodal deep auto-encoder did not outperform their video-only deep auto-encoder, because the bimodal auto-encoder might not have been optimal when only the visual input was provided.

Given the inefficiency of the bimodal auto-encoders proposed in [41], Srivastava et al. [62] used a Deep Boltzmann Machine (DBM), which was first proposed in [59], for AVSR. Like the deep learning models introduced above, the DBM is also a method from the Boltzmann machine family of models, and it has the potential to learn the complex and non-linear representations of the data. Moreover, it can also exploit information from a large amount of unlabelled data for pre-training purposes. The major difference between the DBM and other RBM-based models is shown in Fig. 5. Unlike other RBM-based models, which only employ a top-down approximation inference procedure, the DBM incorporates a bottom-up pass with a top-down feedback. Given that the approximation inference procedure of the DBM has two directions, the DBM model is an undirected model (Fig. 5b), while other RBM-based models are directed (Figs. 4b and 5a). Because of the undirected characteristics of the DBM models, the DBM is more capable of handling uncertainty in the data, and it is more robust to ambiguous inputs [59].

Before applying the DBM model to AVSR, Srivastava et al. [63] first applied the DBM on image and text classification, which is also a multimodal learning task. In their work, the image and text data were trained separately using two single-stream DBMs, and the outputs of these two single-stream DBMs were then merged to train joint representations of the image and text data. As the image and text data are highly correlated, it is difficult for the model proposed in [41] to learn these correlations and produce multimodal representations. In fact, as the approximation inference procedure is directed, the responsibility of the multimodal modelling falls

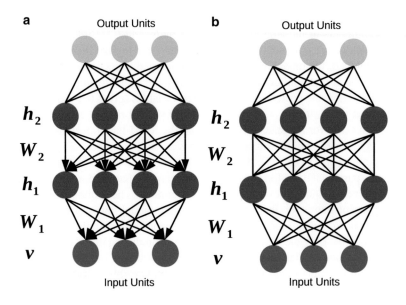

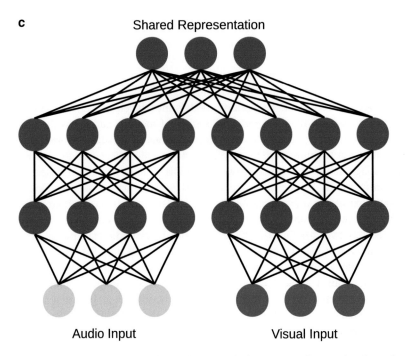

Fig. 5 Different deep models. The *blue* and *orange* circles represent input units, the *red* units represent hidden units, and the *green* circles represent representation units. (**a**): A DBN. (**b**): A DBM. (**c**): A multimodal DBM. When we compare (**a**) with (**b**), one can note that the DBN model is a directed model, while the DBM model is undirected

entirely on the joint layer [63]. In contrast, the model introduced in [63] solved this challenge effectively because the DBM can approximate the learning model both from the top-up pass and the bottom-down feedback, which makes the multimodal modelling responsibility spread out over the entire network [63].

Moreover, as shown in Fig. 5a, the top two layers of the DBN consist of an RBM (which is an undirected model), while the remaining lower layers form a directed generative model. Hence, the directed DBN model is not capable of modelling the missing inputs. Conversely, as the DBM is an undirected generative model and employs a two-way approximate inference procedure, it can be used to generate a missing modality by clamping the observed modality at the inputs and running the standard Gibbs sampler. In [63], the DBM was shown to be capable of generating missing text tags from the corresponding images. Srivastava et al. [62] then used this model for the task of AVSR. Experimental results on the CUAVE [50] and AVLetters [38] datasets showed that the multimodal DBM can effectively combine features across modalities and achieve slightly better results than the video deep auto-encoder proposed in [41]. Although this work demonstrated that the DBM could combine features effectively for speech recognition across audio and visual modalities, the inference of audio from the visual feature was not discussed. However, it provides a method that may be able to solve the problem proposed in [41]-that is, how to generate the missing audio from the video.

Despite these promising results, it should be noted that all of the aforementioned deep learning-based VSR methods have the objective of learning a more informative spatio-temporal descriptor that extracts speech-relevant information directly from the video. However, in order to use deep learning techniques for real-world VSR applications, sequential inference-based approaches, which are widely used by the acoustic speech recognition community, need to be developed.

In terms of acoustic continuous speech recognition, Mohamed et al. [40] developed acoustic phone recognition using a DBN. In this work, MFCCs were used as an input to the DBN. The DBN was pre-trained layer by layer, followed by a fine-tuning process that used 183 target class labels (i.e., three states for each of the 61 phonemes). The output of the DBN represents the probability distribution over possible classes. The probability distribution yielded by the DBN was fed to a Viterbi decoder to generate the final phone recognition results. Inspired by this method, Huang and Kingsbury [27] presented a similar framework for AVSR. Compared with the Hidden Markov Model/Gaussian Mixture Model (HMM/GMM) framework, the DBN achieved a 7 % relative improvement on the audio-visual continuously spoken digit recognition task. This work also presented a mid-level feature fusion method that concatenated the hidden representations from the audio and visual DBN, and the LDA was then used to reduce the dimensionality of the original concatenated hidden representations. At the last stage, the LDA projected representations were used as inputs to train a HMM/GMM model, and achieved a 21 % relative gain over the baseline system. However, using the DBN for visual-only speech recognition did not produce any improvements over the standard HMM/GMM model in [27].

In addition to the RBM-based deep learning techniques introduced above, Vincent et al. [71] proposed a Stacked Denoising Auto-encoder (SDA) based on a new scheme to pre-train a multi-layer neural network. Instead of training the RBM to initialise the hidden units, the hidden units are learned by reconstructing input data from artificial corruption. Paleček [47] explored the possibility of using the auto-encoder to learn useful feature representations for the VSR task. The learned features were further processed by a hierarchical LDA to capture the speech dynamics before feeding them into the HMM for classification. The auto-encoder-learned features produced a 4–8 % improvement in accuracy over the standard DCT feature in the case of isolated word recognition. However, only the single-layer auto-encoder was discussed in their paper [47], suggesting that the superiority of the stacked auto-encoder was not fully analysed. In addition to the conventional SDA, deep bottleneck feature extraction methods based on SDA [16, 58, 73] were extensively used in acoustic speech recognition. Inspired by the deep bottleneck audio features for continuous speech recognition, Sui et al. [70] developed a deep bottleneck feature learning scheme for VSR. This technique was successfully used with the connected word VSR, and it demonstrated superior performance over handcrafted features such as DCT and LBP-TOP [70].

Although RBM-based deep networks and SDA-based methods achieved an impressive performance for various tasks, these techniques did not take the topological structure of the input data into account (e.g., the 2D layout of images and the 3D structure of videos). However, topological information is very important for visual-driven tasks, because a large amount of speech-relevant information is embedded in the topological structure of the video data. Hence, developing a method to explore the topological structure of the input should help to boost VSR performance. The CNN model proposed by Lecun et al. [32] can exploit the spatial correlation that is presented in input images. This model has achieved great success in visual-driven tasks in recent years [33]. Noda et al. [43, 44] developed a lipreading system based on a CNN to recognise isolated Japanese words. In their paper, the CNN was trained using mouth images as input to recognise the phonemes. The parameters of the fully trained CNN were used as features for the HMM/GMM models. The experimental results showed that their proposed CNN-learned features significantly outperformed those acquired by PCA.

A number of deep learning-based methods have achieved promising results in the case of acoustic speech recognition. However, their use in the task of VSR has not yet been explored. For example, deep recurrent neural networks [19] have been recently proposed for acoustic speech recognition. It would be interesting to explore their applications to VSR in future research.

5 Discussion

This chapter provides an overview of some handcrafted, graph-based and deep learning-based visual features that have recently been proposed. To compare the VSR performance achieved by the different visual feature representations, we

Table 1 Summary of the recently proposed multi-speaker and speaker-independent visual-only speech recognition performance on popular and publicly available visual speech corpora

Data corpus	Feature category	Feature extraction methods	Classifier	Accuracy (%)
AVLetters	Hand crafted	ASM [38]	HMM	26.91
		Optical Flow	SVM	32.31
		AAM [38]	HMM	41.9
		MSA [38]	HMM	44.6
		DCT	SVM	53.46
		LBP-TOP [74]	SVM	58.85
	Graph-based	Bakry and Elgammal [2]	SVM	65.64
	Deep learning	Ngiam et al. [41]	SVM	64.4
		Srivastava et al. [62]	SVM	64.7
OuluVS	Hand crafted	LBP-TOP [74]	SVM	62.4
	Graph-based	Ong and Bowden [46]	SVM	65.6
		Zhou et al. [77]	SVM	81.3
		Bakry and Elgammal [2]	SVM	84.84
		Zhou et al. [78]	SVM	85.6
		Pei et al. [51]	SVM	89.7
CUAVE	Hand crafted	DCT [20]	HMM	64
		AAM [48]	HMM	75.7
		Lucey and Sridharan [37]	HMM	77.08
		Visemic AAM [49]	HMM	83
	Deep learning	Ngiam et al. [41]	SVM	66.7
		Srivastava et al. [62]	SVM	69.0

list the performance of these methods for three popular publicly available visual speech corpora in Table 1. The table shows that graph-based and deep learning-based methods generally perform better than handcrafted feature-based approaches. Although some geometric-based handcrafted features [37, 48, 49] achieved more accurate results compared to the graph-based and deep learning-based methods, it is required that the landmarks on the facial area are laboriously labelled beforehand. On this basis, the VSR research community generally recognises that graph-based and deep learning-based methods should be the focus of future research.

Most graph-based and deep learning-based methods have been developed in an attempt to pose lipreading as a classification problem. However, in order to employ VSR for connected and continuous speech applications, the VSR problem should be tackled in a similar way to a speaker-independent acoustic speech recognition task [29]. In terms of continuous speech recognition, instead of extracting holistic visual features from the videos, visual information needs to be represented in a frame-wise manner-that is, the spatio-temporal visual features should be extracted frame by frame, and the temporal dynamic information needs to be captured by the classifiers (e.g., HMM). Given that acoustic modelling for speech recognition using deep learning techniques has been extensively investigated by the speech community, and given that some of these systems have already been commercialised in recent years [26], it is worth investigating whether these methods can be used for VSR.

Another challenge in the area of VSR is that a large-scale and comprehensive data corpus needs to be available. Although there are a large number of data corpora available for VSR research, all of the existing ones cannot satisfy the ultimate goal, which is to build a practical lipreading system that can be used in real-life applications. That is, in order to treat the VSR problem in a way that is similar to continuous speech recognition, which needs to capture the temporal dynamics of the data (e.g., by using HMMs), a large-scale audio-visual data corpus needs to be established, as this will provide visual speech in the same context as audio speech. Currently, popular benchmark corpora such as AVLetters [38], CUAVE [50] and OuluVS [74] are not fully useful because they are limited in both speaker number and speech content. In addition, some large-scale data corpora such as AVTIMIT [23], IBMSR [37], IBMIH [28] are not publicly accessible. Although the publicly available XM2VTSDB [39] has 200 speakers, the speech is limited to simple sequences of isolated word and digit utterances. A large-scale and comprehensive data corpus called AusTalk was recently created [5, 6, 67, 72]. AusTalk is a large 3D audio-visual database of spoken Australian English recorded at 15 different locations in all states and territories of Australia. The contemporary voices of one thousand Australian English speakers of all ages have been recorded in order to capture the variability in their accent, linguistic characteristics and speech patterns. To satisfy a variety of speech-driven tasks, several types of data have been recorded, including isolated words, digit sequences and sentences. Given that the AusTalk data corpus is a relatively new dataset, only a few works have used this data corpus to date [68–70]. A comprehensive review on the availability of data corpora can also be found in [79].

This chapter reviewed the recent advances in the area of visual speech feature representation. One can conclude from this survey that graph-based and deep learning-based feature representations are generally considered state-of-the-art. Instead of directly using handcrafted visual features for the VSR task, handcrafted visual feature extraction methods are widely used during the pre-processing phase before the extraction of visual features that are finally used for graph-based and deep learning techniques. Despite the exciting recent achievements by the VSR community, several challenges still need to be addressed before a system is developed that can fulfil the specifications of real-life applications. We have summarised the major challenges and proposed possible solutions in this chapter.

References

1. T.R. Almaev, M.F. Valstar, Local gabor binary patterns from three orthogonal planes for automatic facial expression recognition, in *Proceedings of Humaine Association Conference on Affective Computing and Intelligent Interaction* (IEEE, Geneva, 2013), pp. 356–361

2. A. Bakry, A. Elgammal, Mkpls: manifold kernel partial least squares for lipreading and speaker identification, in *Proceedings of IEEE Conference on Computer Vision and Pattern Recognition* (IEEE, Washington, 2013), pp. 684–691
3. Y. Bengio, P. Lamblin, D. Popovici, H. Larochelle, Greedy layer-wise training of deep networks. Adv. Neural Inf. Process. Syst. **19**, 153 (2007)
4. C. Bregler, Y. Konig, eigenlips for robust speech recognition, in *Proceedings of IEEE International Conference on Acoustics, Speech, and Signal Processing*, vol. 2 (IEEE, Washington, 1994), pp. 669–672
5. D. Burnham, E. Ambikairajah, J. Arciuli, M. Bennamoun, C.T. Best, S. Bird, A. Butcher, C. Cassidy, G. Chetty, F.M. Cox et al., A blueprint for a comprehensive Australian English auditory-visual speech corpus, in *Selected Proceedings of the 2008 HCSNet Workshop on Designing the Australian National Corpus* (2009), pp. 96–107
6. D. Burnham, D. Estival, S. Fazio, J. Viethen, F. Cox, R. Dale, S. Cassidy, J. Epps, R. Togneri, M. Wagner et al. Building an audio-visual corpus of Australian English: Large corpus collection with an economical portable and replicable black box, in *Proceedings of Twelfth Annual Conference of the International Speech Communication Association* (2011)
7. H.E. Cetingul, Y. Yemez, E. Erzin, A.M. Tekalp, Discriminative analysis of lip motion features for speaker identification and speech-reading. IEEE Trans. Image Process. **15**(10), 2879–2891 (2006)
8. Y. Cheung, X. Liu, X. You, A local region based approach to lip tracking. Pattern Recogn **45**(9), 3336–3347 (2012)
9. N. Dalal, B. Triggs, Histograms of oriented gradients for human detection, in *Proceedings of IEEE Computer Society Conference on Computer Vision and Pattern Recognition*, vol. 1 (IEEE, Washington, 2005), pp. 886–893
10. M. Dantone, J. Gall, G. Fanelli, L. van Gool, Real-time facial feature detection using conditional regression forests, in *Proceedings of International Conference on Computer Vision and Pattern Recognition* (2012)
11. P. Delmas, P. Coulon, V. Fristot, Automatic snakes for robust lip boundaries extraction, in *Proceedings of International Conference on Acoustics, Speech, and Signal Processing*, vol. 6 (IEEE, Washington, 1999), pp. 3069–3072
12. L. Deng, D. Yu, *Deep Learning: Methods and Applications* (Now Publishers, Boston, 2014)
13. S. Dupont, J. Luettin, Audio-visual speech modeling for continuous speech recognition. IEEE Trans. Multimedia **2**(3), 141–151 (2000)
14. A. Elgammal, C.S. Lee, Separating style and content on a nonlinear manifold, in *Proceedings of IEEE Conference on Computer Vision and Pattern Recognition*, vol. I (IEEE, Washington, 2004), pp. 478–485
15. V. Estellers, M. Gurban, J. Thiran, On dynamic stream weighting for audio-visual speech recognition. IEEE Trans. Audio Speech Lang. Process. **20**(4), 1145–1157 (2012)
16. J. Gehring, Y. Miao, F. Metze, A. Waibel, Extracting deep bottleneck features using stacked auto-encoders, in *Proceedings of IEEE International Conference on Acoustics, Speech and Signal Processing* (IEEE, Washington, 2013), pp. 3377–3381
17. Y. Gizatdinova, V. Surakka Feature-based detection of facial landmarks from neutral and expressive facial images. IEEE Trans. Pattern Anal. Mach. Intell. **28**(1), 135–139 (2006)
18. J.N. Gowdy, A. Subramanya, C. Bartels, J. Bilmes, Dbn based multi-stream models for audio-visual speech recognition, in *Proceedings of IEEE International Conference on Acoustics, Speech, and Signal Processing*, vol. 1 (IEEE, Washington, 2004), pp. 993–996
19. A. Graves, Ar. Mohamed, G. Hinton, Speech recognition with deep recurrent neural networks, in *Proceedings of IEEE International Conference on Acoustics, Speech and Signal Processing (ICASSP)* (IEEE, Washington, 2013), pp. 6645–6649
20. M. Gurban, J. Thiran, Information theoretic feature extraction for audio-visual speech recognition. IEEE Trans. Signal Process. **57**(12), 4765–4776 (2009)
21. M. Hayat, M. Bennamoun, S. An, Learning non-linear reconstruction models for image set classification, in *Proceedings of IEEE Conference on Computer Vision and Pattern Recognition* (2014), pp. 1915–1922

22. M. Hayat, M. Bennamoun, S. An, Deep reconstruction models for image set classification. IEEE Trans. Pattern Anal. Mach. Intell. **37**(4), 713–727 (2015)

23. T.J. Hazen, K. Saenko, C.H. La, J.R. Glass, A segment-based audio-visual speech recognizer: Data collection, development, and initial experiments, in *Proceedings of the 6th international conference on Multimodal interfaces* (ACM, New York, 2004), pp. 235–242

24. G. Hinton, Training products of experts by minimizing contrastive divergence. Neural Comput. **14**(8), 1771–1800 (2002)

25. G.E. Hinton, R.R. Salakhutdinov, Reducing the dimensionality of data with neural networks. Science **313**(5786), 504–507 (2006)

26. G. Hinton, L. Deng, D. Yu, G.E. Dahl, Ar. Mohamed, N. Jaitly, A. Senior, V. Vanhoucke, P. Nguyen, T.N. Sainath et al., Deep neural networks for acoustic modeling in speech recognition: the shared views of four research groups. IEEE Signal Process. Mag. **29**(6), 82–97 (2012)

27. J. Huang, B. Kingsbury, Audio-visual deep learning for noise robust speech recognition, in *Proceedings of IEEE International Conference on Acoustics, Speech and Signal Processing (ICASSP)* (IEEE, Washington, 2013), pp. 7596–7599

28. J. Huang, G. Potamianos, J. Connell, C. Neti, Audio-visual speech recognition using an infrared headset. Speech Comm. **44**(1), 83–96 (2004)

29. X. Huang, A. Acero, H.W. Hon et al., *Spoken Language Processing* (Prentice Hall, Englewood Cliffs, 2001)

30. B. Jiang, M.F. Valstar, M. Pantic, Action unit detection using sparse appearance descriptors in space-time video volumes, in *Proceedings of IEEE International Conference on Automatic Face & Gesture Recognition* (IEEE, Washington, 2011), pp. 314–321

31. S. Lankton, A. Tannenbaum, Localizing region-based active contours. IEEE Trans. Image Process. **17**(11), 2029–2039 (2008)

32. Y. LeCun, L. Bottou, Y. Bengio, P. Haffner, Gradient-based learning applied to document recognition. Proc. IEEE **86**(11), 2278–2324 (1998)

33. Y. LeCun, K. Kavukcuoglu, C. Farabet, Convolutional networks and applications in vision, in *Proceedings of 2010 IEEE International Symposium on Circuits and Systems* (IEEE, Washington, 2010), pp. 253–256

34. H. Lee, C. Ekanadham, A.Y. Ng, Sparse deep belief net model for visual area V2, in Proceedings of Adv. Neural Inf. Process. Syst., 873–880 (2008)

35. M. Li, Y. Cheung, Automatic lip localization under face illumination with shadow consideration. Signal Process. **89**(12), 2425–2434 (2009)

36. A. Liew, S. Leung, W. Lau, Lip contour extraction from color images using a deformable model. Pattern Recogn. **35**(12), 2949–2962 (2002)

37. P. Lucey, S. Sridharan, Patch-based representation of visual speech, in *Proceedings of the HCSNet workshop on Use of vision in human-computer interaction*, vol. 56 (Australian Computer Society, Inc., 2006), pp. 79–85

38. I. Matthews, T.F. Cootes, J.A. Bangham, S. Cox, R. Harvey, Extraction of visual features for lipreading. IEEE Trans. Pattern Anal. Mach. Intell. **24**(2), 198–213 (2002)

39. K. Messer, J. Matas, J. Kittler, J. Luettin, G. Maitre, Xm2vtsdb: The extended m2vts database, in *Proceedings of Second International Conference on Audio and Video-based Biometric Person Authentication*, vol. 964, Citeseer (1999), pp. 965–966

40. Ar. Mohamed, G.E. Dahl, G. Hinton, Acoustic modeling using deep belief networks. IEEE Trans. Audio Speech Lang. Process. **20**(1), 14–22 (2012)

41. J. Ngiam, A. Khosla, M. Kim, J. Nam, H. Lee, A.Y. Ng, Multimodal deep learning, in *Proceedings of the 28th International Conference on Machine Learning* (2011), pp. 689–696

42. Q. Nguyen, M. Milgram, T. Nguyen, Multi features models for robust lip tracking, in *Proceedings of International Conference on Control, Automation, Robotics and Vision* (IEEE, Washington, 2008), pp. 1333–1337

43. K. Noda, Y. Yamaguchi, K. Nakadai, H.G. Okuno, T. Ogata, Audio-visual speech recognition using deep learning. Appl. Intell. **42**, 1–16 (2014)

44. K. Noda, Y. Yamaguchi, K. Nakadai, H.G. Okuno, T. Ogata, Lipreading using convolutional neural network, in *Proceedings of INTERSPEECH* (2014), pp. 1149–1153

45. T. Ojala, M. Pietikäinen, D. Harwood, A comparative study of texture measures with classification based on featured distributions. Pattern Recogn. **29**(1), 51–59 (1996)
46. E. Ong, R. Bowden, Learning sequential patterns for lipreading, in Proceedings of the 22nd British Machine Vision Conference (2011), pp. 55.1–55.10
47. K. Paleček, Extraction of features for lip-reading using autoencoders, in *Speech and Computer* (Springer, Berlin, 2014), pp. 209–216
48. G. Papandreou, A. Katsamanis, V. Pitsikalis, P. Maragos, Multimodal fusion and learning with uncertain features applied to audiovisual speech recognition, in *Proceedings of IEEE 9th Workshop on Multimedia Signal Processing* (IEEE, Washington, 2007), pp. 264–267
49. G. Papandreou, A. Katsamanis, V. Pitsikalis, P. Maragos, Adaptive multimodal fusion by uncertainty compensation with application to audiovisual speech recognition. IEEE Trans. Audio Speech Lang. Process. **17**(3), 423–435 (2009)
50. E.K. Patterson, S. Gurbuz, Z. Tufekci, J. Gowdy, Cuave: a new audio-visual database for multimodal human-computer interface research, in *Proceedings of IEEE International Conference on Acoustics, Speech, and Signal Processing*, vol. 2 (IEEE, Washington, 2002), pp. 2017–2020
51. Y. Pei, T.K. Kim, H. Zha, Unsupervised random forest manifold alignment for lipreading, in *Proceedings of IEEE International Conference on Computer Vision* (IEEE, Washington, 2013), pp. 129–136
52. R. Poppe, A survey on vision-based human action recognition. Image Vis. Comput. **28**(6), 976–990 (2010)
53. G. Potamianos, H.P. Graf, E. Cosatto, An image transform approach for hmm based automatic lipreading, in *Proceedings of International Conference on Image Processing* (IEEE, Washington, 1998), pp. 173–177
54. G. Potamianos, C. Neti, G. Gravier, A. Garg, A.W. Senior, Recent advances in the automatic recognition of audiovisual speech. Proc. IEEE **91**(9), 1306–1326 (2003)
55. G. Potamianos, C. Neti, J. Luettin, I. Matthews, Audio-visual automatic speech recognition: an overview. Issues Vis. Audio-Vis. Speech Process. **22**, 23 (2004)
56. M. Ranzato, F.J. Huang, Y.L. Boureau, Y. LeCun, Unsupervised learning of invariant feature hierarchies with applications to object recognition, in *Proceedings of IEEE Conference on Computer Vision and Pattern Recognition* (IEEE, Washington, 2007), pp. 1–8
57. K. Saenko, K. Livescu, J. Glass, T. Darrell, Multistream articulatory feature-based models for visual speech recognition. IEEE Trans. Pattern Anal. Mach. Intell. **31**(9), 1700–1707 (2009)
58. T.N. Sainath, B. Kingsbury, B. Ramabhadran, Auto-encoder bottleneck features using deep belief networks, in *Proceedings of IEEE International Conference on Acoustics, Speech and Signal Processing* (IEEE, Washington, 2012), pp. 4153–4156
59. R. Salakhutdinov, G.E. Hinton, Deep boltzmann machines, in *Proceedings of International Conference on Artificial Intelligence and Statistics* (2009), pp. 448–455
60. P. Scanlon, R. Reilly, Feature analysis for automatic speechreading, in *Proceeding of IEEE Fourth Workshop on Multimedia Signal Processing* (IEEE, Washington, 2001), pp. 625–630
61. X. Shao, J. Barker, Stream weight estimation for multistream audio–visual speech recognition in a multispeaker environment. Speech Comm. **50**(4), 337–353 (2008)
62. N. Srivastava, R. Salakhutdinov, Multimodal learning with deep boltzmann machines. J. Mach. Learn. Res. **15**, 2949–2980 (2014)
63. N. Srivastava, R.R. Salakhutdinov, Multimodal learning with deep boltzmann machines, in *Advances in neural information processing systems* (2012), pp. 2222–2230
64. D. Stewart, R. Seymour, A. Pass, J. Ming, Robust audio-visual speech recognition under noisy audio-video conditions. IEEE Trans. Cybern. **44**(2), 175–184 (2014)
65. J. Su, A. Srivastava, F.D. de Souza, S. Sarkar, Rate-invariant analysis of trajectories on riemannian manifolds with application in visual speech recognition, in *Proceedings of IEEE Conference on Computer Vision and Pattern Recognition* (IEEE, Washington, 2004)
66. J. Su, S. Kurtek, E. Klassen, A. Srivastava et al., Statistical analysis of trajectories on riemannian manifolds: bird migration, hurricane tracking and video surveillance. Ann. Appl. Stat. **8**(1), 530–552 (2014)

67. C. Sui, S. Haque, R. Togneri, M. Bennamoun, A 3D audio-visual corpus for speech recognition, in *Proceedings of Australasian International Conference on Speech Science and Technology* (2012)
68. C. Sui, R. Togneri, S. Haque, M. Bennamoun, Discrimination comparison between audio and visual features, in *Proceedings of the Forty Sixth Asilomar Conference on Signals, Systems and Computers* (IEEE, Washington, 2012), pp. 1609–1612
69. C. Sui, M. Bennamoun, R. Togneri, S. Haque, A lip extraction algorithm using region-based ACM with automatic contour initialization, in *Proceedings of IEEE Workshop on Applications of Computer Vision* (IEEE, Washington, 2013), pp. 275–280
70. C. Sui, R. Togneri, M. Bennamoun, Extracting deep bottleneck features for visual speech recognition, in *Proceedings of IEEE International Conference on Acoustics, Speech, and Signal Processing* (IEEE, Washington, 2015)
71. P. Vincent, H. Larochelle, I. Lajoie, Y. Bengio, P.A. Manzagol, Stacked denoising autoencoders: learning useful representations in a deep network with a local denoising criterion. J. Mach. Learn. Res. **11**, 3371–3408 (2010)
72. M. Wagner, D. Tran, R. Togneri, P. Rose, D. Powers, M. Onslow, D. Loakes, T. Lewis, T. Kuratate, Y. Kinoshita et al., The big Australian speech corpus (the big ASC), in *Proceedings of 13th Australasian International Conference on Speech Science and Technology* (2010), pp. 166–170
73. D. Yu, M.L. Seltzer, Improved bottleneck features using pretrained deep neural networks, in *Proceedings of INTERSPEECH* (2011)
74. G. Zhao, M. Barnard, M. Pietikainen, Lipreading with local spatiotemporal descriptors. IEEE Trans. Multimedia **11**(7), 1254–1265 (2009)
75. G. Zhao, X. Huang, Y. Gizatdinova, M. Pietikäinen, Combining dynamic texture and structural features for speaker identification, in *Proceedings of the 2nd ACM workshop on Multimedia in forensics, security and intelligence* (ACM, 2010), pp. 93–98
76. G. Zhao, T. Ahonen, J. Matas, M. Pietikainen, Rotation-invariant image and video description with local binary pattern features. IEEE Trans. Image Process. **21**(4), 1465–1477 (2012)
77. Z. Zhou, G. Zhao, M. Pietikainen, Towards a practical lipreading system, in *Proceedings of IEEE Conference on Computer Vision and Pattern Recognition* (IEEE, Washington, 2011), pp. 137–144
78. Z. Zhou, X. Hong, G. Zhao, M. Pietikainen, A compact representation of visual speech data using latent variables. IEEE Trans. Pattern Anal. Mach. Intell. **36**(1), 181–187 (2014)
79. Z. Zhou, G. Zhao, X. Hong, M. Pietikäinen, A review of recent advances in visual speech decoding. Image Vis. Comput. **32**(9), 590–605 (2014)

Extended Eye Landmarks Detection for Emerging Applications

Laura Florea, Corneliu Florea, and Constantin Vertan

Abstract In this chapter we focus on the eye landmarking and eye components identification in the framework of emerging psychology-related eye tracking applications. Traditional eye landmarking separates the identification of eye centers and of eye corners and margins, while here we discuss their joint use for face expression analysis in unconstrained environments and precise estimation of non-visual gaze directions, as suggested by the Eye Accessing Cues (EAC) of the Neuro-Linguistic Programming (NLP). Such a system involves a combination of low-level feature extraction, heuristic pre-processing and trained classifiers. The approach is extensively tested across several classical image databases and compared with state of the art traditional methods.

1 Introduction

The exceptional developments in computer vision from the last decade make the automatic analysis of human behavior a goal that seems achievable. An important part of this process is the automatic face and face elements identification and interpretation and a lot of research has been particulary dedicated to eye detection. Eye data and the details of eye movements have numerous applications in face detection, biometric identification, and particularly in human-computer interaction tasks.

One also used to say that eyes are the gate to the soul and witness for various internal cognitive or emotional processes; this observation opened lately a plethora of less-traditional areas of research involving eye detection and tracking. Among these new investigated directions, we note the detection of deception as part of hostile intention perception [2], the estimation of pain intensity via facial expression analysis [3, 4], interpersonal coordination of mother–infant [5], assistance in marketing [6], etc. Recently, literature also reported attempts to interpret more complex

L. Florea (✉) • C. Florea • C. Vertan
Image Processing and Analysis Laboratory (LAPI), University "Politehnica"
of Bucharest, Bucharest, Romania
e-mail: laura.florea@upb.ro; corneliu.florea@upb.ro; constantin.vertan@upb.ro

© Springer International Publishing Switzerland 2016
M. Kawulok et al. (eds.), *Advances in Face Detection and Facial Image Analysis*,
DOI 10.1007/978-3-319-25958-1_15

Fig. 1 The seven classes of Eye Accessing Cues [1]: When eyes are not used for visual tasks, their position can indicate how people are thinking (using visual, auditory or kinesthetic terms) and the mental activity they are doing (remembering, imagining, or having an internal dialogue). Image inspired from [1]

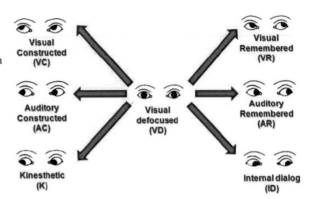

situations, such as dyadic social interactions for the diagnosis and treatment of developmental and behavioral disorders [7] and to experiment within new areas of psychology, as pointed in the recent review by Cohn and De La Torre [8]. The review of Friesen et al. [9] evaluates the social impact of gaze direction and concludes that many opportunities arise upon the understanding of the perceived direction of gaze.

Such an opportunity is offered by the Neuro-Linguistic Programming (NLP) theory, which presents unexplored opportunities for understanding the human patterns of thinking and behavior. One such model is the Eye-Accessing Cue (EAC) from the NLP theory that uses the positions of the iris inside the eye as an indicator of the internal thinking mechanisms of a person. The *direction of gaze*, under the NLP paradigm, can be used to determine the internal representational system employed by a person (see Fig. 1), who, when given a query, may think in visual, auditory or kinesthetic terms, and the mental activity of that person, of remembering, imagining, or having an internal dialogue.

In this chapter we direct the reader's attention to a system [10] that exploits an usual digital video camera (for instance a webcam—as its price makes it widely accessible) to infer eye features and landmarks positions in order to identify the direction of gaze for recognition of the Eye Accessing Cues, which are a potential mean to unravel one's background psychological process.

2 Eye Based Communications in Emergent Applications

The origins of the idea that involuntary eye movements point to inner mechanisms goes back until the nineteenth century [11]. In a review about the perception of interlocutor's gaze, Friesen et al. [9] concluded that "people's eyes convey a wealth of information about their direction of attention and their emotional and mental states". They further note that "eyes and their highly expressive surrounding region can communicate complex mental states such as emotions, beliefs, and desires" and "observing another person's behavior allows the observer to decode and make inferences about a whole range of mental states such as intentions, beliefs, and emotions". Furthermore, the gaze based social mechanisms are specifically decoded by a part of the cortex, namely the *posterior superior temporal sulcus* region, that has been found [12] to responds to the inferred intentionality of social cues.

The underlying expression or the mental process of a person may be enquired by other means than face analysis, such as gaze direction. Liversedge and Findlay [13] showed that saccades parameters are correlated with the underlying cognitive process, namely the duration of fixations and the choice of saccade target emphasize continuities between biological and cognitive descriptions. The connection between underlying emotion, expression and gaze was discussed by Adams and Kleck [14], who proved that, indeed, when gaze direction matches the underlying behavioral intent (approach-avoidance) communicated by an emotional expression, the perception of that emotion is enhanced. Following these findings, a palette of applications based on recording the eye movements have been proposed. Typically developing such applications involves two steps: first, the hypotheses regarding the correlation between eye movements and some behavior pattern or social process is formulated and validated on a series of experiments; next based on the previous found conclusions, practical applications are proposed.

Such a distinct category is the analysis of the reading process with the aim of understanding the learning to read process or how attention correlates with understanding. For instance, Joseph et al. [15] investigated, by means of gaze tracking, insights of children process to *learn to read* and the words frequency impact in sentence reading. Godfroid et al. [16] used eye tracking measurements to test hypothesis concerning words complexity, attention persistence and short/long term memory. In the same line Rayner et al. [17] monitored subjects eye movements while read sentences containing high- or low-predictable target words; their findings showed word predictability (due to contextual constraint) and word length have strong and independent influences on word skipping and fixation durations. Possible application of the eye movement control in teaching are discussed, for instance, in [18].

Moving further, Chun [19] showed that while reading or simply scanning an image, the eye movement may give hints about the observers self build context. This idea was further developed by Bulling and Zander [20] who suggested an application that provides additional information relevant to the context; the various possibilities of the adaptive context are retrieved from analysis of the eye movements.

Another important category is related to the use of eye movements in gaming. For instance Meijering et al. [21] discussed the possibility and show evidence that eye movements are correlated with the plan type used in a strategy game; more precisely forward reasoning (where a player proceed from the initial point to finish) or backward reasoning (where the path from end to start is retrieved) are distinguishable by overlapping the eye movement with the game board. Furthermore, Krejtz et al. [22] investigated the degree of enjoyment when visual cues influences gaming experience and conclude that not only in such a scenario there is no additional cognitive effort but there are many arcade game optimization possible in such a context that would increase the pleasure of users.

2.1 Eye Accessing Cues in Neuro-Linguistic Programming

The Neuro-Linguistic Programming was introduced in the 1970s by Brandler and Grinder [1], as a different model for detecting, understanding and using the patterns that appear between brain, language and body. The NLP theory jumped over intensive and extensive academic investigation and made path very fast into the commercial market. Rigorous investigation was expected after the initial publication and consensus has not been reached yet in the academic world.

The Eye Accessing Cues from the NLP theory are not unanimously accepted, with some of the most recent research on the topic calling for further testing [23]. A recent experiment by Vrânceanu et al. [24], with the scope to gain better insight of the facts, showed that while not 100 % accurate (i.e. universal), the correct apparition rates were higher than random chance, especially between visual, auditory and kinesthetic ways of thinking (corresponding to a separation along the vertical axis of the gaze direction).

2.2 Recognizing Gaze Direction: Premises

The problem of identifying one's direction of gaze is intensively studied in computer vision. These systems may be classified by the position of the recording device as:

1. Head mounted devices (e.g. glasses or head mounted cameras);
2. Stationary and/or remote devices.

The head mounted devices are closer to the eye and because of that they have access to a higher resolution and better precision. But they are rather expensive (their price spans from several thousand dollars, for a professional commercial solution, down to a hundred dollars for the more affordable ones, compared to a few dollars for a normal webcam). Another shortcoming of the head mounted devices that restricts their area of usability is the fact that they are wearable. This may be a distinct indicator that the user is subject to investigation by non-traditional means and it has been showed [25] that voluntary control is exercisable over non-visual eye movements. This is why a stationary webcam is preferable for investigating the eye accessing cues.

Another way of classifying the gaze direction estimation systems may be performed according to the illumination source domain. Here, we may note:

1. Active, infra-red (IR) based illumination;
2. Passive, visible spectrum illumination.

The commercial eye-trackers, which have higher reported precisions, rely on the information from the IR domain. But again we note that this implies a distinct, specialized device (because the IR source is not typically incorporated in webcams). As in the case of wearable eye tracking, the use of specialized recording devices needed by active illumination sources limits the applicability of methods to data that was recorded accordingly. Thus a system having the goal the recognizing the Eye Accessing Cues should be based on a normal digital video camera.

As for the usability of the NLP-EAC hypothesis one can imagine many areas. We will give just two simple examples.

The first example of how can one use the NLP-EAC hypothesis refers to online interviews. Small and medium enterprises may look for employees at a distance and the interview is usually online. In this case the candidate is recorded (sometimes by his/her own device) and, given a query, discrimination between remembering type of activities (looking left) and the constructing ones (looking right) can differentiate experience from creativity. But it is imperative that the interviewed person is not aware that his/hers non-verbal messages are recorded and analyzed. The method described in this chapter requires typical recoding devices for video transmission; thus no distinct means of recording (head mounted camera or/and active light sources) are involved.

The second use-case deals with interactive communication for marketing and training. If the interaction is face-to-face, the meeting may be recorded and analyzed either real-time (with conclusions being shown to the presenter), or before the next session (such that the trainer/seller will ensure that maximum of information reaches his interlocutors). If the communication is online, the restrictions are similar with online interviews: the subject must have access to a recording device which is usually a typical webcam.

3 Databases

3.1 Iris Center Annotated Databases

To set the introductory reports on iris center localization performance, the BioID database[1] is the most popular choice. This database contains 1521 gray-scale, frontal facial images of size 384×286, acquired with frontal illumination conditions in a complex background. The database contains 16 tilted and rotated faces, people that wear eye-glasses and, in very few cases, people that have their eyes closed (i.e. blink) or pose various expressions. The database was released with annotations for iris centers. Being one of the first databases that provided facial annotations, BioID became the most used database for face landmarks localization accuracy tests, even if it provides limited variability and reduced resemblance with real-life cases.

As many methods use the entire eye area as learned template, robustness to face expression should be envisaged, as it induces eye shape changes. In this sense the most appropriate choice would be the Cohn-Kanade database[2] [26]. This database

[1]http://www.bioid.com/downloads/software/bioid-face-database.html.

[2]http://www.pitt.edu/~emotion/ck-spread.htm.

was developed for the study of emotions, contains frontal illuminated portraits and it is challenging through the fact that eyes are in various poses (near-closed, half-open, wide-open).

Further, one should systematically evaluate the robustness of the iris localization methods with respect to lighting and pose changes. Appropriate tests may be conducted onto the Extended Yale Face Database B (B+)[3] [27]. The Extended Yale B database contains 16,128 gray-scale images of 28 subjects, each seen under 576 viewing conditions (9 poses × 64 illuminations). The size of each image is 640×480.

The BioID, Cohn-Kanade and Extended YaleB databases include specific variations as they are acquired under controlled lighting conditions with frontal faces only. In contrast, there are databases like the Labelled Face Parts in the Wild (LFPW) [28] and the Labelled Faces in the Wild[4] (LFW) [29], which are randomly gathered from the Internet and contain large variations in the imaging conditions. While LFPW is annotated with facial point locations, only a subset of about 1500 images is made available and contains high resolution and rather qualitative images. In opposition, the LFW database contains more than 12,000 facial images, having the resolution 250 × 250 pixels, with 5700 individuals that have been collected "in the wild" and vary in pose, lighting conditions, resolution, quality, expression, gender, race, occlusion and make-up. The face landmarks[5] and iris centers are publicly available.

3.2 Iris Center and Eye Landmarks Annotated Databases

To evaluate the performance of the eye landmarking algorithm, four annotated databases are at hand, with publicly available ground-truth: EyeChimera, HPEG, ULM and PUT.

To study the specifics of the EAC detection problem, the *Eye Chimera* Database[6] [24, 30] was developed so that it contains all the seven cues. In generating the database, 40 subjects were asked to move their eyes according to a predefined pattern and their movements were recorded. The movements between consecutive EACs were identified, the first and last frame of each move were selected and labelled with the corresponding EAC tag and eye points were manually marked. In total, the database comprises 1170 frontal face images, grouped according to the seven directions of gaze, with a set of five points marked for each eye: the iris center and four points delimiting the bounding box. Additionally, for more extensive

---

[3] vision.ucsd.edu/~leekc/ExtYaleDatabase/ExtYaleB.html.

[4] The database is available at http://vis-www.cs.umass.edu/lfw/.

[5] At http://blog.gimiatlicho.webfactional.com/?page_id=38.

[6] http://imag.pub.ro/common/staff/cflorea/EyeChimeraReleaseAgreement.pdf.

testing, the still Eye Chimera database was extended with all the consecutive frames that are part of each basic eye movement; this part was named Eye Chimera Sequences.

The HPEG database[7] [31] is given as videos and we have extracted frames with relevant gaze variation, resulting in 233 images (640×480 resolution) of 10 persons, who's eye gaze varies from left to right (no vertical gaze direction is available). The head position includes yaw variations from $-30°$ to $+30°$. The dataset contains two sessions, one in a close-up arrangement while the other with people placed more distantly from the camera. The database comes with annotation related to the head angle, but without landmark positions; these were added in [30] and are available online.

The ULM head and gaze database[8] [32] contains images (1600×1200 pixels) of 20 persons. Variations include gaze direction (left to right), and head pose on both yaw and pitch. The database contains annotation for six eye landmarks: inner eye limits, outer eye limits and pupil centers. Because not all the characters have the images marked and in tests the images with yaw or pitch angles higher in absolute value than $30°$, are excluded; only 335 images have been kept and were used in the current study.

The PUT database[9] [33], is built in a similar manner with ULM. However the marking set is more complete as it contains all ten landmarks envisaged for the eye regions. Overall, it contains slightly more than 1000 annotated images.

4   System Overview

Approaches to the mentioned problem [10, 30, 34, 35] assume a scenario where the image acquisition is done with a single camera with fixed, near-frontal position, under free natural lighting.

The discussed algorithms rely solely on gray-scale images and a coarse-to-fine approach is used for localization, succeeded by gaze direction recognition. The possible schematic of such a system may be followed in Fig. 2.

First, the face bounding box is retrieved; the preferred method is derived from the Viola-Jones algorithm [36]. Given the face bounding box, the image is re-scaled at a fixed size. Next a rough estimation of the iris center should be retrieved. While in some works the iris center is treated as one of the landmarks, yet many of the current solutions make use of the physical particularities of the iris–pupil structure and are specifically optimized for iris center. The rough iris center, even it may be improved

---

[7] http://emotion-research.net/toolbox/toolboxdatabase.2010-02-03.4835728381.

[8] http://www.uni-ulm.de/in/neuroinformatik/mitarbeiter/g-layher/image-databases.html.

[9] https://biometrics.cie.put.poznan.pl.

Fig. 2 The system for recognizing the gaze direction

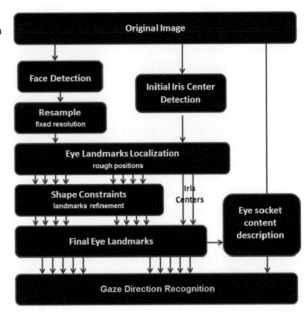

and does vary with respect to gaze direction, is considerable more precise than the face square, thus it acts as a better initialization reference for the position of other landmarks.

The subsequent major step of the procedure refers to the localization of the eye socket landmarks. The prerequisite here are some potential areas deduced using anthropometric criteria and the initial iris position. The process of localization, due to take place in a rather uncommon condition set (gaze variation), usually contains a two step procedure; the initial rough positioning is followed by a refinement step in which case the eye socket landmarks make use of shape constraints.

Once the eye region is determined it is separated and further analyzed for recognizing the gaze direction. While in previous similar solutions this step is referred as gaze (eye) tracking, for the application set envisaged here the outcome is a categorical set of directions (e.g. three or seven directions), thus it seems more appropriate to name it gaze direction recognition.

4.1 Face Detection and Localization

The first step of the system is locating the face bounding box. While many method have been proposed, by far, the most popular is still the boosted cascade of Haar features introduced by Viola and Jones [36]. Currently there exist many public implementations, the OpenCV version being one of the most used.

A recent reevaluation by Mathias et al. [37] showed that, if trained properly, the Vanilla Deformable Part Models [38] reaches top performance. Additional choices for face detection problem may be found on the Face Detection Data Set and Benchmark web page[10] and more recently on the Fine-grained Evaluation on Face detection in the Wild.[11]

5 Iris Center Localization

The problem of iris center localization was well investigated in literature, within a long history, as showed in the review by Song et al. [39]. Methods for eye center (or iris or pupil) localization in passive, remote imaging may approach the problem either as a particular case of pattern recognition application, [40, 41] or by using the physical particularities of the eye, like the high contrast with respect to the neighboring skin [42] or the circular shape of the iris [43]. More recent methods combine the two approaches.

As a general observation, we note that while older solutions [42, 44], tried to estimate also the face position, since the appearance of the Viola-Jones face detection solution [36], eye center search is limited to a subarea within the upper face square. Taking into account the recent advances on the face detection problem, one may truthfully assume that reconsideration of older eye methods may lead to better results than initially stated.

5.1 State of the Art Solutions

In this chapter we will point the attention of the reader to a very fast and robust iris center localization method based on zero-crossing encoded image projection. A list of other methods used for iris centers localization may be retrieved from the review by Song et al. [39] and from the summary presented in Table 1 and following paragraphs.

5.1.1 Projections Based Iris Localization Methods

The same image projections as in the work of Kanade [45] are used to extract information for eye localization in a plethora of methods [46–48]. Feng and Yuen [46] started with a snake based head localization followed by anthropometric reduction (relying on the measurements of Verjak and Stephancic [49]) to the

---

[10]http://vis-www.cs.umass.edu/fddb/results.html.

[11]http://www.cbsr.ia.ac.cn/faceevaluation.

Table 1 Iris center localization methods in remote imaging

Method	Face detection	Features	Machine learning	Public code
Jesorsky et al. [44]	No	Edges	MLP	No
Feng and Yuen [46]	No	VPF	No	No
Zhou and Geng [47]	No	GPF	No	No
Turkan et al. [48]	Yes	EPF	SVM	No
Cristinacce et al [50]	Yes	Image pixels	PRFR	Yes (AAM)
Campadelli et al. [51]	Yes	Haar wavelets	No	No
Hamouz et al. [40]	No	Gabor filters	SVM	No
Niu et al. [52]	Yes	Haar wavelets	AdaBoost	No
Kim et al. [53]	Yes	Image pixels	AdaBoost	No
Asteriadis et al. [41]	Yes	DVF	minDist	No
Valenti and Gevers [43]	Yes	Isophote	Yes	Yes
Asadifard [54]	Yes	Gradient	No	No
Ding and Martinez [55]	Yes	Pixels	AdaBoost/DA	Yes
Timm and Barth [56]	Yes	Image pixels	No	Yes
Kawulok and Szymanek [57]	Yes	Edges	SVM	Yes
Florea et al. [58]	Yes	Image pixels	MLP	Yes

so-called eye-images and introduce the variance projections for localization. The key points of the eye model are the projections particular values, while the conditions are manually crafted.

Zhou and Geng [47] described convex combinations between integral image projections and variance projections that are named generalized projection functions (GPF). These are filtered and analyzed for determining the center of the eye. The analysis is also manually crafted and requires identification of minima and maxima on the computed projection functions. Yet in specific conditions, such as intense expression or side illumination, the eye center does not correspond to a minima or a maxima in the projection functions.

Turkan et al. [48] introduced the edge projections and used them to roughly determine the eye position. Given the eye region, a feature is computed by concatenation of the horizontal and vertical edge image projections. Subsequently, a SVM-based identification of the region with the highest probability is used for marking the eye. Florea et al. [58] also employed a projection based description of the eye region coupled with machine learning; yet this method aims for higher precision and robustness, so multiple projection types are used and simple but efficient dimensionality reduction techniques are employed for speeding up the process.

5.1.2 Pattern Recognition Based Methods

There are many other approaches to the problem of eye localization. Jesorsky et al. [44] proposed a face matching method based on the Hausdorff distance followed by a MLP eye finder. Wu and Zhou [42] even reversed the order of the typical procedure: they used eye contrast specific measures to validate possible face candidates.

Cristinacce et al [50] relied on the Pairwise Reinforcement of Feature Responses algorithm for feature localization. Campadelli et al. [51] used SVM on optimally selected Haar wavelet coefficients.

Hamouz et al. [40] refined with SVM the Gabor filtered faces, for locating ten points of interest; yet the overall approach is different from the face feature fiducial points approach. Niu et al. [52] used an iteratively bootstrapped boosted cascade of classifiers based on Haar wavelets. Kim et al. [53] use multi scale Gabor jets to construct an Eye Model Bunch. Asteriadis et al. [41] used the distance to the closest edge to describe the eye area. Valenti and Gevers [43, 59] used isophote's properties to gain invariance and follow with subsequent filtering with Mean Shift (MS) or nearest neighbor on SIFT feature representation for higher accuracy. Asadifard [54] relied on thresholding the cumulative histogram for segmenting the eyes. Ding and Martinez [55] trained a set of classifiers (with a SVM of with discriminant analysis—DA) to detect multiple face landmarks, including explicitly the pupil center, by using a sliding window approach and test in all possible locations and inter-connect them to estimate the overall shape. Timm and Barth [56] relied their eye localizer on gradient techniques and search for circular shapes. Kawulok and Szymanek [57] fit a multilevel ellipsoid regressed from Hough accumulator planes over the face and the eyes and optimize the localization using SVM.

5.2 Robust Eye Centers Localization with Zero-Crossing Encoded Image Projections

Recently, Florea et al. [58] proposed a framework for the eye centers localization by the joint use of encoding of normalized image projections and a Multi Layer Perceptron (MLP) classifier. This encoding consists in identifying the zero-crossings and extracting the relevant parameters from the resulting modes.

5.2.1   Integral and Edge Image Projections

The integral projections, also named integral projection functions (IPF) or amplitude projections, are tools that have been previously used in face analysis. They appeared as "amplitude projections" [60] or as "integral projections" [45] for face recognition. For a gray-level image sub-window $I(i,j)$ with $i = i_m \ldots i_M$ and $j = j_n \ldots j_N$, the projection on the horizontal axis is the average gray-level along the columns (1), while the vertical axis projection is the average gray-level along the rows (2):

$$P_H(j) = \frac{1}{i_M - i_m + 1} \sum_{i=i_m}^{i_M} I(i,j), \forall j = j_n, \ldots, j_N \tag{1}$$

$$P_V(i) = \frac{1}{j_N - j_n + 1} \sum_{j=j_n}^{j_N} I(i,j), \forall i = i_m, \ldots, i_M \tag{2}$$

To increase the robustness to side illumination, edge projection functions (EPF) could be used to complement the integral ones. To determine them, the classical horizontal and vertical Sobel contour operators are applied, resulting in S_H and S_V, which are combined in the $S(i,j)$ image used to extract edges:

$$S(i,j) = S_H^2(i,j) + S_V^2(i,j) \tag{3}$$

The edge projections are computed on the corresponding image rectangle $I(i,j)$ by replacing I with S in Eqs. (1) and (2).

5.2.2   Fast Computation of Projections

While sums over rectangular image sub-windows may be easily computed using the concept of summed area tables [61] or integral image [36], a fast computation of the integral image projections may be achieved using the *prefix sums* [62] on rows and respectively on columns. A prefix-sum is a cumulative array, where each element is the sum of all elements to the left of it, inclusive, in the original array. They are the 1D equivalent of the integral image, but they definitely precede it.

For the fast computation of image projections, two tables are required: one will hold prefix sums on rows (a table which, for keeping the analogy with the integral image, will be named horizontal 1D integral image) and respectively one vertical 1D integral image that will contain the prefix sums on columns. It should be noted that computation on each row/column is performed separately. Thus, if the image has $M \times N$ pixels, the 1D horizontal integral image, on the column j, $\mathscr{I}_H^j$, is:

$$\mathscr{I}_H^j(i) = \sum_{k=1}^{i} I(k, j) \ , \forall i = 1, \ldots M \tag{4}$$

Thus, the horizontal integral projection corresponding to the rectangle $i = [i_m; i_M] \times [j_n; j_N]$ is:

$$P_H(j) = \frac{1}{i_M - i_m + 1} \left(\mathscr{I}_H^j(i_M) - \mathscr{I}_H^j(i_m - 1) \right) \tag{5}$$

Using the oriented integral images, the determination of the integral projections functions on all sub-windows of size $K \times L$ in an image of $M \times N$ pixels requires one pass through the image and $2 \times M \times N$ additions, $2 \times (M-K) \times (N-L)$ subtractions and two circular buffers of $(K+1) \times (N+1)$ locations, while the classical determination requires $2 \times K \times L \times (M-K) \times (N-L)$ additions. Hence, the time to extract the projections associated with a sub-window, where many sub-windows are considered in an image, is greatly reduced.

The edge projections require the computation of the oriented integral images over the Sobel edge image, $S(i, j)$ which needs to be found on the areas of interest.

5.2.3 Encoding and ZEP Feature

To reduce the complexity (and computation time), the projections are compressed using a zero-crossing based encoding technique. After ensuring that the projections values are in a symmetrical range with respect to zero, one will describe, independently, each interval between two consecutive zero-crossings. Such an interval is called an *epoch* and for its description three parameters are considered (as presented in Fig. 3):

- *Duration*—the number of samples in the epoch;
- *Amplitude*—the maximal signed deviation of the signal with respect to 0;
- *Shape*—the number of local extremes in the epoch.

The proposed encoding is similar with the TESPAR (Time-Encoded Signal Processing and Recognition) technique [63] that is used in the representation and recognition of 1D, band-limited, speech signals. Depending on the problem specifics, additional parameters of the epochs may be considered (e.g. the difference between the highest and the lowest mode from the given epoch). Further extensions are at hand if an epoch is considered to be the approximation of a probability density function and the extracted parameters are the statistical moments of the said distribution. In such a case the *shape* parameter corresponds to the number of modes of the distribution.

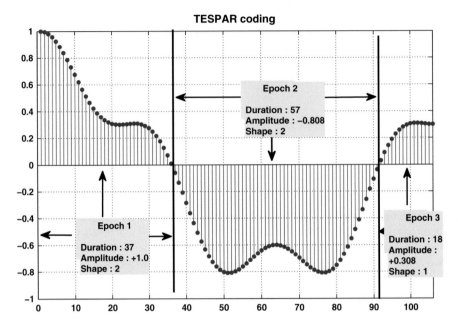

Fig. 3 Example of 1D signal (vertical projection of an eye crop) and the associated encoding. There are three epochs, each encoded with three parameters. The associated code is: $[37, +1, 2; 57, -0.808, 2; 18, 0.308, 1]$

The reason for choosing this specific encoding is twofold. First, the determination of the zero-crossings and the computation of the parameters can be performed in a single pass through the target 1D signal, and, secondly, the epochs have specific meaning when describing the eye region.

Given an image sub-window, the ZEP feature is determined by the concatenation of four encoded projections as described in the following:

1. Compute both the integral and the edge projection functions (P_H, P_V, E_H, E_V);
2. Independently *normalize* each projection within a symmetrical interval. For instance, each of the projections is normalized to the $[-1; 1]$ interval. This will normalize the amplitude of the projection;
3. Encode each projection as described; allocate for each projection a maximum number of epochs;
4. Normalize all other (i.e. duration and shape) encoding parameters;
5. Form the final Zero-crossing based Encoded image Projections (ZEP) feature by concatenation of the encoded projections. Given an image rectangle, the ZEP feature consists of the epochs from all the four projections: (P_H, P_V, E_H, E_V).

Image projections are simplified representations of the original image, each of them carrying specific information; the encoding simplifies even more the image representation. The normalization of the image projections, and thus of the epochs amplitudes, ensures independence of the ZEP feature with respect to uniform

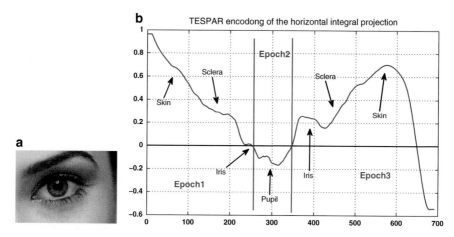

Fig. 4 Horizontal image projection function from a typical eye patch: (**a**) eye crop, (**b**) the integral projection on the eye crop and the physical features marked. The *vertical lines* mark the zero crossing that are typically found on all eye examples

variation of the illumination. The normalization with respect to the number of elements in the image sub-window leads to partial scale invariance: horizontal projections are invariant to stretching on the vertical direction and vice versa. The scale invariance property of the ZEP feature is achieved by completely normalizing the encoded durations to a specific range (e.g. the encoded horizontal projection becomes invariant to horizontal stretching after duration normalization).

In [47] it was noted that image projections in the eye region have a specific sequence of relative minima and maxima assigned with to skin (relative minimum), sclera (relative maximum), iris (relative minimum), etc.

Considering a rectangle from the eye region including the eyebrow (as showed in Fig. 4a), the associated integral projections have specific epochs, as showed in Fig. 4b. The particular succession of positive and negative modes is precisely encoded by the this technique. On the horizontal integral projection there will be a large (one-mode) epoch that is assigned to skin, followed by an epoch for sclera, a triple mode, negative, epoch corresponding to the eye center and another positive epoch for the sclera and skin. On the vertical integral projection, one expects a positive epoch above the eyebrow, followed by a negative epoch on the eyebrow, a positive epoch between the eyebrow and eye, a negative epoch (with three modes) on the eye and a positive epoch below the eye.

The ZEP feature, due to invariance properties already discussed, achieves consistent performance under various stresses and is able to discriminate among eyes (patches centered on pupil) and non-eyes (patches centered on locations at a distance from the pupil center).

Fig. 5 The work flow of the rough iris center localization algorithm

5.2.4 Rough Iris Center Localization

The block schematic of the initial rough eye center localization algorithm is summarized in Fig. 5. The main problem when employing projections for iris localization is their susceptibility to lateral illumination. In such a case, to avoid loosing accuracy, Florea et al. [58] proposed a preemptive detection of the lateral illumination case by observing that for a near-frontal placed light source the distribution of intensity should be symmetrical with respect to the nose area. This is a simple example of the *face relighting* problem and further details may be retrieved from [64].

The conceptual steps of the actual eye localization procedure are: preprocessing, machine learning and postprocessing.

A simple preprocessing is applied for each eye candidate region to accelerate the localization process. Wu and Zhou [42] noted that the pupil center is significantly darker than the surrounding; thus the pixels that are too bright with respect to the eye region (and are not plausible to be pupil centers) are discarded. The "too bright" characteristic is encoded as gray-levels higher than a percentage from the maximum value of the eye region. In the lateral illumination case, this threshold is higher due to the deep shadows that can be found on the skin area surrounding the eye.

In the area of interest, using a sliding window, all plausible locations are investigated by ZEP+MLP. To further accelerate the algorithm only some of the values should be further considered [58]. The potential positions are recorded in a separate image which is further post-processed for eye center extraction. Further attention needs to be given to the confusion between eye and eyebrows; a possible procedure is a pre-segmentation and to locate the iris one will look only to the lower darker region [10].

For the frontal illumination case, in the case of training with L2 distance as objective, one expects a symmetrical shape around the true eye center. Thus the final eye location is taken as the weighted center of mass of the previously selected eye regions. For the lateral illumination, the binary trained MLP is supposed to localize the area surrounding the eye center and the final eye center is the geometrical center of the rectangle circumscribed to the selected region. In both cases, the specific way of selecting the final eye center is able to deal with holes (caused by reflections or glasses) in the eye region.

The training of the machine learning system should be performed with crops of eyes and non-eyes. The positive examples are to be taken near the eye ground truth while the patches corresponding to the negative examples should also overlap with the true eye but to a lesser degree. This choice forces the machine learning to give importance in precise discrimination of patches centered on the iris and patches centered elsewhere.

5.2.5 Iris Center Refinement

If challenged by the gaze direction variation, the rough method does not perform very well and further refinement is necessary [30].

To improve the performance of some initial iris center localization method one may consider a small, centered region of interest (ROI) (e.g. of size $\frac{d_{eye}}{5} \times \frac{d_{eye}}{5}$, where d_{eye} is the inter-ocular distance) around the reported eye center. The improvement of the iris centers, as well as the detection of the eye socket limits require position and intensity priors and template matching. These are identical with eye landmarks localization procedure and will be discussed in the next paragraphs.

5.3 Evaluation of the Iris Center Localization Methods

5.3.1 Evaluation on the BioID, Cohn-Kanade, Yale B+ and LFW Databases

The iris centers localization performance is typically evaluated according to the stringent localization criterion [44]. The eyes are considered to be correctly determined if the specific localization error ϵ, defined in Eq. (6) is smaller than a predefined value.

$$\epsilon = \frac{\max\{\varepsilon_L, \varepsilon_R\}}{D_{eye}} \qquad (6)$$

In the equation above, ε_L is the Euclidean distance between the ground truth left eye center and determined left eye center, ε_R is the corresponding value for the right eye, while D_{eye} is the distance between the ground truth eyes centers. Typical error thresholds are $\epsilon = 0.05$ corresponding to eyes centers found inside the true pupils, $\epsilon = 0.1$ corresponding to eyes centers found inside the true irises, and $\epsilon = 0.25$ corresponding to eyes centers found inside the true sclera. This criterion identifies the worst case scenario.

To give an initial overview of the problem in state of the art, we report in Table 2 the results of multiple methods performance on the BioID database.

Considering as most important criterion the accuracy at $\epsilon < 0.05$, it should be noted that Timm and Barth [56] and Valenti and Gevers [59] provide the highest accuracy. Yet, the best performance achieved by a variation of the method described in [59], namely Val.&Gev.+SIFT contains a tenfold testing scheme, thus using nine

Table 2 Comparison with state of the art (listed in chronological appearance) in terms of localization accuracy on the BioID database

Method	Accuracy		
	$\epsilon < 0.05$	$\epsilon < 0.1$	$\epsilon < 0.25$
Florea et al. [58]	70.46	91.94	98.89
Jesorsky et al. [44]	40.0	79.00	91.80
Wu and Zhou [42]	10.0*	54.00*	93.00*
Zhou and Geng [47]	47.7	74.5	97.9
Cristinacce et al. [50]	55.00*	96.00	98.00
Campadalli et al. [51]	62.00	85.20	96.10
Hamouz et al. [40]	59.00	77.00	93.00
Turkan et al. [48]	19.0*	73.68	*99.46*
Kroon et al. [65]	65.0	87.0	98.8
Asteriadis et al. [41]	62.0*	89.42	96.0
Asadifard et al. [54]	47.0	86.0	96.0
Timm and Barth [56]	82.50	*93.40*	98.00
Val.&Gev. [59]+MS	81.89	87.05	98.00
Val.&Gev. [59]+SIFT[†]	*86.09*	91.67	97.87

The correct localization presents results reported by authors; values marked with "*" were inferred from authors plot. While Zhou [47] reports only the value for $\varepsilon < 0.25$, the rest is reported by Ciesla and Koziol [66]. The method marked with † relied on a tenfold training/testing scheme, thus, at a step, using nine parts of the BioID database for training

parts of the BioID database for training. Furthermore, taking into account that BioID database was used for more that 10 years and provides limited variation, it has been concluded [28, 67] that other tests are also required to validate a method.

Valenti and Gevers [59] provide results on other datasets and made public the associated code for their baseline system (Val.&Gev.+MS), which is not database dependent. Timm and Barth [56] do not provide results on any other database except BioID or source code, yet there is publicly available[12] code developed with author involvement. Thus, in continuation, we will compare the method from [58] against these two on other datasets. Additionally, we include the comparison against the eye detector developed by Ding and Martinez [55] which has also been trained and tested on other databases, thus is not BioID dependent.

As mentioned in the introduction, the purpose of this chapter is to investigate emergent application with potential on behavior inference. Thus we will report eye localization performance with respect to facial expressions, as real-life cases with fully opened eyes looking straight are rare. We tested the performance of various methods on the Cohn-Kanade database [26]. This database was developed for the study of emotions, contains frontal illuminated portraits and it is challenging

---

[12]http://thume.ca/projects/2012/11/04/simple-accurate-eye-center-tracking-in-opencv/.

Table 3 Percentage of correct eye localization on the Cohn-Kanade database

Method	Type	Accuracy		
		$\epsilon < 0.05$	$\epsilon < 0.1$	$\epsilon < 0.25$
Florea et al. [58]	Neutral	76.0	99.0	100
	Apex	71.9	95.7	100
	Total	73.9	97.3	100
Valenti and Gevers [59]	Neutral	46.0	95.7	99.6
	Apex	35.1	92.4	98.8
	Total	40.6	94.0	99.2
Timm and Barth [56]	Neutral	66.0	95.4	99.0
	Apex	61.4	85.1	93.4
	Total	63.7	90.2	96.2
Ding and Martinez [55]	Neutral	14.3	75.9	100
	Apex	11.8	72.8	100
	Total	13.1	74.4	100

We report results of the methods form [55, 56, 58, 59] on the neutral poses, expression apex and overall. We marked with italics the best achieved performance for each accuracy criterion and respectively for each image type

Table 4 Comparative results on the Extended YaleB database. We marked with bold letters the best performance for each accuracy category

Method	$\epsilon < 0.05$	$\epsilon < 0.1$	$\epsilon < 0.25$
Florea et al. [58]	*39.9*	*67.3*	97.3
Valenti and Gevers [59]	37.8	66.6	**98.5**
Timm and Barth [56]	20.1	34.5	51.5
Ding and Martinez [55]	19.7	47.8	58.6

through the fact that eyes are in various poses (near-closed, half-open, wide-open). We tested only on the neutral pose and on the expression apex image from each example.

We note that solutions that try to fit a circular or a symmetrical shape over the iris, like the ones from [59] or [56], and thus, performs well on open eyes, do encounter significant problems when facing eyes in expressions (as it is shown in Table 3). Taking into account the achieved results, which are comparable on neutral pose and expression apex images, it is to be seen which method performs very well under such complex conditions. Results indicate that [58] achieved higher accuracy when compared with rest of the method tested.

A systematic evaluation of the robustness of iris center localization algorithms with respect to lighting and pose changes may be done using the Extended Yale Face Database B (B+) [27]. Here, by a small margin the best performance is obtained by the method from [58], followed closely by the one from [59]. The methods proposed by Timm and Barth [56] and respectively Ding and Martinez [55] have greater susceptibility to errors due to uneven illumination (Table 4).

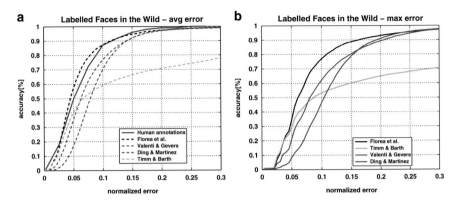

416

L. Florea et al.

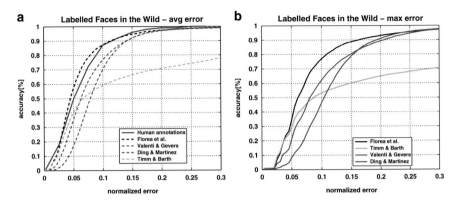

Fig. 6 Results achieved on the LFW database: (**a**) average error and (**b**) maximum error as imposed by the stringent criterion—Eq. (6). With *dashed blue line* is the average error for human evaluation

The difficulty of localizing features on the LFW database is certified by the performance of human evaluation error as reported in [67]. While the ground truth is taken as the average of human markings for each point normalized to inter-ocular distance, human evaluation error is considered as the averaged displacement of the one marker. Regarding the achieved results, the ZEP based method [58] provides the most accurate results, as one can see in Fig. 6. The improvement is by almost 50 % at $\epsilon < 0.05$ compared with [59] and from [56] and with more over [55].

5.3.2 Evaluation on Databases with Gaze Variability

The main purpose of this chapter is to discuss the potential application of computer vision methods for the communication by gaze induced methods; thus we will evaluate the performance of various methods on databases illustrating gaze variability. In Table 5, for comparison purposes, we report the performance the method from [58] and of its refinement from [30]. Additional we report the results achieved with the method of Sun et al. [68]. This localizes face landmarks; yet from their searched set, of interest for the current problem are only the iris centers; thus we report their performance specifically to the iris center section.

The highest accuracy is achieved by the deep convolution network based system [68], with very high results especially for medium accuracy ($\varepsilon = 0.1$). Yet, multistage method from [30] clearly outperforms it for high accuracy ($\varepsilon = 0.05$), which is critical in achieving high EAC recognition rate.

Table 5 Iris center localization accuracy measured as percentage of the inter-ocular distance (stringent criterion [44])

Method	Database					
	Eye Chimera		HPEG		ULM	
	$\varepsilon = 0.05$ (%)	$\varepsilon = 0.1$ (%)	$\varepsilon = 0.05$ (%)	$\varepsilon = 0.1$ (%)	$\varepsilon = 0.05$ (%)	$\varepsilon = 0.1$ (%)
Florea et al. [58]	42.7	68.9	53.5	78.8	32.5	74.1
Timm and Barth [56]	33.6	67.8	46.9	82.0	61.2	85.2
Valenti and Gevers [59]	16.1	50.5	24.6	55.7	50.1	77.0
Sun et al. [68]	59.9	91.2	54.2	90.3	60.8	93.2
Florea et al. [30]	65.3	78.7	71.1	83.2	76.6	92.7

Fig. 7 The five eye landmarks searched

6 Eye Landmarking

In order to locate properly the position of the eye for further processing while determining the direction of gaze, one needs to identify the eye landmarks. The typical set of eye landmarks is showed in Fig. 7. While there have been developed methods focusing only on the eye, most of the state of the art methods are general face landmarking methods. An overview of some of the most relevant such methods may be followed in Table 6.

Facial landmarking originates in the classical holistic approaches of Active Shape Models (ASMs) [69], Active Appearance Models (AAMs) [70] and Elastic Graph Matching [71]. Active Shape Models describe an eigenspace of the geometrical shape having as vertices the landmarks, while the AAM complement the information with pixel values from shape interior.

Building on the ASMs/AAMs versatility, a multitude of extensions appeared. For instance in [72], a 2D profile model and a denser point set are used. In [73], higher independence between the facial components is encouraged while the actual fitting step is further optimized by a Viterbi optimization process.

In the later period, the ASM underlying holistic fitting switched to independent models built on top of local part detectors, to form the so-called Constrained Local Models (CLMs) [74], or to a combination of local shape models and PCA-based global shapes, as in [75]. The CLM model was extended with full voting from a random forest in [76] or by probabilistic interpretation for optimization of the shape parameters in [77].

Another direction assumes independent point localizer followed by aggregation of location. This is a rich class of methods, including some of the most recent and accurate solutions. Thus Valstar et al. [78] complemented the SVM regressed feature point location with conditional MRF to keep the estimates globally consistent. Belhumeur et al. [28] proposed a Bayesian model combining the outputs of the local detectors (formed by SVM classifier trained over SIFT features) with a consensus of non-parametric global models for part locations; Zhu and Ramanan [79] relied on a connected set of local templates described with HOG. Dantone et al. [67] constructed multiple random forrest that use image patches as input conditioned by the head pose to estimate in real time a set of face landmarks. Yu et al. [80] used 3D deformable shape models to iteratively fit over 2D data and identify without respect to pose the facial landmarks positions.

Martinez et al. [81] trained SVM regressors with selected Local Binary Patterns to formulate initial predictions that are further iteratively aggregated for improving accuracy. In [68], the relation between fiducial points is encoded directly in the localization system, which is based on deep convolutional networks. Florea et al. [30] used a multi stage method for precisely fitting the landmarks in the eye are. In the next subsection we will detail this method as it is focussed on the eyes.

6.1 Multi-level Eye Landmark Localization

6.1.1 Position and Intensity Priors

During training, a bounding box is computed on the range of each feature location within the region found by the face detector, as in the classical CLM [74]. For each candidate landmark, its *position prior* ic constructed as a probability map spanning its region of interest (ROI), such that each position is given the likelihood of being close to the ROI center. This is in fact the two-dimensional histogram of the positions.

The position prior is further denoted as $p_1(i, j)$ and an example for the left eye outer corner position prior map is shown in Fig. 8a.

While for eye center localization it is common to investigate only the darkest pixels [82], this idea may be extended to all landmarks with appropriate conditions (such as considering that the inner eye corner is darker than most of its neighboring pixels while the upper limit of the eye is brighter than most of its neighboring pixels). Thus, for each candidate landmark, its *intensity prior* is constructed as a probability map spanning its region of interest, such that each position is given the probability of occurrence of its corresponding graylevel within the ROI. This prior is denoted as $p_2(i, j)$ and is exemplified in Fig. 8b for the case of the outer corner of the left eye.

Table 6 Face landmarks localization methods in remote imaging

Method	Face detection	Features	Learning	Public code
Cootes et al. [69]—ASM	No	Edges	Yes	Yes
Cootes et al. [70]—AAM	No	Edges and pixels	Yes	Yes
Leung et al. [71]—EGM	No	VPF	No	No
Cristinacce et al. [74]—CLM	Yes	Edge	Yes	Yes
Tresadern et al. [75]	Yes	Edge+pixels	MRF	No (AAM)
Saragih et al. [77]—PAAM	Yes	Edge+pixels	Yes	Yes
Milborrow and Nichols [72]—STASM	Yes	Edges	Yes	Yes
Valstar et al. [78]—Borman	Yes	Image pixels	SVR+MRF	Yes
Zhu and Ramanan [79]	Yes	pHoG	Bagged trees	Yes
Belhumeur et al. [28]	Yes	SIFT	SVM	No
Dantone et al. [67]	Yes	Pixels	RF	No
Martinez et al. [81]—Lear	Yes	Gradient	SVR	Yes
Sun et al. [68]	Yes	Pixels	CNN	Yes
Yu et al. [80]	Yes	HoG	EM	Yes
Florea et al. [30]	Yes	IPF+EPF	MLP	Yes

The learning column signals that method parameters are regressed on a training database; we nominate when a specific machine learning system is employed

6.1.2 Template Matching

It is typical in the landmark localization [28, 30, 67] that the bulk of the search to be performed by a template matching procedure. In such a case, the problem is to determine for each pixel, within a reasonable neighborhood of an initial landmark, the probability of that pixel being the true landmark. A typical procedure implies considering consecutive sub-windows in the search area. Each sub-window is centered in the investigated location and it is represented in a descriptor space; a machine learning system is then trained to determine how likely is for the sub-window to be centered on the true position of the landmark.

Similar with the iris center localization procedure, given a rectangular sub-windows centered in the truth landmark position and a Multi-Layer Perceptron (MLP) is trained with the integral and edge projections within that image sub-window. The definition of the IPF and EPF were presented in Sect. 5.

To ensure a better robustness to illumination variation, each of the projections should be independently normalized to the $[-128, 127]$ range. Each sub-window, for

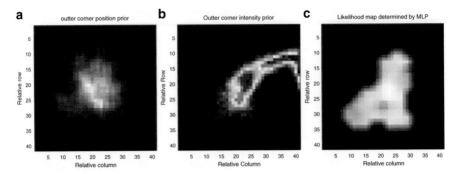

Fig. 8 (**a**) Position prior map p_1 for left eye outer corner, gathered from all training images; (**b**) Intensity prior map p_2 for left eye outer corner; (**c**) Likelihood Map determined by the MLP given the actual data, p_3 for a specific example of left eye outer corner. Image taken from [30]. Copyright of the authors

each landmark, is thus described by a feature vector formed by the concatenation of the four projections: horizontal/vertical, integral/edge. Thus, the length of the features is $2 \cdot W + 2 \cdot H = 120$.

A MLP is separately trained for each landmark (with feature vectors from the sub-windows as input) to output the Euclidian distance between the center of the input sub-window and the true landmark position (thus performing regression). Thus, given an original image and a landmark to localize, the complement of the output of the MLP, $c(i,j)$ is the landmark likelihood map $p_3(i,j)$, with $p_3(i,j) = 1 - c(i,j)$. A MLP with one hidden layer with 30 neurons is good choice [30]. A typical p_3 map is showed in Fig. 8c for the location of the outer left eye corner.

Alternatively, Vranceanu et al. [10] feed the concatenated image projection into a logistic regressor to determine only one dimension of the landmarks in order to obtain the bounding box of the eye.

6.1.3   Initial Landmarks Estimates

To fuse the information from intensity and position priors with template matching data, the final probability map is:

$$p_{123}(i,j) = \alpha \cdot p_1(i,j) + \beta \cdot p_2(i,j) + \gamma \cdot p_3(i,j) \tag{7}$$

where α, β and γ weight the confidence in each type of map; their values are to be deduced on the training database independently for each landmark, for each eye. Florea et al. [30] suggested the following empirical values $\alpha = \beta = 0.25$, $\gamma = 0.5$. The *weighted center of mass* of the p_{123} probability map for each landmark represents an accurate initial estimate of that landmark.

6.1.4 Shape Constraints

Even if gaze varies across the tested cases, the relative position of the eye remains stable enough, thus the eye socket shape may be further constrained. To achieve this, either the Constrained Local Model [74] or the Global Models Locally Constrained [30] are at hand. The second solution, instead of gathering local constrains and maximizing the global output, for each landmark (locally), iteratively uses the global shape to construct a local constraint. As eye landmarks are investigated in terms of gaze variation, the iris center position has very large variations, thus it is not included in the shape. For each face there will be two shapes, one for each eye, refined independently; a shape contains only four points, namely the limits of the eye socket.

Distinct shapes for the left and the right eye are envisaged. In ASM [69], given s sets of points $\{\mathbf{x}_i\}$, after alignment (centering all shapes in origin and aligning them by a specific rule), one may compute the mean shape $\bar{\mathbf{x}}$, and the projection matrix on a PCA-reduced space $\mathbf{P}$. Any of the input shapes is given as:

$$\mathbf{x} \approx \bar{\mathbf{x}} + \mathbf{Pb} \tag{8}$$

The $\mathbf{P}$ matrix is made by the first t eigenvectors of the covariance matrix of all the shapes $\{\mathbf{x}_i\}$ in the training database. Thus, the transformation vector is found as:

$$\mathbf{b} = \mathbf{P}^T (\mathbf{x} - \bar{\mathbf{x}}) \tag{9}$$

Given a large number of shapes $\{\mathbf{x}_i\}$, one may compute a histogram for the transformation vector, $\mathbf{b}$. More precisely, for each shape, $\{\mathbf{x}_i\}$ in the database, using Eq. (9), one will obtain a t-dimensional transformation vector $\mathbf{b}_i$.

Since the dimensionality reduction is performed by means of decorrelation (i.e. PCA), and the histogram is Gaussian-like, the histogram of $\mathbf{b}$ may be approximated with t-dimensional independent Gaussian distributions with $\mathbf{0}$-mean and $\Sigma = diag(\lambda_i)$ covariance matrix, where λ_i is the shapes eigenvalue on the dimension i, in the original shape space.

Given an initial shape $\mathbf{x}'$ and keeping all the landmarks fixed with the exception of the current one (which is to be improved), if one assumes that this landmark is at (i, j) location, a new value, $\mathbf{b}'$ will be obtained. Given the original histogram of $\mathbf{b}$, the newly obtained value, $\mathbf{b}'$, is back-propagated into a probability value, which will be named $p_s(i, j)|\mathbf{x}'$.

Thus, for each location in the searched area and for each of the landmarks, $p_s(i, j)|\mathbf{x}$ the probability to have the landmark at position (i, j) given the remainder of the shape $\mathbf{x}$ to its initial position is computed. Practical choices are $t = 2$ and the alignment of the shapes such that to have a horizontal outer-inner axis [30].

Taking into account that at any step, for each landmark, an estimate of the shape is available by computing the weighted center of mass from Eq. (7) and observing that some landmarks (e.g. eye outer corners) are more reliable than others (upper and lower boundaries), by keeping all points fixed with the exception of the least reliable, the likelihood of various positions for the current landmark is built. The order of the landmarks is with respect to their reliability. Then, the procedure is repeated iteratively for each landmark N_{it} times (typically $N_{it} = 3$).

The *final landmark position* is taken as the weighted center of mass of the convex combination, $p_F(i,j)$, between the initial stages and the shape fitting likelihood:

$$p_F(i,j) = \delta \cdot p_{123}(i,j) + (1 - \delta) \cdot p_S(i,j)|\mathbf{x}, \tag{10}$$

where $\delta = 0.75$ was experimentally chosen.

6.2 Evaluation

6.2.1 Evaluation Procedure

For multiple landmarks the proximity measure [74] m_e is the common measure. It is computed as:

$$m_e = \frac{1}{t \cdot D_{eye}} \sum_{i=1}^{t} \varepsilon_i \tag{11}$$

In Eq. (11), ε_i are the point to point errors for each individual landmark location and t is the number of feature points searched (ten points in the current considered case, thus the measure being further referred as m_{e10}). Again, the interest is in obtaining a higher accuracy for low threshold values.

6.2.2 Eye Landmarks Localization

For this specific task, we compare the performance of the methods from [30, 78–81]. The accuracies are shown in Figs. 9 and 10. We note that the method from [81] does not locate the iris centers, thus in this case we take into account only eight landmarks and all methods are trained outside the tested databases.

The results show that mostly the best performance is achieved by Florea et al. [30]. The largest difference is on the Eye-Chimera database, where other methods, that are general face landmarking methods, significantly under-perform due to the significant variation of gaze on both horizontal and vertical directions.

7 Gaze Direction Recognition

In computer vision, extensive research was done in the field of detecting the direction of gaze [83, 84], by means of so-called eye trackers. Usually, *eye tracking* technology relies on measuring reflections of the infrared/near-infrared light on the

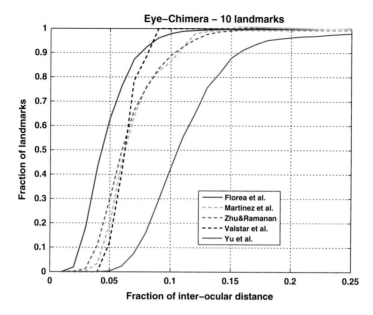

Fig. 9 Eye landmarks localization performance on the Eye-Chimera database of several methods: Valstar et al. [78], Zhu-Ramanan [79], Martinez et al. [81], Yu et al. [80] and Florea at al. [30] on the Eye-Chimera, HPEG and ULM databases

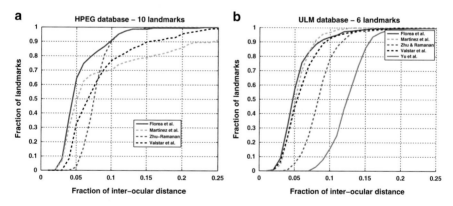

Fig. 10 Eye landmarks localization performance of several methods: Valstar et al. [78], Zhu-Ramanan [79], Martinez et al. [81], Yu et al. [80] and Florea at al. [30] on the HPEG (**a**) and ULM (**b**) databases

eye: the first Purkinje image (P1) is the reflection from the outer surface of the cornea, while the fourth (P4) is the reflection from the inner surface of the lens; these two images form a vector that is used to compute the angular orientation of the eye, in the so-called "dual Purkinje" method [84]. An example of such an eye tracking systems is, for instance, found in the work of Yoo and Chung [85] that relies on two cameras and four infrared sources to achieve high accuracy.

A method relying on a head mounted device with visible spectrum illumination is found in the work of Pires et al. [86] who extracted the iris contour followed by a Hough transform to detect the iris center and, respectively, by the localization of the eye corners contours; the head-mounted device, permits high resolution for the eye image thus extending the range of a wearable eye-tracking for sport. Due to reasons detailed in the previous subsection, we will avoid both the IR-based and, respectively, the head mounted category of solutions.

The alternative is to develop non-intrusive, low-cost techniques that directly measure the gaze direction, such as the approaches used in [34, 82, 87–89]. Wang et al. [82] selected recursive nonparametric discriminant features from a topographic image feature pool to train an Adaboost that locates the eye direction. Hansen and Pece [87] modelled the eye contour as an ellipse and use Expectation-Maximization to locally fit the actual contour. Cadavid et al. [88] trained a Support Vector Machine with spectrally projected eye region images to identify the direction of gaze. Heyman et al. [89] used correlation-based methods (more precisely the so-called Canonical Correlation Analysis) to match the new eye data with marked data and to find the direction of gaze. Wolf et al. [34] applied the eye landmark localizer provided by Everingham and Zisserman [90] to initialize the fit of the eye double parabola model. We note that all these methods first localize eye landmarks and subsequently analyze the identified eye regions.

Florea et al. [30] used the eye landmarks and the interior of the eye shape to directly estimate the gaze into seven possible directions. The same set of directions, which matches the NLP identified direction, was also searched by the methods form [10, 35]. Radlak et al. [35] employed Hybrid Image Projections functions as defined by Zhou and Geng [47] followed by either a SVM or a random forest. Vranceanu et al. [10] complemented the projection based information with the geometrical location of four segments extracted from the eye region.

Also in the direction of gaze recognition in terms of EAC-NLP we note the work of Diamantopoulos [91] which used a head mounted device. Taking into account that Laeng and Teodorescu [25] showed that, even for non-visual tasks, voluntary control affects eye movement, we may conclude that they explore the theme only from a computer vision perspective, without direct practical applications. Furthermore, the head mounted device has the un-realistic advantage of being closer to the eye and, thus, of having access to higher resolution and more precisely located eye image patches. For images with high resolution, the method implied by Pires et al. [86] (iris contour detection followed by Hough transform for circles) works very well. However, for the lower resolution images, which are associated with remote

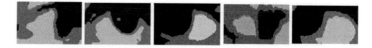

Fig. 11 Separation (segmentation) of the eye components using K-Means

acquisition devices, the contours in the eye region are no longer sharp and the accumulation in the Hough transform, very often, points to wrong locations (eye socket or brow).

Once the eye bounding box has been delimited by means of the eye landmarks, the specific EAC is retrievable by analyzing the positions of different eye components. The natural choice is to analyze the position of the iris inside the eye bounding box. Yet as eye localizers are imperfect [10], especially when challenged by gaze variation, to improve the accuracy, a separation of the components of the eye within the bounding box and use their relative position as indicators of the EAC, improves the landmark achievable performance.

7.1 Separating the Eye Components

7.1.1 Segmentation

The segmentation is a well known problem and many solutions have been proposed through the years. For the specific problem of the eye components separation for gaze direction estimation, it should be required that the segmentation of the eye components allows a good EAC recognition in a reasonable amount of time. According to the tests performed by Vranceanu et al. [10], the best compromise is achievable using a *K-Means* segmentation (Fig. 11).

7.1.2 Post-processing and Classification

The eye area, given by the detected landmarks is normalized to a standard size and position. The coordinates of each of the resulting eye components' centers of mass in the normalized bounding box and the average luminance are used as features describing the eye. To improve the region separation resulted from segmentation, the integral projections functions (IPF) complements the bounding box in a variation of the Appearance Models. Therefore, for a more general description inside the bounding box, the vertical and horizontal integral and edge projections are added as features for the classifier, next to the segmented regions center of mass and landmarks.

Table 7 Influence of the number of regions on the EAC recognition rate, RR (%), when simple K-Means segmentation is used (without additional projection information). We have marked with bold letters the best results for each EAC case

Regions no.	$C = 2$	$C = 3$	$C = 4$
RR (%) (7 EAC classes)	50.73	62.08	**64.56**
RR (%) (3 EAC classes)	62.77	77.28	**82.32**

In order to recognize the seven EAC classes, the feature vector is composed by:

- $3 \times C$ elements (which correspond to the centers of mass coordinates and the average luminance for each of the C regions);
- the concatenated horizontal and vertical integral and edge projections;
- landmarks

Various classification methods are considered and, as the number of features is small, the Logistic Classification [92] was found [10] to give good results.

7.2 Results

While computing the EAC recognition rate, two scenarios are evaluated: the seven-case and the three-case. The complete seven EACs set contains all the situations described by the NLP theory and presented in Fig. 1. Additionally, as the vertical direction of gaze is harder to identify [84], one may consider only three cases assigned to: looking forward (center), looking left and looking right; in terms of EACs, here, the focus is on the type of mental activity, while the representational systems are merged together. This particular test is relevant for the interview scenario, where, when given a query, if the subject remembers the solution, it indicates experience in the field, while if he/she constructs the answers, it points to creativity.

7.2.1 Segmentation Influence

Eye region segmentation is an important step for the accurate recognition of the EAC and a critical aspect is the number of classes, C, in which the input data should be divided. As can be seen in Table 7 a larger number of regions increases the EAC recognition rate; therefore, the eye space should be in fact divided in four regions corresponding to all the eye components present in the bounding box: the iris and the sclera, the eyelashes and the surrounding skin area.

Table 8 Individual recognition rate for each EAC case on the still Eye Chimera database

VD	VR	VC	AR	AC	ID	K
88.62	74.66	80.00	71.83	61.43	71.76	80.43

The acronyms for the EACs are presented in Fig. 1

Fig. 12 Automatic recognition examples: correct (*green arrow*) and false (*red arrow*). Image taken from [10]. Copyright by Springer

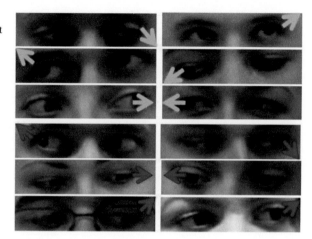

7.2.2 EAC Recognition

The final solution that gives the best results for EAC recognition consists of using iris-oriented K-Means segmentation together with projection information and landmarks. Vrancenu et al. [10] showed that the extra use of the integral projections in the feature vector leads to an improvement of approx. $+5\%$ in the recognition rate.

Furthermore, the recognition rates for each individual EAC are presented in Table 8. It can be seen that a higher confusion rate appears vertically, between eyes looking to the same side. In a NLP interpretation, this corresponds to a better separability between the internal activities and a poorer separability between representational systems. Visual examples of correct and false recognitions are shown in Fig. 12 and it should be noted that even for a human observer it is difficult, in some cases, to correctly classify the direction of gaze.

As said, one intuitive way to recognize the EAC is to use the coordinates of eye fiducial points. Thus, we consider as relevant several foremost such methods. First, the BoRMaN algorithm [78] can be employed for detecting the eye bounding box and a good iris center localization can be obtained using the maximum isophote algorithm presented in [43]. The eye landmarking method proposed in [30] also provides the required points for such an analysis. Finally, using the landmarking technique proposed in [79], out of a larger number of detected fiducial points, the points delimiting the eye and the iris center can be selected for the EAC analysis.

Comparative results are presented in Table 9. As one can see the best results are retrieved by the methods from [10]. There, the refined bounding box (or more precisely its height) is necessary to differentiate between looking down and looking

L. Florea et al.

Table 9 Recognition rate (%) on the still Eye Chimera database for the three EAC cases scenario (when the focus is on the type of mental activity) and for the seven EAC cases scenario (the complete EAC set)

Bounding box method	Iris detection method	RR (%)	
		7 classes	3 classes
Manual	Manual	73.98	94.52
Manual	Valenti [43]	32.30	36.40
BoRMaN [78]	Valenti [43]	32.00	33.12
Zhu [79]	Zhu [79]	39.21	45.57
Zhu [79]	Sun [68]	47.25	68.15
Florea [30]	Florea [30]	48.64	78.57
Radlak [35]	Kawulok [57]	49.47	77.00
Vranceanu [10]	Vranceanu [10]	77.54	89.92

Table 10 EAC recognition rates (%) of various solutions, when information from both eyes is used

Method	RR (%)	[43] + [78]	[79]	[10] (1 eye)	[10] (2 eyes)
Still Eye	7 classes	39.83	43.29	77.54	83.08
Chimera	3 classes	55.73	63.01	89.92	95.21

elsewhere, while the iris position inside it, actually defines the direction of gaze. The pre-processing step removes the eye-lashes as it interferes with the iris separation from the rest of the eye components. The integral projections functions added in the post-processing step supplement the information used by the classifier for the EACs recognition. Due to these facts, when all seven EAC classes are considered, the algorithm from [10] surpasses the upper limit of a point-based analysis, which is obtained when only the five manual markings are used.

Both Eyes Information In order to further improve the detection rate, information from both eyes can be concatenated in the feature vectors. Vranceanu et al. [10] showed (also in Table 10) that this leads to an improvement of approx. +6 % in the detection rate, in both the three cases scenario as well as for the complete EAC set.

Other Databases For a thorough evaluation, we report results on other databases, where the eye cues are partially represented. Since these databases are not designed for an EAC-NLP application, each poses different challenges and are somewhat incomplete from the EAC point of view. The HPEG database does contain all seven EACs, but in a small number, the UUlm database contains only three of the eye cues: Visual Defocus (VD—looking straight), Auditory Remember (AR—looking center-right) and Auditory Constructed (AC—looking center-left). The PUT database contains all seven cues but disproportionably represented. Furthermore, all three databases have a considerable head pose variation.

Comparative results can be seen in Table 11. Although the results vary considerably across databases, the eye components based method [10] offers the best results

Table 11 EAC recognition rate (%) computed on ULM, HPEG and PUT databases

Method	RR (%)	[43] + [78]	[79]	[10] (1 eye)	[10] (2 eyes)
HPEG	7 classes	18.52	31.82	43.71	50.00
	3 classes	29.34	49.15	68.54	75.17
ULM	7 classes	40.63	29.37	23.57	29.29
	3 classes	41.28	44.39	70.35	80.89
PUT	7 classes	11.01	31.11	55.68	62.18
	3 classes	13.18	44.11	63.76	71.43

in all scenarios. While testing on the ULM database, we looked also for seven cases, and any output different from the correct one is marked as an error; to make the test more relevant to the work, we ignored that, for this specific test, only three possible outputs could exist.

8 Discussion and Conclusions

The purpose of this research was to discuss some potential alternatives for an automatic solutions that recognize the direction of gaze in images that contain a frontal face. Such a solution would facilitate advances in areas such as non-conventional teaching, gaming industry and, at last but not at least, for the Eye Accessing Cues.

Eye Accessing Cues are a hypothesis from the Neuro-Lingvistic-theory and they have been only partially validated; more precisely, it has been found that correlation greater than random chance is possible. These results, which are in line with most prior art on the topic, in fact, motivates large scale intensive tests to find the truth behind. If validated, the direction of gaze may be used for better understanding of the mental patterns of a person. We nominated two applications: on-line interviews and interactive presentation.

From a computer vision point of view, while several efficient approaches were investigated, the best results for the recognition of the direction of gaze were reported by Vranceanu et al. [10]. There, consecutively, the face square, iris center, face landmarks and eye components are extracted.

The results on specifically built Eye Chimera database show that the method from Vranceanu et al. [10] surpasses in accuracy some of the most efficient state of the art methods for detecting landmarks and implicitly eye points. It was also shown that it surpasses the pure eye landmarking techniques proving that a region-based solution provides better accuracy than a point-only based approach. These findings were confirmed on other databases too.

Finally, using the video (sequence) part of the Eye-Chimera database, it was proven that, when dealing with sequences, the recognition rate can be increased by considering the temporal redundancy and the correlation between consecutive frames. This observation, together with the low computational cost, offers potential for an implementation where eye cues are detected, tracked and interpreted for mass applications.

The current maximum reported performance reached 77.54 % accuracy on the Eye-Chimera database, which is assumed to be the most relevant for the EAC-NLP theme. This means that in average 1 out of 4 cases is mis-labelled. Thus, if the goal is to validate the hypothesis behind the EAC-NLP than the existing solution is reliable only for automatic initialization of the annotations followed by manual verification. Yet even for this case automatic solution brings speed-ups up to 10×. If the EAC-NLP hypothesis are validated, than the described application may work in real cases as an estimator, with the observation that it needs to be applied independently in consecutive cases, and the overall conclusion needs to be manually validated.

Some additional issues remain for further investigation and development. First, the hypothesis of the Eye Accessing Cues needs to be fully confirmed and most likely bounded. Secondly the EACs are related to non-visual tasks and, therefore, separation between visual and non-visual tasks is required. In normal conditions, the difference between voluntary eye movements (as for seeing something) and involuntary ones (as part of non-verbal communication) is retrievable by the analysis of duration and amplitude [83] as non-visual movements are shorter and with smaller amplitude. However, in both visual memory related task [25] as in the NLP theory, the actual difference between visual and non-visual tasks is achieved by integrating additional information about the person specific activities. More precisely, the Eye Accessing Cues are expected to appear following specific predicates (such as immediately after a question marked by "How?" or "Why?"). Thus, for a complete autonomous solution, the labels required for segmenting the video in visual and non-visual tasks should be inferred from an analysis of the audio channel, that should complement the visual data. To the moment, a completely functional system would be the one where the trainer/interviewer marks the beginning and the end of the non-visual period, as he is aware of the nature of communication.

Concluding, the gaze direction estimation from passive remotely acquired image is an interesting area with many un-explored development directions.

Acknowledgements This work was partially supported by the Romanian Sectoral Operational Programme Human Resources Development 2007–2013 through the European Social Fund Financial Agreements POSDRU/159/1.5/S/134398 (Knowledge).

References

1. R. Bandler, J. Grinder, *Frogs into Princes: Neuro Linguistic Programming* (Real People Press, Moab, 1979)
2. P. Tsiamyrtzis, J. Dowdall, D. Shastri, I.T. Pavlidis, M.G. Frank, P. Ekman, Imaging facial physiology for the detection of deceit. Int. J. Comput. Vis. **71**, 197–214 (2007)

3. A.B. Ashraf, S. Lucey, J.F. Cohn, T. Chen, Z. Ambadar, K.M. Prkachin, P. Solomon, The painful face – pain expression recognition using active appearance models. Image Vis. Comput. **27**, 1788–1796 (2009)
4. C. Florea, L. Florea, C. Vertan, Learning pain from emotion: transferred hot data representation for pain intensity estimation, in *Proceedings of European Conference on Computer Vision Workshop on ACVR* (2014)
5. D.S. Messinger, M.H. Mahoor, S.M. Chow, J. Cohn, Automated measurement of facial expression in infant-mother interaction: a pilot study. Infancy **14**(3), 285–305 (2009)
6. D. McDuff, R.E. Kaliouby, R. Picard, Predicting online media effectiveness based on smile responses gathered over the internet, in *IEEE Face and Gesture* (2013), pp. 1–8
7. J. Rehg, G. Abowd, A. Rozga et al., Decoding children's social behavior, in *Proceedings of Computer Vision and Pattern Recognition* (2013), pp. 3414–3421
8. J.F. Cohn, F. De la Torre, Automated face analysis for affective computing, in *The Oxford Handbook of Affective Computing* (Oxford University Press, Oxford, 2014)
9. A. Frischen, A.P. Bayliss, S.P. Tipper, Gaze cueing of attention. Psychol. Bull. **133**, 694–724 (2007)
10. R. Vranceanu, C. Florea, L. Florea, C. Vertan, Gaze direction estimation by component separation for recognition of eye accessing cues. Mach. Vis. Appl. **26**(2–3), 267–278 (2015)
11. W. James, *The Principles of Psychology* (Harvard University Press, Cambridge, 1890)
12. L. Nummenmaa, A. Calder, Neural mechanisms of social attention. Trends Cogn. Sci. **13**, 135–43 (2009)
13. S. Liversedge, J. Findlay, Saccadic eye movements and cognition. Trends Cogn. Sci. **4**(1), 6–14 (2000)
14. R. Adams, R.E. Kleck, Effects of direct and averted gaze on the perception of facially communicated emotion. Emotion **5**, 3–11 (2005)
15. H. Joseph, K. Nation, S.P. Liversedge, Using eye movements to investigate word frequency effects in children's sentence reading. Sch. Psychol. Rev. **42**, 207–222 (2013)
16. A. Godfroid, F. Boers, A. Housen, An eye for words: gauging the role of attention in incidental l2 vocabulary acquisition by means of eye-tracking. Stud. Second Lang. Acquis. **35**, 483–517 (2013)
17. K. Rayner, T.J. Slattery, D. Drieghe, S.P. Liversedge, Eye movements and word skipping during reading: effects of word length and predictability. J. Exp. Psychol. Hum. Percept. Perform. **37**, 514–528 (2011)
18. K. Rayner, B.R. Foorman, C.A. Perfetti, D. Pesetsky, M.S. Seidenberg, How psychological science informs the teaching of reading. Psychol. Sci. Public Interest **2**, 31–74 (2001)
19. M.M. Chun, Contextual cueing of visual attention. Trends Cogn. Sci. **4**, 170–178 (2000)
20. A. Bulling, T. Zander, Cognition-aware computing. IEEE Trans. Pervasive Comput. **13**, 80–83 (2014)
21. B. Meijering, H. van Rijn, N.A. Taatgen, R. Verbrugge, What eye movements can tell about theory of mind in a strategic game. PLoS One **7**(9) (2012) doi:10.1371/journal.pone.0045961
22. K. Krejtz, C. Biele, D. Chrzastowski, A. Kopacz, A. Niedzielska, P. Toczyski, A. Duchowski, Gaze-controlled gaming: immersive and difficult but not cognitively overloading, in *Proceedings of the ACM International Joint Conference on Pervasive and Ubiquitous Computing: Adjunct Publication* (2014), pp. 1123–1129
23. J. Sturt, S. Ali, W. Robertson, D. Metcalfe, A. Grove, C. Bourne, C. Bridle, Neurolinguistic programming: systematic review of the effects on health outcomes. Br. J. Gen. Pract. **62**, 757–764 (2012)
24. R. Vranceanu, C. Florea, L. Florea, C. Vertan, NLP EAC recognition by component separation in the eye region, in *Proceedings of Computer Analysis and Image Processing* (2013), pp. 225–232
25. B. Laeng, D.S. Teodorescu, Eye scanpaths during visual imagery reenact those of perception of the same visual scene. Cogn. Sci. **26**, 207–231 (2002)
26. T. Kanade, J.F. Cohn, Y. Tian, Comprehensive database for facial expression analysis, in *IEEE Face and Gesture* (2000), pp. 46–53

27. K. Lee, J. Ho, D. Kriegman, Acquiring linear subspaces for face recognition under variable lighting. IEEE Trans. Pattern Anal. Mach. Intell. **27**, 684–698 (2005)

28. P. Belhumeur, D. Jacobs, D. Kriegman, N. Kumar, Localizing parts of faces using a consensus of exemplars, in *Proceedings of Computer Vision and Pattern Recognition* (2011), pp. 545–552

29. G. Huang, M. Ramesh, T. Berg, E. Learned-Miller, Labeled faces in the wild: a database for studying face recognition in unconstrained environments. Technical report, University of Massachusetts, 2007

30. L. Florea, C. Florea, R. Vranceanu, C. Vertan, Can your eyes tell me how you think? A gaze directed estimation of the mental activity, in *Proceedings of British Machine Vision Conference* (2013)

31. S. Asteriadis, D. Soufleros, K. Karpouzis, S. Kollias, A natural head pose and eye gaze dataset, in ACM Workshop on Affective Interaction in Natural Environments (2009), pp. 1–4

32. U. Weidenbacher, G. Layher, P. Strauss, H. Neumann, A comprehensive head pose and gaze database, in *IET International Conference on Intelligent Environments* (2007), pp. 455–458

33. A. Kasiński, A. Florek, A. Schmidt, The PUT face database. Image Process. Commun. **13**, 59–64 (2008)

34. L. Wolf, Z. Freund, S. Avidan, An eye for an eye: a single camera gaze-replacement method, in *Proceedings of Computer Vision and Pattern Recognition* (2010), pp. 817–824

35. K. Radlak, M. Kawulok, B. Smolka, N. Radlak, Gaze direction estimation from static images, in *Proceedings of IEEE Multimedia Signal Processing* (2014), pp. 1–4

36. P. Viola, M. Jones, Robust real-time face detection. Int. J. Comput. Vis. **57**, 137–154 (2004)

37. M. Mathias, R. Benenson, M. Pedersoli, L.V. Gool, Face detection without bells and whistles, in *Proceedings of the European Conference on Computer Vision*, vol. 8692 (2014), pp. 720–735

38. P. Felzenszwalb, R. Girshick, D. McAllester, D. Ramanan, Object detection with discriminatively trained part-based models. Pattern Recogn. Lett. **19**, 899–906 (2010)

39. F. Song, X. Tan, S. Chen, Z. Zhoub, A literature survey on robust and efficient eye localization in real-life scenarios. Br. J. Gen. Pract. **46**, 3157–3173 (2013)

40. M. Hamouz, J. Kittlerand, J.K. Kamarainen, P. Paalanen, H. Kalviainen, J. Matas, Feature-based affine-invariant localization of faces. IEEE Trans. Pattern Anal. Mach. Intell. **27**, 643–660 (2005)

41. S. Asteriadis, N. Nikolaidis, I. Pitas, Facial feature detection using distance vector fields. Pattern Recogn. **42**, 1388–1398 (2009)

42. J. Wu, Z.H. Zhou, Efficient face candidates selector for face detection. Pattern Recogn. **36**, 1175–1186 (2003)

43. R. Valenti, T. Gevers, Accurate eye center location and tracking using isophote curvature, in *Proceedings of Computer Vision and Pattern Recognition* (2008), pp. 1–8

44. O. Jesorsky, K. Kirchberg, R. Frischholz, Robust face detection using the Hausdorff distance, in *Proceedings of International Conference on Audio- and Video-Based Biometric Person Authentication* (2001), pp. 90–95

45. T. Kanade, Picture processing by computer complex and recognition of human faces. Technical Report, Kyoto University, Department of Information Science, 1973

46. G.C. Feng, P.C. Yuen, Variance projection function and its application to eye detection for human face recognition. Pattern Recogn. Lett. **19**, 899–906 (1998)

47. Z. Zhou, Projection functions for eye detection. Pattern Recogn. **37**, 1049–1056 (2004)

48. M. Turkan, M. Pardas, A.E. Cetin, Edge projections for eye localization. Opt. Eng. **47**, 047–054 (2008)

49. M. Verjak, M. Stephancic, An anthropological model for automatic recognition of the male human face. Ann. Hum. Biol. **21**, 363–380 (1994)

50. D. Cristinacce, T. Cootes, I. Scott, A multi-stage approach to facial feature detection, in *Proceedings of British Machine Vision Conference* (2004), pp. 277–286

51. P. Campadelli, R. Lanzarotti, G. Lipori, Precise eye localization through a general-to-specific model definition, in *Proceedings of British Machine Vision Conference*, **I**, 187–196 (2006)

52. Z. Niu, S. Shan, S. Yan, X. Chen, W. Gao, 2D cascaded adaboost for eye localization, in *Proceedings of International Conference of Pattern Recognition* (2006), pp. 1216–1219
53. S. Kim, S.T. Chung, S. Jung, D. Oh, J. Kim, S. Cho, World Academy of Science, Engineering and Technology, in *WASET*, vol. 21 (World Academy of Science, Engineering and Technology, 2007), pp. 483–487
54. M. Asadifard, J. Shanbezadeh, Automatic adaptive center pupil detection using face detection and CDF analysis, in *Proceedings of International Multimedia Conference of Engineers and Computer Scientist* (2010), pp. 130–133
55. L. Ding, A.M. Martinez, Features versus context: an approach for precise and detailed detection and delineation of faces and facial features. IEEE Trans. Pattern Anal. Mach. Intell. **32**, 2022–2038 (2010)
56. F. Timm, E. Barth, Accurate eye centre localisation by means of gradients, in *Proceedings of International Conference on Computer Theory and Applications* (2011), pp. 125–130
57. M. Kawulok, J. Szymanek, Precise multi-level face detector for advanced analysis of facial images. IET Image Process. **6**, 95–103 (2012)
58. C. Florea, L. Florea, C. Vertan, Robust eye centers localization with zero-crossing encoded image projections. Pattern Anal. Applic. 1–17 (2015), DOI:10.1007/s10044-015-0479-x, http://dx.doi.org/10.1007/s10044-015-0479-x
59. R. Valenti, T. Gevers, Accurate eye center location through invariant isocentric patterns. IEEE Trans. Pattern Anal. Mach. Intell. **34**, 1785–1798 (2012)
60. H.C. Becker, W.J. Nettleton, P.H. Meyers, J.W. Sweeney, C.M. Nice, Digital computer determination of a medical diagnostic index directly from chest X-ray images. IEEE Trans. Biomed. Eng. **11**, 62–72 (1964)
61. F. Crow, Summed-area tables for texture mapping. Proc. SIGGRAPH **18**, 207–212 (1984)
62. G.E. Blelloch, Prefix sums and their applications. synthesis of parallel algorithms. Technical report, University of Massachusetts, 1990
63. R.A. King, T.C. Phipps, Shannon, TESPAR and approximation strategies. Comput. Secur. **18**, 445–453 (1999)
64. X. Chen, H. Wu, X. Jin, Q. Zhao, Face illumination manipulation using a single reference image by adaptive layer decomposition. IEEE Trans. Image Processing **22**(11), 4249–4259 (2013)
65. B. Kroon, A. Hanjalic, S.M. Maas, Eye localization for face matching: is it always useful and under what conditions, in *Proceedings of International Conference on Content-Based Image and Video Retrieval* (2008), pp. 379–387
66. M. Ciesla, P. Koziol, Eye pupil location using webcam. CoRR, (2012) http://arxiv.org/abs/1202.6517
67. M. Dantone, J. Gall, G. Fanelli, L.V. Gool, Real-time facial feature detection using conditional regression forests, in *Proceedings of Computer Vision and Pattern Recognition* (2012), pp. 2578–2585
68. Y. Sun, X. Wang, X. Tang, Deep convolutional network cascade for facial point detection, in *Proceedings of Computer Vision and Pattern Recognition* (2013), pp. 3476–3483
69. T. Cootes, C. Taylor, D. Cooper, J. Graham, Active shape models - their training and application. Comput. Vis. Image Underst. **61**, 38–59 (1995)
70. T.F. Cootes, G.J. Edwards, C.J. Taylor, Active appearance models. IEEE Trans. Pattern Anal. Mach. Intell. **23**, 681–685 (2001)
71. T. Leung, M. Burl, P. Perona, Finding faces in cluttered scenes using random labeled graph matching, in *Proceedings of International Conference on Computer Vision* (1995), pp. 637–644
72. S. Milborrow, F. Nicolls, Locating facial features with an extended active shape model, in *Proceedings of European Conference on Computer Vision* (2008), pp. 504–513
73. V. Le, J. Brandt, Z. Lin, L. Bourdev, T.S. Huang, Interactive facial feature localization, in *Proceedings of European Conference on Computer Vision* (2012), pp. 679–692
74. D. Cristinacce, T. Cootes, Feature detection and tracking with constrained local models, in *Proceedings of British Machine Vision Conference* (2006), pp. 929–938

75. P. Tresadern, H. Bhaskar, S. Adeshina, C. Taylor, T. Cootes, Combining local and global shape models for deformable object matching, in *Proceedings of British Machine Vision Conference* (2009)
76. T. Cootes, M.C. Ionita, C. Lindner, P. Sauer, Robust and accurate shape model fitting using random forest regression voting, in *Proceedings of European Conference on Computer Vision* (2012)
77. J. Saragih, S. Lucey, J. Cohn, Deformable model fitting by regularized landmark mean-shift. Int. J. Comput. Vis. **91**, 200–215 (2011)
78. M. Valstar, T. Martinez, X. Binefa, M. Pantic, Facial point detection using boosted regression and graph models, in *Proceedings of Computer Vision and Pattern Recognition* (2010), pp. 2729–2736
79. X. Zhu, D. Ramanan, Face detection, pose estimation, and landmark localization in the wild, in *Proceedings of Computer Vision and Pattern Recognition* (2012), pp. 2879–2886
80. X. Yu, J. Huang, S. Zhang, W. Yan, D.N. Metaxas, Pose-free facial landmark fitting via optimized part mixtures and cascaded deformable shape model, in *Proceedings of International Conference on Computer Vision* (2013), pp. 1944–1951
81. B. Martinez, M.F. Valstar, X. Binefa, M. Pantic, Local evidence aggregation for regression based facial point detection. IEEE Trans. Pattern Anal. Mach. Intell. **35**, 1149–1163 (2013)
82. P. Wang, M.B. Green, Q. Ji, J. Wayman, Automatic eye detection and its validation, in *IEEE Workshop on FRGC, Computer Vision and Pattern Recognition* (2005), p. 164
83. A. Duchowski, *Eye Tracking Methodology: Theory and Practice* (Springer, Berlin, 2007)
84. D. Hansen, J. Qiang, In the eye of the beholder: a survey of models for eyes and gaze. IEEE Trans. Pattern Anal. Mach. Intell. **32**, 478–500 (2010)
85. D. Yoo, M. Chung, A novel non-intrusive eye gaze estimation using cross-ratio under large head motion. Comput. Vis. Image Underst. **98**, 25–51 (2005)
86. B. Pires, M. Hwangbo, M. Devyver, T. Kanade, Visible-spectrum gaze tracking for sports, in *WACV* (2013)
87. D. Hansen, A. Pece, Eye tracking in the wild. Comput. Vis. Image Underst. **98**, 182–210 (2005)
88. S. Cadavid, M. Mahoor, D. Messinger, J. Cohn, Automated classification of gaze direction using spectral regression and support vector machine, in *Proceedings of Affective Computing and Intelligent Interaction* (2009), pp. 1–6
89. T. Heyman, V. Spruyt, A. Ledda, 3d face tracking and gaze estimation using a monocular camera, in *Proceedings of International Conference on Positioning and Context-Awareness* (2011), pp. 23–28
90. M. Everingham, A. Zisserman, Regression and classification approaches to eye localization in face images, in *IEEE Face and Gesture* (2006), pp. 441–446
91. G. Diamantopoulos, Novel eye feature extraction and tracking for non-visual eye-movement applications. Ph.D. thesis, University of Birmingham, 2010
92. S. le Cessie, J. van Houwelingen, Ridge estimators in logistic regression. Appl. Stat. **41**, 191–201 (1992)